SOLUTIONS MANUAL TO ACCOMPANY FUNDAMENTALS OF MATRIX ANALYSIS WITH APPLICATIONS

SOLUTIONS MANUAL TO ACCOMPANY FUNDAMENTALS OF MATRIX ANALYSIS WITH APPLICATIONS

EDWARD BARRY SAFF
Vanderbilt University

ARTHUR DAVID SNIDER
University of South Florida

For general information on our other products and services please contact our Customer Care
Department with the U.S. at 877-762-2974, outside the U.S. at 317-572-3993 or fax 317-572-4002.

Wiley also publishes its books in a variety of electronic formats. Some content that appears in print,
however, may not be available in electronic format.

Library of Congress Cataloging-in-Publication data has been applied for

Set in 11/14pt Times by SPi Global, Pondicherry, India

Printed in the United States of America.

1   2016

ACKNOWLEDGMENTS

Special thanks to Viktor Maymeskul, Samuel Garrett, Rares Rasdeaconu, and Oleksandr Vlasiuk for contributions to this work.

CONTENTS

1

SYSTEMS OF LINEAR ALGEBRAIC EQUATIONS

1.1 LINEAR ALGEBRAIC EQUATIONS

1. (a) Solving the last equation for x_4 yields $x_4 = 4/4 = 1$. We substitute this value of x_4 into the third equation and solve for x_3 to get

$$x_3 + 2(1) = -1 \quad \Rightarrow \quad x_3 = -3.$$

With these values of x_3, the second equation becomes

$$3x_2 + 2(-3) + (1) = 2 \quad \Rightarrow \quad 3x_2 = 7 \quad \Rightarrow \quad x_2 = \frac{7}{3}.$$

For x_1, we substitute the values of x_2 and x_4 into the first equation.

$$2x_1 + \left(\frac{7}{3}\right) - (1) = -2 \quad \Rightarrow \quad 2x_1 = -\frac{10}{3} \quad \Rightarrow \quad x_1 = -\frac{5}{3}.$$

Therefore, the answer is: $x_1 = -5/3$, $x_2 = 7/3$, $x_3 = -3$, and $x_4 = 1$.

Solutions Manual to Accompany Fundamentals of Matrix Analysis with Applications, First Edition. Edward Barry Saff and Arthur David Snider.

(b) Solving the first equation for x_1, we obtain $x_1 = 3$. We now substitute this value of x_1 into the second equation and solve for x_2.

$$-2(3) - 3x_2 = -12 \quad \Rightarrow \quad -3x_2 = -6 \quad \Rightarrow \quad x_2 = 2.$$

The third equation then becomes

$$(3) + (2) + x_3 = 5 \quad \Rightarrow \quad x_3 = 0.$$

Therefore, the answer is $x_1 = 3$, $x_2 = 2$, and $x_3 = 0$.

(c) From the third equation, we immediately get $x_4 = 1$. This value, when substituted into the first equation, yields

$$-x_3 + 2(1) = 1 \quad \Rightarrow \quad -x_3 = -1 \quad \Rightarrow \quad x_3 = 1.$$

From the fourth equation, we obtain

$$x_2 + 2(1) + 3(1) = 5 \quad \Rightarrow \quad x_2 = 0.$$

Finally, from the second equation, we conclude that

$$4x_1 + 2(0) + (1) = -3 \quad \Rightarrow \quad x_1 = -1.$$

So, the solution is $x_1 = -1$, $x_2 = 0$, $x_3 = 1$, and $x_4 = 1$.

(d) The third equation implies that $x_1 = 1$. Then, the first equation says

$$2(1) + x_2 = 3 \quad \Rightarrow \quad x_2 = 1.$$

We now substitute these values of x_1 and x_2 into the second equation to get

$$3(1) + 2(1) + x_3 = 6 \quad \Rightarrow \quad x_3 = 1.$$

The solution to this problem is $x_1 = x_2 = x_3 = 1$.

3. To eliminate x_1 from the first and second equations, we subtract from the first equation the third equation multiplied by 3 and subtract from

the second equation the third equation multiplied by 2, resp. Thus we get an equivalent system

$$
\begin{array}{rrcr}
-x_2 & & = & -2 \\
-9x_2 & -x_3 & = & -16 \\
x_1 \quad +3x_2 & +x_3 & = & 3.
\end{array}
$$

From the first equation, we get $x_2 = 2$. Making the back substitutions into the second and third equations yields

$$
\begin{array}{rcrcr}
-9(2) - x_3 = -16 & \Rightarrow & -x_3 = 2 & \Rightarrow & x_3 = -2 \\
x_1 + 3(2) + (-2) = 3 & \Rightarrow & x_1 = -1.
\end{array}
$$

The solution is $x_1 = -1$, $x_2 = 2$, and $x_3 = -2$.

5. We eliminate x_1 and x_4 from the first and the the last equations by subtracting from them the second equation. We also eliminate x_1 from the third equation by subtracting from it twice the second equation. The new system is

$$
\begin{array}{rrcr}
x_2 \quad +x_3 & & = & 1 \\
x_1 & +x_4 & = & 0 \\
2x_2 \quad -x_3 & -x_4 & = & 6 \\
2x_2 \quad -x_3 & & = & 0
\end{array}
$$

Next, we eliminate x_2 from the third and fourth by subtracting from them the first equation multiplied by 2. This gives

$$
\begin{array}{rrcr}
x_2 \quad +x_3 & & = & 1 \\
x_1 & +x_4 & = & 0 \\
-3x_3 & -x_4 & = & 4 \\
-3x_3 & & = & -2
\end{array}
$$

We now can go with the back substitution. The last equation gives $x_3 = 2/3$. With this value, we find x_1 and x_4 from the first and third equations, resp.

$$x_2 + \left(\frac{2}{3}\right) = 1 \quad \Rightarrow \quad x_2 = \frac{1}{3}$$

$$-3\left(\frac{2}{3}\right) - x_4 = 4 \quad \Rightarrow \quad -x_4 = 6 \quad \Rightarrow \quad x_4 = -6.$$

Finally, the second equation says $x_1 = -x_4 = 6$. so, the answer is $x_1 = 6$, $x_2 = 1/3$, $x_3 = 2/3$, and $x_4 = -6$.

7. To eliminate x_1 from the second equation, we multiply the first equation by $0.987/0.123$ and subtract the result from the second one. Thus, we get

$$
\begin{aligned}
0.123x_1 \quad +0.456x_2 &= \quad 0.789 \\
-3.005x_2 &= -6.010
\end{aligned}
$$

From the second equation, we get $x_2 = (-6.010)/(-3.005) = 2.000$. Substituting this value into the first equation, we get

$$0.123x_1 + 0.456(2) = 0.789 \quad \Rightarrow \quad 0.123x_1 = -0.123$$
$$\Rightarrow \quad x_1 = -1.000$$

The solution, rounded to three decimal places, is $x_1 = -1.000$, $x_2 = 2.000$. The number of arithmetic operations required is 3 divisions, 3 multiplications, and 3 additions.

9. Following the notations in Problem 8, the coefficients in Problem 7 are

$$a = 0.123, \ b = 0.456, \ c = 0.789,$$
$$d = 0.987, \ e = 0.654, \ f = 0.321.$$

Applying the formulas given in Problem 8, we obtain

$$x = \frac{(0.789)(0.654) - (0.456)(0.321)}{(0.123)(0.654) - (0.456)(0.987)} = -1.000;$$
$$y = \frac{(0.123)(0.321) - (0.789)(0.987)}{(0.123)(0.654) - (0.456)(0.987)} = 2.000.$$

The number of arithmetic operations required is 2 divisions, 6 multiplications, and 6 additions.

11. To eliminate x_1 from the second equation, we subtract from it the first equation multiplied by $0.987/0.123$. Similarly, we eliminate x_1 from the third equation using the factor of $0.333/0.123$, and obtain

$$\begin{aligned} 0.123x_1 \quad +0.456x_2 \quad +0.789x_3 &= \quad 0.111 \\ -3.005x_2 \quad -6.010x_3 &= -0.446 \\ -1.790x_2 \quad -2.913x_3 &= \quad 0.587 \end{aligned}$$

We subtract from the third equation the second one multiplied by $(1.790/3.005)$:

$$\begin{aligned} 0.123x_1 \quad +0.456x_2 \quad +0.789x_3 &= \quad 0.111 \\ -3.005x_2 \quad -6.010x_3 &= -0.446 \\ 0.667x_3 &= \quad 0.853 \end{aligned}$$

Going from the third equation up, using the back substitution we find that

$$x_3 = \frac{0.853}{0.667} = 1.279;$$
$$x_2 = \frac{-0.446 + 6.010(1.279)}{-3.005} = -2.410;$$
$$x_1 = \frac{0.111 - 0.456(-2.410) - 0.789(1.279)}{0.123} = 1.633.$$

The number of arithmetic operations required is 6 divisions, 11 multiplications, and 11 additions.

13. (a) From the third equation we find that $x_3 = 5.16/1.42 \approx 3.6338$ (rounded to four decimal places). Substituting this value into the second equation, we get

$$x_2 = \frac{1.11 - 1.34(3.6338)}{2.73} \approx -1.3770$$

Finally, we use the values of x_2 and x_3 to find x_1 from the first equation.

$$x_1 = \frac{-4.22 - 7.29(-1.3770) + 3.21(3.6338)}{1.23} \approx 14.2137.$$

Rounding the results to two decimal places, the answer is

$$x_1 = 14.21, \ x_2 = -1.38, \ x_3 = 3.63.$$

The total number of arithmetic operations required is 3 divisions, 3 multiplications, and 3 additions.

(b) We start from the fourth equation to find x_4. Substituting its value into the third equation, we evaluate x_3, and so on. These computations give

$$x_4 = \frac{-1}{0.250} = -4.0000;$$

$$x_3 = \frac{1 - 0.888(-4)}{0.999} = 4.5565;$$

$$x_2 = \frac{-1 - 0.222(4.5565) - 0.333(-4)}{0.111} = -6.1220;$$

$$x_1 = \frac{1 - 0.333(-6.1220) - 0.250(4.5565) - 0.200(-4)}{0.500}$$
$$= 5.3990.$$

Rounding the results to three decimal places yields

$$x_1 = 5.399, \ x_2 = -6.122, \ x_3 = 4.557, \ x_4 = -4.000.$$

The number of arithmetic operations required is 4 divisions, 6 multiplications, and 6 additions.

15. No, the suggested procedure is inconsistent with the Gauss elimination rules. First of all, these rules require, on each step, the elimination of one of the variables from all but one equations. One can do some *preliminary steps* adding to an equation a multiple of another equation, but on each such step the *new system* must be considered.

In the problem, let's follow steps. Subtracting the second equation from the first, we get a *new system*:

$$
\begin{array}{rrrr}
-x_1 & +x_2 & +x_3 & = 0 \\
2x_1 & +x_2 & +x_3 & = 6 \\
x_1 & +x_2 & +3x_3 & = 6
\end{array}
$$

Subtracting the third equation from the second, yields a *new system*:

$$
\begin{array}{rrrl}
-x_1 & +x_2 & +x_3 & = 0 \\
x_1 & & -2x_3 & = 0 \\
x_1 & +x_2 & +3x_3 & = 6
\end{array}
$$

Performing the last step suggested in the problem, we obtain

$$
\begin{array}{rrrl}
-x_1 & +x_2 & +x_3 & = 0 \\
x_1 & & -2x_3 & = 0 \\
2x_1 & & +2x_3 & = 6
\end{array}
$$

After these preliminary steps (that were not really necessary), we can now go with Gauss elimination procedure. Adding the second equation to the third yields

$$
\begin{array}{rrrl}
-x_1 & +x_2 & +x_3 & = 0 \\
x_1 & & -2x_3 & = 0 \\
3x_1 & & & = 6
\end{array}
$$

We can now use the back substitution method to solve the system.

$$
x_1 = \frac{6}{3} = 2; \quad x_3 = \frac{x_1}{2} = 1; \quad x_2 = x_1 - x_3 = 1.
$$

17. In the "derivation", a standard mistake was made: the equation $x^2 = (y-1)^2$ *does not* conclude that $x = y - 1$. The correct conclusion is $x = \pm(y-1)$. With the sign "+", we get the answer given; i.e. $x = 1$, $y = 2$. Choosing the sign "−", we get

$$
-(y-1) + 2y = 5 \quad \Rightarrow \quad y = 4 \quad \Rightarrow \quad x = -(4-1) = -3.
$$

The figure below indicates these *two* solutions – the points of intersection of the graphs of the equations.

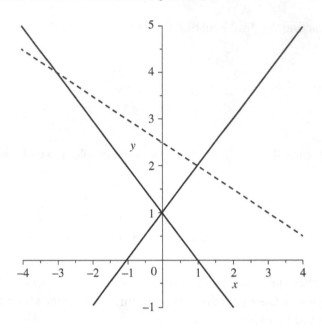

19. The coordinates of both points must satisfy the equation $ax + by = c$. Thus, we have a system of two linear equations with three unknowns – a, b, and c:

$$ax_1 + by_1 = c$$
$$ax_2 + by_2 = c.$$

Since there is always a line passing through two given points, this system is consistent. If the triple (a_0, b_0, c_0) is a solution, multiplying both equations by $k \neq 0$, we get an equivalent system

$$(ka_0) x_1 + (kb_0) y_1 = (kc_0)$$
$$(ka_0) x_2 + (kb_0) y_2 = (kc_0).$$

Thus, the triple (ka_0, kb_0, kc_0) also satisfies the conditions and, actually, determines the same line because

$$a_0 x + b_0 y = c_0 \quad \Leftrightarrow \quad (ka_0) x + (kb_0) y = (kc_0).$$

Thus, the equation of a line is determined up to a non-zero constant multiple.

21. Since the points $(x_0, y_0), \ldots, (x_n, y_n)$ are on the graph $y = a_0 + \cdots + a_n x^n$, they must satisfy the equation. Substituting the coordinates of these points into the equation, yield the system of $(n + 1)$ equations with $(n + 1)$ unknowns – $a_0, \ldots, a_n$.

$$
\begin{aligned}
a_0 \;+a_1 x_0 \;+\cdots+\; a_n x_0^n &= y_0 \\
a_0 \;+a_1 x_1 \;+\cdots+\; a_n x_1^n &= y_1 \\
&\;\;\vdots \\
a_0 \;+a_1 x_n \;+\cdots+\; a_n x_n^n &= y_n
\end{aligned}
$$

23. Multiplying the second equation by $(1 + i)$ we get

$$
(1 + i)x_1 + (1 + i)ix_2 + (1 + i)(2 - i)x_3 = (1 + i)^2
$$
$$
\Leftrightarrow \qquad (1 + i)x_1 + (i - 1)x_2 + (3 + i)x_3 = 2i.
$$

We eliminate now x_1 from the first equation by subtracting the above result from it (keeping the original second equation unchanged). Thus we get

$$
\begin{aligned}
(3 - i)x_2 \qquad -(3 + i)x_3 &= \quad 0 \\
x_1 \qquad\quad +ix_2 \qquad\quad +(2 - i)x_3 &= 1 + i \\
(-1 + 2i)x_2 \;+(-2 - 3i)x_3 &= \quad 0
\end{aligned}
$$

To eliminate x_2 from the first equation, we multiply it by $(-1 + 2i)$ and then subtract from it the third equation multiplied by $(3 - i)$. Thus, we get

$$
[-(3 + i)(-1 + 2i) - (-2 - 3i)(3 - i)] x_3 = 0
$$
$$
\Rightarrow \qquad (9 + 7i)x_3 = 0,
$$

so that we have a new system

$$
\begin{aligned}
-(9 + 7i)x_3 &= \quad 0 \\
x_1 \qquad\quad +ix_2 \qquad\quad +(2 - i)x_3 &= 1 + i \\
(-1 + 2i)x_2 \;+(-2 - 3i)x_3 &= \quad 0
\end{aligned}
$$

From the first equation, we find that $x_3 = 0$. Substituting this value into the third equation yields $x_2 = 0$. With zero values of x_2 and x_3, the second equation says that $x_1 = 1 + i$. Therefore, the solution is

$$
x_1 = 1 + i, \; x_2 = 0, \; x_3 = 0.
$$

25. If x and y are integers, then $6x$ and $4y$ are *even* numbers. Thus, their difference, $6x - 4y$, is *even*, and so cannot be equal to an *odd* number 9.

27. The first equation is equivalent to

$$3(2x + 4y) = 3(3) \qquad \Rightarrow \qquad 6x + 12y = 9.$$

In the integers modulo 6, we have the sides as

$$(6x + 12y) \mod 6 = 6(x + 2y) \mod 6 = 0, \qquad 9 \mod 6 = 3.$$

Thus, the equation is *inconsistent* in integers modulo 6, and so is the system.

Concerning a solution in integers modulo 7, there are different ways to go. One of them is the following. First, we get a system that is equivalent to the given system by multiplying the first equation by 3 and the second equation by 2.

$$
\begin{aligned}
6x &+ 12y = 9 \\
6x &+ 4y = 6
\end{aligned}
$$

Subtracting the second equation from the first one yields $8y = 3$. Rewriting this equation in integers modulo 7 (since $8 \mod 7 = 1$) as $y \mod 7 = 3$, the back substitution into the second equation in the original system yields

$$3x = 3 - 2y \qquad \Rightarrow \qquad (3x) \mod 7 = (3 - 2y)$$
$$\mod 7 = 4 \mod 7.$$

Solving for x, we get $x \mod 7 = 6$. Therefore, in integers modulo 7, the solution to the given system is

$$x \mod 7 = 6, \ y \mod 7 = 3.$$

29. To simplify computations, we note that, for large n, the total number of each arithmetic operation can be approximated with a good *relative accuracy* by

Additions: $\dfrac{(n-1)n(2n+5)}{6} \approx \dfrac{n \cdot n \cdot (2n)}{6} = \dfrac{n^3}{3}$;

Multiplications: $\dfrac{(n-1)n(2n+5)}{6} \approx \dfrac{n \cdot n \cdot (2n)}{6} = \dfrac{n^3}{3}$;

Divisions: $\dfrac{n(n+1)}{2} \approx \dfrac{n^2}{2}$.

Thus, the total number $N(n)$ of arithmetic operations can be approximated as

$$N(n) \approx \frac{n^3}{3} + \frac{n^3}{3} + \frac{n^2}{2} \approx \frac{2n^3}{3}.$$

For the "Thermal stress", $n = 10,000 = 10^4$. Therefore, using the performance of computers, we get that computers require approximately

Typical PC: $\dfrac{2\left(10^4\right)^3 / 3}{5 \times 10^9} = \dfrac{2 \times 10^{12}}{15 \times 10^9} \approx 133.33$ (sec.)

Tianhe -2: $\dfrac{2\left(10^4\right)^3 / 3}{3.38 \times 10^{16}} \approx \dfrac{10^{12}}{5.07 \times 10^{16}} \approx 1.97 \times 10^{-5}$ (sec.)

to solve the system. Performing similar computations for other systems, we fill in the following table.

Model	n	Typical PC	Tianhe-2
Thermal Stress	10^4	133.333	1.97×10^{-5}
American Sign Language	10^5	1.33×10^5	0.020
Chemical Plant Modeling	3×10^5	3.59×10^6	0.532
Mechanics of Composite Materials	10^6	1.33×10^8	19.723
Electromagnetic Modeling	10^8	1.33×10^{14}	1.97×10^7
Computation Fluid Dynamics	10^9	1.33×10^{17}	1.97×10^{10}

1.2   MATRIX REPRESENTATION OF LINEAR SYSTEMS AND THE GAUSS-JORDAN ALGORITHM

1. We subtract the first row the second row multiplied by 2.

$$
\begin{bmatrix} 2 & 1 & \vdots & 8 \\ 1 & -3 & \vdots & -3 \end{bmatrix}
\quad \Rightarrow \quad
\begin{bmatrix} 2 & 1 & \vdots & 8 \\ 0 & -7 & \vdots & -14 \end{bmatrix}
$$

Therefore, $x_2 = (-14)/(-7) = 2$, and the back substitution yields

$$2x_1 + (2) = 8 \quad \Rightarrow \quad x_1 = 3.$$

The solution is $x_1 = 3, x_2 = 2$.

3. We perform elementary row operations to reduce the given augmented coefficient matrix to upper triangular form.

$$
\begin{bmatrix} 4 & 2 & 2 & \vdots & 8 & 0 \\ 1 & -1 & 1 & \vdots & 4 & 2 \\ 3 & 2 & 1 & \vdots & 2 & 0 \end{bmatrix}
\quad \rho_2 \leftrightarrow \rho_1 \quad
\begin{bmatrix} 1 & -1 & 1 & \vdots & 4 & 2 \\ 4 & 2 & 2 & \vdots & 8 & 0 \\ 3 & 2 & 1 & \vdots & 2 & 0 \end{bmatrix}
$$

$$
\begin{array}{c} \rho_2 - 4\rho_1 \to \rho_2 \\ \rho_3 - 3\rho_1 \to \rho_3 \end{array}
\quad
\begin{bmatrix} 1 & -1 & 1 & \vdots & 4 & 2 \\ 0 & 6 & -2 & \vdots & -8 & -8 \\ 0 & 5 & -2 & \vdots & -10 & -6 \end{bmatrix}
$$

$$
6\rho_3 - 5\rho_2 \to \rho_3
\quad
\begin{bmatrix} 1 & -1 & 1 & \vdots & 4 & 2 \\ 0 & 6 & -2 & \vdots & -8 & -8 \\ 0 & 0 & -2 & \vdots & -20 & 4 \end{bmatrix}
$$

Evaluating x_3 from the third equation and making back substitutions we obtain the following solutions:

For the first system,

$$x_3 = \frac{-20}{-2} = 10, \ x_2 = \frac{-8 + 2(10)}{6} = 2, \ x_1 = 4 + (2) - (10) = -4.$$

For the second system,

$$x_3 = \frac{4}{-2} = -2, \ x_2 = \frac{-8 + 2(-2)}{6} = -2,$$
$$x_1 = 2 + (-2) - (-2) = 2.$$

5. For an $m \times n$ matrix $\mathbf{A}$, let's find a formula for the address of the element a_{ij}, $1 \leq i \leq m$, $1 \leq j \leq n$. This element is in the j's column. Therefore, the number of preceding columns is $(j - 1)$. Each of them contain m elements, and so the number of preceding elements is $m(j - 1)$. The element a_{ij} is the i's element in the j's column, and so it has the number

$$\# \text{ of } a_{ij} = m(j - 1) + i, \quad 1 \leq i \leq m, \quad 1 \leq j \leq n.$$

This is the answer to the part (c).
(a) We have $m = 7$ and $n = 8$. Using the formula derived for the part (c) yields

$$\# \text{ of } a_{1,7} = 7(7 - 1) + 1 = 43; \quad \# \text{ of } a_{7,1} = 7(1 - 1) + 7 = 7;$$
$$\# \text{ of } a_{5,5} = 7(5 - 1) + 5 = 33; \quad \# \text{ of } a_{7,8} = 7(8 - 1) + 7 = 56.$$

(d) Let's first find the answer to this part, and then apply it in the part (b). From the formula in the part (c), we have

$$\frac{\# \text{ of } a_{ij} - i}{m} = j - 1 \quad \Leftrightarrow \quad j = \frac{\# \text{ of } a_{ij} - i + m}{m}.$$

Since j is a natural number and $1 \leq i \leq m$, we can eliminate the (unknown) i from this equation by writing

$$j = \left\lceil \frac{\# \text{ of } a_{ij}}{m} \right\rceil,$$

where $\lceil \cdot \rceil$ means the "ceiling function", i.e., $\lceil x \rceil$ is the smallest integer that is not less than x. Once j is found, we can compute the index i using

$$i = \# \text{ of } a_{ij} - m(j - 1).$$

(b) Here, $m = 5$, $n = 12$. With these values, the formulas in the part (d) give

$$\#4: \qquad j = \left\lceil \frac{4}{5} \right\rceil = 1, \quad i = 4 - 5(1 - 1) = 4 \Rightarrow a_{4,1};$$

$$\#20: \qquad j = \left\lceil \frac{20}{5} \right\rceil = 4, \quad i = 20 - 5(4 - 1) = 5 \Rightarrow a_{5,4};$$

$$\#50: \qquad j = \left\lceil \frac{50}{5} \right\rceil = 10, \quad i = 50 - 5(10 - 1) = 5 \Rightarrow a_{5,10}.$$

7. (a) Following the Gauss-Jordan elimination procedure, we proceed as follows.

$$\left[\begin{array}{ccc:c} 3 & 8 & 3 & 7 \\ 2 & -3 & 1 & -10 \\ 1 & 3 & 1 & 3 \end{array} \right]$$

$$(1/3)\rho_1 \to \rho_1 \qquad \left[\begin{array}{ccc:c} 1 & 8/3 & 1 & 7/3 \\ 2 & -3 & 1 & -10 \\ 1 & 3 & 1 & 3 \end{array} \right]$$

$$\begin{array}{c} \rho_2 - 2\rho_1 \to \rho_2 \\ \rho_3 - \rho_1 \to \rho_3 \end{array} \qquad \left[\begin{array}{ccc:c} 1 & 8/3 & 1 & 7/3 \\ 0 & -25/3 & -1 & -44/3 \\ 0 & 1/3 & 0 & 2/3 \end{array} \right]$$

$$(-3/25)\rho_2 \to \rho_2 \qquad \left[\begin{array}{ccc:c} 1 & 8/3 & 1 & 7/3 \\ 0 & 1 & 3/25 & 44/25 \\ 0 & 1/3 & 0 & 2/3 \end{array} \right]$$

$$\begin{array}{c} \rho_1 - (8/3)\rho_2 \to \rho_1 \\ \rho_3 - (1/3)\rho_2 \to \rho_3 \end{array} \qquad \left[\begin{array}{ccc:c} 1 & 0 & 17/25 & -177/75 \\ 0 & 1 & 3/25 & 44/25 \\ 0 & 0 & -1/25 & 2/25 \end{array} \right]$$

$$(-25)\rho_3 \to \rho_3 \quad \begin{bmatrix} 1 & 0 & 17/25 & \vdots & -177/75 \\ 0 & 1 & 3/25 & \vdots & 44/25 \\ 0 & 0 & 1 & \vdots & -2 \end{bmatrix}$$

$$\begin{matrix} \rho_1 - (17/25)\rho_3 \to \rho_1 \\ \rho_2 - (3/25)\rho_3 \to \rho_2 \end{matrix} \quad \begin{bmatrix} 1 & 0 & 0 & \vdots & -1 \\ 0 & 1 & 0 & \vdots & 2 \\ 0 & 0 & 1 & \vdots & -2 \end{bmatrix}$$

Therefore, the answer is $x_1 = -1$, $x_2 = 2$, and $x_3 = -2$.

(b) This problem, actually, contains three systems of linear equations with the same coefficient matrix but different right-hand side columns. They are

$$\begin{bmatrix} 1 \\ 0 \\ 0 \end{bmatrix}, \quad \begin{bmatrix} 0 \\ 1 \\ 0 \end{bmatrix}, \quad \begin{bmatrix} 0 \\ 0 \\ 1 \end{bmatrix}.$$

So, can proceed with the Gauss-Jordan elimination process for the coefficient matrix keeping a track of the right-hand sides on each step. For convenience, we first rearrange the equations.

$$\begin{bmatrix} 1 & 2 & 1 & \vdots & 1 & 0 & 0 \\ 1 & 3 & 2 & \vdots & 0 & 1 & 0 \\ 1 & 0 & 1 & \vdots & 0 & 0 & 1 \end{bmatrix} \quad \begin{matrix} \rho_3 \to \rho_1 \\ \rho_1 \to \rho_2 \\ \rho_2 \to \rho_3 \end{matrix} \quad \begin{bmatrix} 1 & 0 & 1 & \vdots & 0 & 0 & 1 \\ 1 & 2 & 1 & \vdots & 1 & 0 & 0 \\ 1 & 3 & 2 & \vdots & 0 & 1 & 0 \end{bmatrix}$$

Then we proceed as follows:

$$\begin{matrix} \rho_2 - \rho_1 \to \rho_2 \\ \rho_3 - \rho_1 \to \rho_3 \end{matrix} \quad \begin{bmatrix} 1 & 0 & 1 & \vdots & 0 & 0 & 1 \\ 0 & 2 & 0 & \vdots & 1 & 0 & -1 \\ 0 & 3 & 1 & \vdots & 0 & 1 & -1 \end{bmatrix}$$

$$2\rho_3 - 3\rho_2 \to \rho_3 \quad \begin{bmatrix} 1 & 0 & 1 & \vdots & 0 & 0 & 1 \\ 0 & 2 & 0 & \vdots & 1 & 0 & -1 \\ 0 & 0 & 2 & \vdots & -3 & 2 & 1 \end{bmatrix}$$

$$
\begin{array}{c}
(1/2)\rho_2 \to \rho_2 \\
(1/2)\rho_3 \to \rho_3
\end{array}
\left[
\begin{array}{ccc:ccc}
1 & 0 & 1 & 0 & 0 & 1 \\
0 & 1 & 0 & 1/2 & 0 & -1/2 \\
0 & 0 & 1 & -3/2 & 1 & 1/2
\end{array}
\right]
$$

$$
\rho_1 - \rho_3 \to \rho_1
\left[
\begin{array}{ccc:ccc}
1 & 0 & 0 & 3/2 & -1 & 1/2 \\
0 & 1 & 0 & 1/2 & 0 & -1/2 \\
0 & 0 & 1 & -3/2 & 1 & 1/2
\end{array}
\right]
$$

Therefore, the solutions to the given systems, respectively, are

$$
x_1 = \frac{3}{2}, \quad x_2 = \frac{1}{2}, \quad x_3 = -\frac{3}{2};
$$
$$
x_1 = -1, \quad x_2 = 0, \quad x_3 = 1;
$$
$$
x_1 = \frac{1}{2}, \quad x_2 = -\frac{1}{2}, \quad x_3 = \frac{1}{2}.
$$

9. Let us find out which approach, the standard one or the suggested modified one, is *more* efficient (meaning *less* time consuming). In other words, we have to compare the number of arithmetic operations, N_s and N_m, resp.

   Intuitively, the modified approach should be more efficient:
   In the standard way, given $1 \le j \le n$, to eliminate x_{ij}, $1 \le i \le n$, $i \ne j$, from the ith column, we have to perform arithmetics with entries on *all* rows ρ_i. In the modified procedure, in obtaining an upper triangular matrix, we deal only with the rows ρ_i, $i > j$. It is, approximately, twice less time consuming. Going then from the upper triangular coefficient matrix to a diagonal one, as described, essentially requires arithmetics performed *only* with the last column of the augmented matrix for the rows ρ_i, $1 \le i < j$. These arguments suggest that $N_m \le N_s$.

   As an example, let us compute the number of arithmetic operations used in Problem 9(a), where we followed the standard Gauss-Jordan elimination process: $N_s = 27$. (Computations with obvious zero results are not counted.)

   Using the modified way, we get $N_m = 20$. (Check it!)

   Rigorous computations show that, for n large, the suggested modification of the Gauss-Jordan elimination procedure can save approximately 30% of the time spent for solving an $n \times n$ linear system vs the standard procedure.

1.3   THE COMPLETE GAUSS ELIMINATION ALGORITHM

1. To simplify computations, we rearrange the equations first.

$$
\begin{bmatrix}
1 & 1 & 1 & 1 & \vdots & 1 \\
2 & 2 & -1 & 1 & \vdots & 0 \\
1 & 0 & 0 & 1 & \vdots & 0 \\
1 & 2 & -1 & 1 & \vdots & 0
\end{bmatrix}
\begin{array}{l}
\rho_1 \rightarrow \rho_2 \\
\rho_2 \rightarrow \rho_4 \\
\rho_3 \rightarrow \rho_1 \\
\rho_4 \rightarrow \rho_3
\end{array}
\begin{bmatrix}
1 & 0 & 0 & 1 & \vdots & 0 \\
1 & 1 & 1 & 1 & \vdots & 1 \\
1 & 2 & -1 & 1 & \vdots & 0 \\
2 & 2 & -1 & 1 & \vdots & 0
\end{bmatrix}
$$

$$
\begin{array}{l}
\rho_2 - \rho_1 \rightarrow \rho_2 \\
\rho_3 - \rho_1 \rightarrow \rho_3 \\
\rho_4 - 2\rho_1 \rightarrow \rho_4
\end{array}
\begin{bmatrix}
1 & 0 & 0 & 1 & \vdots & 0 \\
0 & 1 & 1 & 0 & \vdots & 1 \\
0 & 2 & -1 & 0 & \vdots & 0 \\
0 & 2 & -1 & -1 & \vdots & 0
\end{bmatrix}
$$

$$
\begin{array}{l}
\rho_4 - \rho_3 \rightarrow \rho_4 \\
\rho_3 - 2\rho_2 \rightarrow \rho_3
\end{array}
\begin{bmatrix}
1 & 0 & 0 & 1 & \vdots & 0 \\
0 & 1 & 1 & 0 & \vdots & 1 \\
0 & 0 & -3 & 0 & \vdots & -2 \\
0 & 0 & 0 & -1 & \vdots & 0
\end{bmatrix}
$$

From the third and fourth equations we find that

$$
x_3 = \frac{-2}{-3} = \frac{2}{3}, \quad x_4 = \frac{0}{-1} = 0.
$$

Substituting these values in the first two equations, we obtain

$$
x_1 + (0) = 0 \Rightarrow x_1 = 0, \quad x_2 + \left(\frac{2}{3}\right) = 1 \Rightarrow x_2 = \frac{1}{3}.
$$

Therefore, the solution is

$$x_1 = 0, \ x_2 = \frac{1}{3}, \ x_3 = \frac{2}{3}, \ x_4 = 0.$$

3. We follow the Gauss elimination procedure with back substitution.

$$\begin{bmatrix} 1 & 1 & 1 & 1 & \vdots & 1 \\ 2 & 2 & -1 & 1 & \vdots & 0 \\ -1 & -1 & 0 & 1 & \vdots & 0 \\ 0 & 0 & 3 & -4 & \vdots & 11 \end{bmatrix}$$

$$\begin{matrix} \rho_2 - 2\rho_1 \to \rho_2 \\ \rho_3 + \rho_1 \to \rho_3 \end{matrix} \quad \begin{bmatrix} 1 & 1 & 1 & 1 & \vdots & 1 \\ 0 & 0 & -3 & -1 & \vdots & -2 \\ 0 & 0 & 1 & 2 & \vdots & 1 \\ 0 & 0 & 3 & -4 & \vdots & 11 \end{bmatrix}$$

$$\begin{matrix} \rho_4 + \rho_2 \to \rho_4 \\ -\rho_2 \to \rho_2 \end{matrix} \quad \begin{bmatrix} 1 & 1 & 1 & 1 & \vdots & 1 \\ 0 & 0 & 3 & 1 & \vdots & 2 \\ 0 & 0 & 1 & 2 & \vdots & 1 \\ 0 & 0 & 0 & -5 & \vdots & 9 \end{bmatrix}$$

$$\rho_2 - 3\rho_3 \to \rho_2 \quad \begin{bmatrix} 1 & 1 & 1 & 1 & \vdots & 1 \\ 0 & 0 & 0 & -5 & \vdots & -1 \\ 0 & 0 & 1 & 2 & \vdots & 1 \\ 0 & 0 & 0 & -5 & \vdots & 9 \end{bmatrix}$$

$$\rho_4 - \rho_2 \to \rho_4 \quad \begin{bmatrix} 1 & 1 & 1 & 1 & \vdots & 1 \\ 0 & 0 & 0 & -5 & \vdots & -1 \\ 0 & 0 & 1 & 2 & \vdots & 1 \\ 0 & 0 & 0 & 0 & \vdots & 10 \end{bmatrix}$$

The last equation is *inconsistent*, so is the original system.

5. Performing elementary row operations yields

$$
\begin{bmatrix}
1 & 3 & 1 & \vdots & 2 \\
3 & 4 & -1 & \vdots & 1 \\
1 & -2 & -3 & \vdots & 1
\end{bmatrix}
\begin{matrix}
\rho_2 - 3\rho_1 \to \rho_2 \\
\rho_3 - \rho_1 \to \rho_3
\end{matrix}
\begin{bmatrix}
1 & 3 & 1 & \vdots & 2 \\
0 & -5 & -4 & \vdots & -5 \\
0 & -5 & -4 & \vdots & -1
\end{bmatrix}
$$

$$
\rho_3 - \rho_2 \to \rho_3 \quad
\begin{bmatrix}
1 & 3 & 1 & \vdots & 2 \\
0 & -5 & -4 & \vdots & -5 \\
0 & 0 & 0 & \vdots & 4
\end{bmatrix}
$$

The system is *inconsistent* because of inconsistency of the third equation.

7. This augmented matrix defines three systems of linear equations in four unknowns with the same coefficient matrix **A** and right-hand side vectors b_1, b_2, and b_3, where

$$
A = \begin{bmatrix}
1 & -1 & -2 & 3 \\
1 & 2 & 1 & -2 \\
2 & 1 & -1 & 1
\end{bmatrix}, \quad b_1 = \begin{bmatrix} 1 \\ 1 \\ 2 \end{bmatrix},
$$

$$
b_2 = \begin{bmatrix} 0 \\ 0 \\ 0 \end{bmatrix}, \quad b_3 = \begin{bmatrix} 1 \\ -1 \\ 1 \end{bmatrix}.
$$

We perform the Gauss elimination procedure with back substitution.

$$
\begin{bmatrix}
1 & -1 & -2 & 3 & \vdots & 1 & 0 & 1 \\
1 & 2 & 1 & -2 & \vdots & 1 & 0 & -1 \\
2 & 1 & -1 & 1 & \vdots & 2 & 0 & 1
\end{bmatrix}
$$

$$
\begin{matrix}
\rho_2 - \rho_1 \to \rho_2 \\
\rho_3 - 2\rho_1 \to \rho_3
\end{matrix}
\begin{bmatrix}
1 & -1 & -2 & 3 & \vdots & 1 & 0 & 1 \\
0 & 3 & 3 & -5 & \vdots & 0 & 0 & -2 \\
0 & 3 & 3 & -5 & \vdots & 0 & 0 & -1
\end{bmatrix}
$$

$$\rho_3 - \rho_2 \rightarrow \rho_3 \quad \begin{bmatrix} 1 & -1 & -2 & 3 & \vdots & 1 & 0 & 1 \\ 0 & 3 & 3 & -5 & \vdots & 0 & 0 & -2 \\ 0 & 0 & 0 & 0 & \vdots & 0 & 0 & 1 \end{bmatrix}$$

In the first system, the augmented matrix is

$$\begin{bmatrix} 1 & -1 & -2 & 3 & \vdots & 1 \\ 0 & 3 & 3 & -5 & \vdots & 0 \\ 0 & 0 & 0 & 0 & \vdots & 0 \end{bmatrix}$$

The last equation is redundant. Taking $x_3 = s$ and $x_4 = t$ as free variables, we obtain from the second equation

$$x_2 = \frac{5x_4 - 3x_3}{3} = \frac{5t - 3s}{3}.$$

Back substitution of x_2, x_3, and x_4 into the first equation yields

$$x_1 = 1 + x_2 + 2x_3 - 3x_4 = 1 + \frac{5t - 3s}{3} + 2s - 3t = \frac{3 - 4t + 3s}{3}.$$

In the second system, we have

$$\begin{bmatrix} 1 & -1 & -2 & 3 & \vdots & 0 \\ 0 & 3 & 3 & -5 & \vdots & 0 \\ 0 & 0 & 0 & 0 & \vdots & 0 \end{bmatrix}$$

Eliminating again the redundant third equation and letting $x_3 = s$ and $x_4 = t$ be free variables, we conclude that

$$x_2 = \frac{5x_4 - 3x_3}{3} = \frac{5t - 3s}{3};$$

$$x_1 = x_2 + 2x_3 - 3x_4 = \frac{5t - 3s}{3} + 2s - 3t = \frac{3s - 4t}{3}.$$

The third system,

$$\begin{bmatrix} 1 & -1 & -2 & 3 & \vdots & 1 \\ 0 & 3 & 3 & -5 & \vdots & -2 \\ 0 & 0 & 0 & 0 & \vdots & 1 \end{bmatrix}$$

is inconsistent because of the last equation that says

$$(0)x_1 + (0)x_2 + (0)x_3 + (0)x_4 = 1,$$

which is wrong for any values of the variables involved.

9. We perform elementary row operations to reduce the given augmented coefficient matrix to upper trapezoidal form.

$$\begin{bmatrix} 1 & -1 & 2 & 0 & 0 & \vdots & 1 \\ 2 & -2 & 4 & 1 & 0 & \vdots & 5 \\ 3 & -3 & 6 & -1 & 1 & \vdots & -2 \end{bmatrix} \quad \begin{array}{l} \rho_2 - 2\rho_1 \to \rho_2 \\ \rho_3 - 3\rho_1 \to \rho_3 \end{array}$$

$$\begin{bmatrix} 1 & -1 & 2 & 0 & 0 & \vdots & 1 \\ 0 & 0 & 0 & 1 & 0 & \vdots & 3 \\ 0 & 0 & 0 & -1 & 1 & \vdots & -5 \end{bmatrix}$$

From the last two equations, we obtain

$$x_4 = 3, \quad -(3) + x_5 = -5 \quad \Rightarrow \quad x_5 = -2.$$

The first equation is a linear equation with three unknowns, x_1, x_2, and x_3. We can choose any two of them, whose values are *arbitrary*, and then solve the equation for the other variable. For example,

$$x_2 = s, \ x_3 = t, \quad \Rightarrow \quad x_1 - s + 2(t) = 1 \quad \Rightarrow \quad x_1 = 1 + s - 2t.$$

Thus, a solution to the given system can be written in the form

$$x_1 = 1 + s - 2t, \ x_2 = s, \ x_3 = t, \ x_4 = 3, \ x_5 = -2,$$

where s and t are *any* real numbers.

11. There are three systems of linear equations in this problem. All of them have the same *coefficient* matrix appeared in Example 3 but different right-hand sides vectors (columns). The first column is that given in Example 3 with the answer

$$x_1 = -\frac{2}{3} + 2s, \; x_2 = s, \; x_3 = \frac{1}{3} - 2s, \; x_4 = -\frac{1}{3}, \; x_5 = \frac{1}{3}.$$

To solve the other two problems, all that we need is to follow the elementary row operations in Example 3 with the matrix representing the right-hand side columns of the remaining equations. Thus we get

$$
\begin{bmatrix}
1 & 0 \\
1 & 0 \\
1 & 0 \\
0 & 0 \\
0 & 0
\end{bmatrix}
\quad \rho_1 \leftrightarrow \rho_3 \quad
\begin{bmatrix}
1 & 0 \\
1 & 0 \\
1 & 0 \\
0 & 0 \\
0 & 0
\end{bmatrix}
$$

$$
(-2/3)\rho_1 + \rho_2 \rightarrow \rho_2 \quad
\begin{bmatrix}
1 & 0 \\
1/3 & 0 \\
1 & 0 \\
0 & 0 \\
0 & 0
\end{bmatrix}
$$

$$
\rho_2 \leftrightarrow \rho_3 \quad
\begin{bmatrix}
1 & 0 \\
1 & 0 \\
1/3 & 0 \\
0 & 0 \\
0 & 0
\end{bmatrix}
\quad
\begin{matrix}
\rho_2 + \rho_4 \rightarrow \rho_4 \\
-\rho_2 + \rho_5 \rightarrow \rho_5
\end{matrix}
\quad
\begin{bmatrix}
1 & 0 \\
1 & 0 \\
1/3 & 0 \\
1 & 0 \\
-1 & 0
\end{bmatrix}
$$

$$
\begin{matrix}
(1/2)\rho_3 + \rho_4 \rightarrow \rho_4 \\
(-1/2)\rho_3 + \rho_5 \rightarrow \rho_5
\end{matrix}
\quad
\begin{bmatrix}
1 & 0 \\
1 & 0 \\
1/3 & 0 \\
7/6 & 0 \\
-7/6 & 0
\end{bmatrix}
$$

$$
-\rho_4 + \rho_5 \rightarrow \rho_5 \quad
\begin{bmatrix}
1 & 0 \\
1 & 0 \\
1/3 & 0 \\
7/6 & 0 \\
-7/3 & 0
\end{bmatrix}
$$

Thus, the new augmented matrix for the second and the third systems is

$$
\begin{bmatrix}
3 & 6 & 6 & 0 & 0 & \vdots & 1 & 0 \\
0 & 2 & 1 & 1 & 0 & \vdots & 1 & 0 \\
0 & 0 & 0 & 2 & 2 & \vdots & 1/3 & 0 \\
0 & 0 & 0 & 0 & 3 & \vdots & 7/6 & 0 \\
0 & 0 & 0 & 0 & 0 & \vdots & -7/3 & 0
\end{bmatrix}
$$

The first system here is inconsistent because the fifth equation is:

$$
(0)x_1 + (0)x_2 + (0)x_3 + (0)x_4 + (0)x_5 = 0 \neq -\frac{7}{3}.
$$

In the second system, the fifth equation is redundant. From the fourth equation we find that $x_5 = 0$. The back substitution into the third equation gives $x_4 = 0$. Thus, the first two equations give the system

$$
3x_1 + 6x_2 + 6x_3 = 0
$$
$$
2x_2 + x_3 = 0
$$

We can choose x_2 to be a free parameter: $x_2 = s$. Then $x_3 = -2x_2 = -2s$, and the back substitution into the first equation gives us

$$
x_1 = -2x_2 - 2x_3 = -2(s) - 2(-2s) = 2s.
$$

We can write the solution as a column vector in the form

$$
\begin{bmatrix} x_1 \\ x_2 \\ x_3 \\ x_4 \\ x_5 \end{bmatrix} = \begin{bmatrix} 2s \\ s \\ -2s \\ 0 \\ 0 \end{bmatrix},
$$

where s is any number.

13. Starting with the original matrix, we perform the following elementary row operations.

$$2\rho_1 + \rho_2 \rightarrow \rho_2 \qquad \begin{bmatrix} 1 & 1 & 1 & \vdots & 1 \\ 0 & 9 & 6 & \vdots & 2+\alpha \\ 0 & 3 & 2 & \vdots & 2 \end{bmatrix}$$

$$\rho_3 - (1/3)\rho_2 \rightarrow \rho_3 \qquad \begin{bmatrix} 1 & 1 & 1 & \vdots & 1 \\ 0 & 9 & 6 & \vdots & \alpha+2 \\ 0 & 0 & 0 & \vdots & (4-\alpha)/3 \end{bmatrix}$$

If $\alpha \neq 4$, then the last equation is inconsistent, and so the system is inconsistent.

Assume now that $\alpha = 4$. Then the last equation becomes redundant, and we come up with a system of two linear equations in three unknowns, whose augmented matrix is

$$\begin{bmatrix} 1 & 1 & 1 & \vdots & 1 \\ 0 & 9 & 6 & \vdots & 6 \end{bmatrix}$$

Letting $x_3 = s$ be a free parameter, we obtain

$$x_2 = \frac{6-6s}{9} = \frac{2-2s}{3}; \ x_1 = 1 - x_2 - x_3$$
$$= 1 - \frac{2-2s}{3} - s = \frac{1-s}{3}.$$

Thus, the system has infinitely many solutions if $\alpha = 4$.

15. Performing

$$\begin{bmatrix} 2 & 1 & 2 & \vdots & 1 \\ 2 & 2 & \alpha & \vdots & 1 \\ 4 & 2 & 4 & \vdots & 1 \end{bmatrix} \qquad \rho_3 - 2\rho_1 \rightarrow \rho_3 \qquad \begin{bmatrix} 2 & 1 & 2 & \vdots & 1 \\ 2 & 2 & \alpha & \vdots & 1 \\ 0 & 0 & 0 & \vdots & -1 \end{bmatrix}$$

The new third equation does not involve α, and it is inconsistent. Therefore, the system has no solution for any α.

17. Since, in Example 3, the values of $x_4 = -1/3$ and $x_5 = 1/3$ are uniquely determined, their back substitution into the last augmented matrix yields

$$\begin{bmatrix} 1 & 2 & 2 & \vdots & 0 \\ 0 & 2 & 1 & \vdots & 1/3 \end{bmatrix} \quad \rho_1 - \rho_2 \to \rho_1 \quad \begin{bmatrix} 1 & 0 & 1 & \vdots & -1/3 \\ 0 & 2 & 1 & \vdots & 1/3 \end{bmatrix}$$

If $x_1 = s$ is a free parameter, then the first equation yields

$$x_3 = -\frac{1}{3} - x_1 = -\frac{1}{3} - s = -\frac{1 + 3s}{3}.$$

From the second equation, we then obtain

$$2x_2 = \frac{1}{3} - x_3 = \frac{1}{3} + \frac{1 + 3s}{3} \quad \Rightarrow \quad x_2 = \frac{2 + 3s}{6}.$$

Therefore, in terms of the free parameter x_1, the answer in Example 3 is

$$x_1 = s, \ x_2 = \frac{2 + 3s}{6}, \ x_3 = -\frac{1 + 3s}{3}, \ x_4 = -\frac{1}{3}, \ x_5 = \frac{1}{3}.$$

19. (a) This is a system of three equations with four unknowns. The coefficient matrix is in upper trapezoidal form with non-zero entries on the main diagonal. Taking x_4 as a *free* parameter, we can find the value of x_3 from the last equation, and then use the back substitutions to find x_2 and x_1. Therefore, the system has infinitely many solutions.

(b) The last equation is redundant. Removing it from the augmented matrix, we obtain an upper triangular coefficient matrix whose pivot elements equal to 1. Therefore, the system has a unique solution.

(c) No solution because the last equation is inconsistent.

(d) The last equation is redundant. We can find x_5 from the fourth equation and, performing back substitution, x_4 from the third equation. Substituting these values into the first two equations, we will get a system of two equations with three unknowns, x_1, x_2, and x_3, whose augmented matrix is upper trapezoidal form with non-zero pivot elements in each row. Taking x_3 as a *free* parameter, we than can solve the system for x_2 and x_1 in terms of x_3. Therefore, the given system has infinitely many solutions.

1.4 ECHELON FORM AND RANK

1. This matrix is in a row echelon form (follows from the definition). The rank of this matrix is 1 because it has just one nonzero row.

3. This matrix is in a row echelon form. Since this matrix has *no* zero rows, its rank is 3.

5. This matrix is in a row echelon form. No zero row. Therefore, the rank equals the number of rows, i.e., 2.

7. This augmented matrix is in a row echelon form having

$$\text{rank} \, [\mathbf{A}|\mathbf{b}] = \text{rank} \, [\mathbf{A}] = 2.$$

Therefore, the system is consistent.

Since the number of unknowns (the number of columns) is $3 > 2$, we conclude that the system has infinitely many solutions.

9. This augmented matrix is in a row echelon with

$$\text{rank} \, [\mathbf{A}|\mathbf{b}] = \text{rank} \, [\mathbf{A}] = 1.$$

Thus, the system is consistent.

Since the number of unknowns (the number of columns) is $3 > 1$, we conclude that the system has infinitely many solutions.

11. This augmented matrix is in a row echelon;

$$\text{rank} \, [\mathbf{A}|\mathbf{b}] = \text{rank} \, [\mathbf{A}] = 4.$$

Thus, the system is consistent.

Since the number of unknowns (the number of columns), 4 , is the same as the rank, we conclude that the system has a unique solution.

13. (a) This homogeneous system has fewer equations (2) than unknowns (3). Thus, for any α, it has infinitely many solutions.

(b) The number of equations (3) in this homogeneous system equals the number of unknowns. Thus, we have to compute the rank of the coefficient matrix and find the value(s) of α, for which the rank is less than 3. In this and only in this case, the system will have nontrivial solutions.

$$\begin{bmatrix} 1 & 1 & 1 \\ \alpha & 4 & 3 \\ 4 & 3 & 2 \end{bmatrix} \quad \begin{matrix} \rho_2 - \alpha\rho_1 \rightarrow \rho_2 \\ \rho_3 - 4\rho_1 \rightarrow \rho_3 \end{matrix} \quad \begin{bmatrix} 1 & 1 & 1 \\ 0 & 4-\alpha & 3-\alpha \\ 0 & -1 & -2 \end{bmatrix}$$

$$\rho_2 \leftrightarrow \rho_3 \quad \begin{bmatrix} 1 & 1 & 1 \\ 0 & -1 & -2 \\ 0 & 4-\alpha & 3-\alpha \end{bmatrix}$$

$$\rho_3 + (4-\alpha)\rho_2 \rightarrow \rho_3 \quad \begin{bmatrix} 1 & 1 & 1 \\ 0 & -1 & -2 \\ 0 & 0 & \alpha-5 \end{bmatrix}$$

The rank of this matrix is less than 3 (namely, equals to 2) if and only if

$$\alpha - 5 = 0 \quad \Leftrightarrow \quad \alpha = 5.$$

15. We have to verify that

$$\# \text{ columns } \mathbf{A}^{\text{ech}} = \text{rank} \left[\mathbf{A}^{\text{ech}} | \mathbf{b}\right] + \# \text{ free parameters.}$$

(a) The given coefficient matrix is in a row echelon form, whose rank, 3. equals the number of unknowns (number of columns in it). Thus, this system has a unique solution, and so the number of free parameters is 0. The equation to be verified becomes

$$3 \overset{?}{=} 3 + 0$$

with the obvious answer.

(b) The coefficient matrix is in a row echelon form. It has 6 columns (unknowns) and rank 3. x_6 is uniquely defined from the third equation: $x_6 = 2$. The back substitution into the first two equations results two equations with 5 unknowns:

$$\left[\begin{array}{ccccc:c} 3 & 5 & 0 & 0 & 0 & 0 \\ 0 & 0 & 1 & 3 & 2 & -10 \end{array}\right]$$

In the second equation, we have 2 free parameters; for example, $x_4 = v$ and $x_5 = w$ give $x_3 = -10 - 3v - 2w$. These values, when substituted into the first equation, say that

$$3x_1 + 5x_2 = 0.$$

One of the variables, say x_2, can be taken as a third free parameter, $3u$. Then, the solution to the given system is

$$x_1 = -5u, \ x_2 = 3u, \ x_3 = -10 - 3v - 2w,$$
$$x_4 = v, \ x_5 = w, \ x_6 = 2.$$

The total number of free parameters, u, v, and w, is three. The question then is about whether or not the equality

$$6 \overset{?}{=} 3 + 3$$

holds, with the obvious positive answer.

(c) The coefficient matrix is in a row echelon form. It has 6 columns (unknowns) and rank 3. x_6 is uniquely defined from the third equation: $x_6 = 2$. The back substitution into the first two equations results two equations with 5 unknowns:

$$\left[\begin{array}{ccccc:c} 3 & 0 & 1 & 0 & 0 & 2 \\ 0 & 0 & 1 & 0 & 0 & -10 \end{array}\right]$$

From the second equation, we have $x_3 = -10$, and the back substitution into the first equation gives $x_1 = 4$. Therefore, we get

$$x_1 = 4, \ x_3 = -10, \ x_6 = 2,$$

and three other variables (x_2, x_4, and x_5) can take any values – they are free parameters. Since $6 = 3+3$, the equality in question holds.

17. Let $\mathbf{x} = [x_j]$, $\mathbf{y} = [y_j]$, and $\mathbf{A} = [a_{ij}]$. Since $\mathbf{x}$ and $\mathbf{y}$ are solutions to the homogeneous system $[\mathbf{A}|\mathbf{0}]$, for any $1 \leq i \leq n$ we have

$$\sum_{j=1}^{n} a_{ij}x_j = 0, \qquad \sum_{j=1}^{n} a_{ij}y_j = 0.$$

Summing these two equations, we get

$$\sum_{j=1}^{n} a_{ij}x_j \pm \sum_{j=1}^{n} a_{ij}y_j = \sum_{j=1}^{n} (a_{ij}x_j \pm a_{ij}y_j) = \sum_{j=1}^{n} a_{ij} (x_j \pm y_j) = 0.$$

These equations tell us that the n-by-1 column vectors $[x_j \pm y_j]$ are solutions to $[\mathbf{A}|\mathbf{0}]$. According to the addition/subtraction rules for vectors, these vectors are $\mathbf{x} \pm \mathbf{y}$.

19. Let $\mathbf{x} = [x_j]$ and $\mathbf{A} = [a_{ij}]$. Since $\mathbf{x}$ is a solution to the homogeneous system $[\mathbf{A}|\mathbf{0}]$, for any $1 \leq i \leq n$ we have

$$\sum_{j=1}^{n} a_{ij}x_j = 0.$$

Multiplying this equation by c yields

$$c\sum_{j=1}^{n} a_{ij}x_j = \sum_{j=1}^{n} ca_{ij}x_j = \sum_{j=1}^{n} a_{ij} (cx_j) = 0.$$

This equations tell us that the n-by-1 column vector $[cx_j]$ is a solution to $[\mathbf{A}|\mathbf{0}]$. According to the scalar multiplication rule for vectors, this vector equals to $c\mathbf{x}$.

21. Let $\mathbf{x} = [x_j]$, $\mathbf{y} = [y_j]$, and $\mathbf{A} = [a_{ij}]$, $\mathbf{b} = [b_i]$. Since $\mathbf{x}$ and $\mathbf{y}$ are solutions to the system $[\mathbf{A}|\mathbf{b}]$, for any $1 \leq i \leq n$ we have

$$\sum_{j=1}^{n} a_{ij}x_j = b_i, \qquad \sum_{j=1}^{n} a_{ij}y_j = b_i.$$

Subtracting these two equations, we get

$$\sum_{j=1}^{n} a_{ij}x_j - \sum_{j=1}^{n} a_{ij}y_j = b_i - b_i = 0.$$

On the other hand,

$$\sum_{j=1}^{n} a_{ij}x_j - \sum_{j=1}^{n} a_{ij}y_j = \sum_{j=1}^{n} (a_{ij}x_j - a_{ij}y_j) = \sum_{j=1}^{n} a_{ij}(x_j - y_j).$$

These equations tell us that the n-by-1 column vector $[x_j - y_j]$ is a solution to the homogeneous system $[A|0]$. According to the subtraction rule for vectors, this vector is $x - y$.

23. (a) We will verify the first and the third equation. Others are similar.

For the first equation, we use **KCL** saying that, in a circuit, the *algebraic* sum of all currents at any junction node is 0. At the left node, we see one *in*coming current, I_2, and two *out*going currents, I_1 and I_3. Thus, we have the equation

$$I_2 - I_1 - I_3 = 0,$$

which is equivalent to the give equation.

For the third equation, we use **KVL** saying that the combined voltage around *any* loop in a circuit is 0. In the loop 1, we have the following voltage inputs: the current I_3 passing through the resistant $R = 5(\Omega)$ and the voltage source $E = 5(V)$. Taking into account the orientation of the loop, we get

$$-(5)I_3 + 5 = 0 \quad \Leftrightarrow \quad 5I_3 = 5.$$

(b) In the matrix form, the system is

$$
\left[
\begin{array}{cccccc:c}
1 & -1 & 1 & 0 & 0 & 0 & 0 \\
0 & 1 & 0 & -1 & -1 & 1 & 0 \\
0 & 0 & 5 & 0 & 0 & 0 & 5 \\
0 & 0 & 0 & 3 & 0 & 0 & 1 \\
0 & 0 & 0 & 0 & 0 & 2 & 6 \\
0 & 0 & 5 & 0 & 0 & -2 & 0
\end{array}
\right]
$$

We perform elementary row operations to reduce this matrix to a row echelon form.

$$\begin{bmatrix} 1 & -1 & 1 & 0 & 0 & 0 & \vdots & 0 \\ 0 & 1 & 0 & -1 & -1 & 1 & \vdots & 0 \\ 0 & 0 & 5 & 0 & 0 & 0 & \vdots & 5 \\ 0 & 0 & 0 & 3 & 0 & 0 & \vdots & 1 \\ 0 & 0 & 0 & 0 & 0 & 2 & \vdots & 6 \\ 0 & 0 & 5 & 0 & 0 & -2 & \vdots & 0 \end{bmatrix}$$

$$\rho_6 - \rho_3 \rightarrow \rho_6 \qquad \begin{bmatrix} 1 & -1 & 1 & 0 & 0 & 0 & \vdots & 0 \\ 0 & 1 & 0 & -1 & -1 & 1 & \vdots & 0 \\ 0 & 0 & 5 & 0 & 0 & 0 & \vdots & 5 \\ 0 & 0 & 0 & 3 & 0 & 0 & \vdots & 1 \\ 0 & 0 & 0 & 0 & 0 & 2 & \vdots & 6 \\ 0 & 0 & 0 & 0 & 0 & -2 & \vdots & -5 \end{bmatrix}$$

$$\rho_6 + \rho_5 \rightarrow \rho_6 \qquad \begin{bmatrix} 1 & -1 & 1 & 0 & 0 & 0 & \vdots & 0 \\ 0 & 1 & 0 & -1 & -1 & 1 & \vdots & 0 \\ 0 & 0 & 5 & 0 & 0 & 0 & \vdots & 5 \\ 0 & 0 & 0 & 3 & 0 & 0 & \vdots & 1 \\ 0 & 0 & 0 & 0 & 0 & 2 & \vdots & 6 \\ 0 & 0 & 0 & 0 & 0 & 0 & \vdots & 1 \end{bmatrix}$$

The last equation is inconsistent and, therefore, the system is inconsistent.

(c) Changing the 5V battery to 6V battery affects only the loop 1 (the third equation). Thus we have

$$\begin{bmatrix} 1 & -1 & 1 & 0 & 0 & 0 & \vdots & 0 \\ 0 & 1 & 0 & -1 & -1 & 1 & \vdots & 0 \\ 0 & 0 & 5 & 0 & 0 & 0 & \vdots & 6 \\ 0 & 0 & 0 & 3 & 0 & 0 & \vdots & 1 \\ 0 & 0 & 0 & 0 & 0 & 2 & \vdots & 6 \\ 0 & 0 & 5 & 0 & 0 & -2 & \vdots & 0 \end{bmatrix}$$

$$\rho_6 - \rho_3 + \rho_5 \rightarrow \rho_6 \qquad \begin{bmatrix} 1 & -1 & 1 & 0 & 0 & 0 & \vdots & 0 \\ 0 & 1 & 0 & -1 & -1 & 1 & \vdots & 0 \\ 0 & 0 & 5 & 0 & 0 & 0 & \vdots & 6 \\ 0 & 0 & 0 & 3 & 0 & 0 & \vdots & 1 \\ 0 & 0 & 0 & 0 & 0 & 2 & \vdots & 6 \\ 0 & 0 & 0 & 0 & 0 & 0 & \vdots & 0 \end{bmatrix}$$

The last equation is redundant. The rank of the coefficient matrix is 5, the number of unknowns is 6. Therefore, there are infinitely many solutions with $6 - 5 = 1$ free parameter.

From the equations three, four, and five we find, resp.,

$$I_3 = \frac{6}{5}, \quad I_4 = \frac{1}{3}, \quad I_6 = 3.$$

Substituting these values into the first two equations, we obtain

$$\begin{aligned} I_1 - I_2 &= -6/5 \\ I_2 - I_5 &= -8/3 \end{aligned} \quad (\rho_1 + \rho_2 \rightarrow \rho_1) \quad \begin{aligned} I_1 - I_5 &= -58/15 \\ I_2 - I_5 &= -8/3 \end{aligned}$$

Taking $I_5 = t$ as a free parameter, we get the solution

$$I_1 = -\frac{58}{15} + t, \ I_2 = -\frac{8}{3} + t, \ I_3 = \frac{6}{5}, \ I_4 = \frac{1}{3}, \ I_5 = t, \ I_6 = 3.$$

(d) The parametrized solution shows that increasing I_5 by t units increases I_1 and I_2 by the same amount, but does not affect the currents I_3, I_4, and I_6.

25. (a) For the jth spring, according to the Hooke's Law, we have $F_j = k_j x_j$. Since the system is in equilibrium, we have

$$F_1 + F_2 + F_3 = 650 \quad \Rightarrow \quad 5000x_1 + 10000x_2 + 20000x_3 = 650$$
$$\Rightarrow \quad x_1 + 2x_2 + 4x_3 = 0.13.$$

The torque equilibrium gives another equation:

$$F_1 = F_3 \quad \Rightarrow \quad 5000x_1 = 20000x_3 \quad \Rightarrow \quad x_1 - 4x_3 = 0.$$

Finally, from the geometry considerations, we get

$$x_2 = \frac{x_1 + x_3}{2} \quad \Rightarrow \quad x_1 - 2x_2 + x_3 = 0.$$

We arrange these three equations in the order convenient for Gauss elimination.

$$
\begin{array}{rrrcr}
x_1 & & -4x_3 & = & 0 \\
x_1 & -2x_2 & +x_3 & = & 0 \\
x_1 & +2x_2 & +4x_3 & = & 0.13
\end{array}
$$

with the augmented matrix

$$
\left[
\begin{array}{rrr:r}
1 & 0 & -4 & 0 \\
1 & -2 & 1 & 0 \\
1 & 2 & 4 & 0.13
\end{array}
\right]
$$

$$
\begin{array}{c}
p_2 - p_1 \to p_2 \\
p_3 - p_1 \to p_3
\end{array}
\left[
\begin{array}{rrr:r}
1 & 0 & -4 & 0 \\
0 & -2 & 5 & 0 \\
0 & 2 & 8 & 0.13
\end{array}
\right]
$$

$$
p_3 + p_2 \to p_3
\left[
\begin{array}{rrr:r}
1 & 0 & -4 & 0 \\
0 & -2 & 5 & 0 \\
0 & 0 & 13 & 0.13
\end{array}
\right]
$$

Therefore, the solution is

$$x_3 = \frac{0.13}{13} = 0.01, \quad x_2 = \frac{5x_3}{2} = 0.025, \quad x_1 = 4x_3 = 0.04.$$

(b) The equilibrium conditions for forces say that

$$\begin{matrix} F_1 & & -F_3 & = & 0 \\ F_1 & +F_2 & +F_3 & = & 650 \end{matrix} \quad \Leftrightarrow \quad \begin{bmatrix} 1 & 0 & -1 & \vdots & 0 \\ 1 & 1 & 1 & \vdots & 650 \end{bmatrix}$$

$$\rho_2 - \rho_1 \rightarrow \rho_2 \qquad \begin{bmatrix} 1 & 0 & -1 & \vdots & 0 \\ 0 & 1 & 2 & \vdots & 650 \end{bmatrix}$$

The coefficient matrix has rank 2 but the number of columns (unknowns) is 3. Thus, the system has infinitely many solutions.

1.5   COMPUTATIONAL CONSIDERATIONS

1. This system has the augmented matrix

$$\begin{bmatrix} 0.003 & 59.14 & \vdots & 59.17 \\ 5.291 & -6.130 & \vdots & 46.78 \end{bmatrix}$$

First, let us try to perform the Gauss elimination procedure with this matrix, i.e., without pivoting.

$$\begin{bmatrix} 0.003 & 59.14 & \vdots & 59.17 \\ 5.291 & -6.130 & \vdots & 46.78 \end{bmatrix}$$

$$\frac{\rho_1}{0.003} \rightarrow \rho_1 \qquad \begin{bmatrix} 1 & 1.971 \times 10^4 & \vdots & 1.972 \times 10^4 \\ 5.291 & -6.130 & \vdots & 46.78 \end{bmatrix}$$

$$\rho_2 - 5.291\rho_1 \to \rho_2 \quad \begin{bmatrix} 1 & 1.971 \times 10^4 & \vdots & 1.972 \times 10^4 \\ 0 & -1.043 \times 10^5 & \vdots & -1.039 \times 10^4 \end{bmatrix}$$

$$\rho_2/(-1.043 \times 10^5) \to \rho_2 \quad \begin{bmatrix} 1 & 1.971 \times 10^4 & \vdots & 1.972 \times 10^4 \\ 0 & 1 & \vdots & 9.962 \times 10^{-2} \end{bmatrix}$$

Therefore, the solution obtained is

$$x_2 = 9.962 \times 10^{-2}, \quad x_1 = 1.776 \times 10^4.$$

With pivoting, the augmented matrix is

$$\begin{bmatrix} 5.291 & -6.130 & \vdots & 46.78 \\ 0.003 & 59.14 & \vdots & 59.17 \end{bmatrix}$$

Performing the Gauss elimination yields

$$\begin{bmatrix} 5.291 & -6.130 & \vdots & 46.78 \\ 0.003 & 59.14 & \vdots & 59.17 \end{bmatrix}$$

$$\frac{\rho_1}{5.291} \to \rho_1 \quad \begin{bmatrix} 1 & -1.159 & \vdots & 8.841 \\ 0.003 & 59.14 & \vdots & 59.17 \end{bmatrix}$$

$$\rho_2 - 0.003\rho_1 \to \rho_2 \quad \begin{bmatrix} 1 & -1.159 & \vdots & 8.841 \\ 0 & 59.14 & \vdots & 59.14 \end{bmatrix}$$

Therefore,

$$x_2 = \frac{59.14}{59.14} = 1.000, \quad x_1 = 8.841 + 1.159(1.000) = 10.00.$$

The comparison of the answers obtained without and with pivoting indicate that the former is absolutely inaccurate (due to the machine specifications), and the latter gives the accurate answer.

3. The system satisfying the given conditions is

$$
\begin{aligned}
0.004x_1 \quad +2.8x_2 &= 252.8 \\
x_1 \quad +x_2 &= 290
\end{aligned}
$$

or, in the matrix form

$$
\begin{bmatrix}
0.004 & 2.8 & \vdots & 252.8 \\
1 & 1 & \vdots & 290
\end{bmatrix}
$$

We reduce this matrix to a row echelon form keeping in mind that the machine retains only 3 significant digits after each computation.

$$
\begin{bmatrix}
0.004 & 2.8 & \vdots & 252.8 \\
1 & 1 & \vdots & 290
\end{bmatrix}
$$

$$
\rho_2 - \rho_1/0.004 \rightarrow \rho_2 \qquad
\begin{bmatrix}
0.004 & 2.8 & \vdots & 252.8 \\
0 & -699 & \vdots & -62900
\end{bmatrix}
$$

Therefore,

$$
x_2 = \frac{-62900}{-699} = 89.9, \quad x_1 = \frac{252.8 - 2.8(89.9)}{0.004} = 270.
$$

With pivoting, we deal with the augmented matrix

$$
\begin{bmatrix}
1 & 1 & \vdots & 290 \\
0.004 & 2.8 & \vdots & 252.8
\end{bmatrix}
$$

$$
\rho_2 - (0.004)\rho_1 \rightarrow \rho_2 \qquad
\begin{bmatrix}
1 & 1 & \vdots & 290 \\
0 & 2.79 & \vdots & 251
\end{bmatrix}
$$

Therefore,

$$
x_2 = \frac{251}{2.79} = 89.9, \quad x_1 = 290 - 89.9 = 200.
$$

5. (a) We start with the augmented matrix corresponding to the first two systems given in this problem.

$$
\begin{bmatrix}
1 & 0 & 0 & 0 & 1 & \vdots & 1 & \vdots & 0 \\
-1 & 1 & 0 & 0 & 1 & \vdots & 1 & \vdots & 0 \\
-1 & -1 & 1 & 0 & 1 & \vdots & 1 & \vdots & 0 \\
-1 & -1 & -1 & 1 & 1 & \vdots & 1 & \vdots & 0 \\
-1 & -1 & -1 & -1 & 1 & \vdots & 1 & \vdots & s
\end{bmatrix}
$$

and perform the following elementary row operations following the Gauss elimination procedure.

$$
\begin{bmatrix}
1 & 0 & 0 & 0 & 1 & \vdots & 1 & \vdots & 0 \\
-1 & 1 & 0 & 0 & 1 & \vdots & 1 & \vdots & 0 \\
-1 & -1 & 1 & 0 & 1 & \vdots & 1 & \vdots & 0 \\
-1 & -1 & -1 & 1 & 1 & \vdots & 1 & \vdots & 0 \\
-1 & -1 & -1 & -1 & 1 & \vdots & 1 & \vdots & s
\end{bmatrix}
$$

$$
\begin{matrix}
\rho_2 + \rho_1 \rightarrow \rho_2 \\
\rho_3 + \rho_1 \rightarrow \rho_3 \\
\rho_4 + \rho_1 \rightarrow \rho_4 \\
\rho_5 + \rho_1 \rightarrow \rho_5
\end{matrix}
\quad
\begin{bmatrix}
1 & 0 & 0 & 0 & 1 & \vdots & 1 & \vdots & 0 \\
0 & 1 & 0 & 0 & 2 & \vdots & 2 & \vdots & 0 \\
0 & -1 & 1 & 0 & 2 & \vdots & 2 & \vdots & 0 \\
0 & -1 & -1 & 1 & 2 & \vdots & 2 & \vdots & 0 \\
0 & -1 & -1 & -1 & 2 & \vdots & 2 & \vdots & s
\end{bmatrix}
$$

$$
\begin{matrix}
\rho_3 + \rho_2 \rightarrow \rho_3 \\
\rho_4 + \rho_2 \rightarrow \rho_4 \\
\rho_5 + \rho_2 \rightarrow \rho_5
\end{matrix}
\quad
\begin{bmatrix}
1 & 0 & 0 & 0 & 1 & \vdots & 1 & \vdots & 0 \\
0 & 1 & 0 & 0 & 2 & \vdots & 2 & \vdots & 0 \\
0 & 0 & 1 & 0 & 4 & \vdots & 4 & \vdots & 0 \\
0 & 0 & -1 & 1 & 4 & \vdots & 4 & \vdots & 0 \\
0 & 0 & -1 & -1 & 4 & \vdots & 4 & \vdots & s
\end{bmatrix}
$$

$$
\begin{array}{ccccc}
\rho_4 + \rho_3 \rightarrow \rho_4 \\
\rho_5 + \rho_3 \rightarrow \rho_5
\end{array}
\left[
\begin{array}{ccccc|c|c}
1 & 0 & 0 & 0 & 1 & 1 & 0 \\
0 & 1 & 0 & 0 & 2 & 2 & 0 \\
0 & 0 & 1 & 0 & 4 & 4 & 0 \\
0 & 0 & 0 & 1 & 8 & 8 & 0 \\
0 & 0 & 0 & -1 & 8 & 8 & s
\end{array}
\right]
$$

$$
\rho_5 + \rho_4 \rightarrow \rho_5
\left[
\begin{array}{cccccc|c|c}
1 & 0 & 0 & 0 & 1 & 1 & 0 \\
0 & 1 & 0 & 0 & 2 & 2 & 0 \\
0 & 0 & 1 & 0 & 4 & 4 & 0 \\
0 & 0 & 0 & 1 & 8 & 8 & 0 \\
0 & 0 & 0 & 0 & 16 & 16 & s
\end{array}
\right]
$$

(b) The pattern is clear. Since the last column in the original augmented matrix is the sum of two preceding rows, the largest value in the third system in n-row case will occur in the position $(n, n + 3)$ and equals to $2^{n-1} + s$.

(c) From the row echelon form obtained in part (a) (for $n = 5$) we conclude that, for the first system,

$$
\begin{aligned}
x_5 &= 16/16 = 1, \\
x_4 &= 8 - 8(1) = 0, \\
x_3 &= 4 - 4(1) = 0, \\
x_2 &= 2 - 2(1) = 0, \\
x_1 &= 1 - (1) = 0.
\end{aligned}
$$

For the second system,

$$
\begin{aligned}
x_5 &= s/16, \\
x_4 &= -8(x_5) = -s/2, \\
x_3 &= -4(x_5) = -s/4, \\
x_2 &= -2(x_5) = -s/8, \\
x_1 &= -x_5 = -s/16.
\end{aligned}
$$

For the third system, the value of x_4 will be the sum of values of this variable in the first two systems; namely, $x_4 = 0 + (-s/2) = -s/2$.

2

MATRIX ALGEBRA

2.1 MATRIX MULTIPLICATION

1. (a)

$$\mathbf{A} + \mathbf{B} = \begin{bmatrix} (2) + (2) & (1) + (3) \\ (3) + (-1) & (5) + (0) \end{bmatrix} = \begin{bmatrix} 4 & 4 \\ 2 & 5 \end{bmatrix}.$$

(b)

$$3\mathbf{A} - \mathbf{B} = \begin{bmatrix} 3(2) - (2) & 3(1) - (3) \\ 3(3) - (-1) & 3(5) - (0) \end{bmatrix} = \begin{bmatrix} 4 & 0 \\ 10 & 15 \end{bmatrix}.$$

3. (a)

$$(\mathbf{AB})_{1,1} = a_{1,1}b_{1,1} + a_{1,2}b_{2,1} + a_{1,3}b_{3,1}$$
$$= (2)(-1) + (4)(5) + (0)(4) = 18;$$
$$(\mathbf{AB})_{1,2} = a_{1,1}b_{1,2} + a_{1,2}b_{2,2} + a_{1,3}b_{3,2}$$
$$= (2)(3) + (4)(2) + (0)(5) = 14;$$
$$(\mathbf{AB})_{1,3} = a_{1,1}b_{1,3} + a_{1,2}b_{2,3} + a_{1,3}b_{3,3}$$
$$= (2)(0) + (4)(1) + (0)(1) = 4.$$

Solutions Manual to Accompany Fundamentals of Matrix Analysis with Applications,
First Edition. Edward Barry Saff and Arthur David Snider.
© 2016 John Wiley & Sons, Inc. Published 2016 by John Wiley & Sons, Inc.

Similarly, we find other entries of **AB**.

$$\mathbf{AB} = \begin{bmatrix} 18 & 14 & 4 \\ 16 & 20 & 4 \\ 15 & 23 & 4 \end{bmatrix}.$$

(b)

$$(\mathbf{BA})_{1,1} = b_{1,1}a_{1,1} + b_{1,2}a_{2,1} + b_{1,3}a_{3,1}$$
$$= (-1)(2) + (3)(1) + (0)(2) = 1;$$
$$(\mathbf{BA})_{1,2} = b_{1,1}a_{1,2} + b_{1,2}a_{2,2} + b_{1,3}a_{3,2}$$
$$= (-1)(4) + (3)(1) + (0)(1) = -1;$$
$$(\mathbf{BA})_{1,3} = b_{1,1}a_{1,3} + b_{1,2}a_{2,3} + b_{1,3}a_{3,3}$$
$$= (-1)(0) + (3)(3) + (0)(3) = 9.$$

Similarly, we find other entries of **BA**.

$$\mathbf{BA} = \begin{bmatrix} 1 & -1 & 9 \\ 14 & 23 & 9 \\ 15 & 22 & 18 \end{bmatrix}.$$

(c)

$$\left(\mathbf{A}^2\right)_{1,1} = a_{1,1}a_{1,1} + a_{1,2}a_{2,1} + a_{1,3}a_{3,1}$$
$$= (2)(2) + (4)(1) + (0)(2) = 8;$$
$$\left(\mathbf{A}^2\right)_{1,2} = a_{1,1}a_{1,2} + a_{1,2}a_{2,2} + a_{1,3}a_{3,2}$$
$$= (2)(4) + (4)(1) + (0)(1) = 12;$$
$$\left(\mathbf{A}^2\right)_{1,3} = a_{1,1}a_{1,3} + a_{1,2}a_{2,3} + a_{1,3}a_{3,3}$$
$$= (2)(0) + (4)(3) + (0)(3) = 12.$$

Similarly, we find other entries of $\mathbf{A}^2$.

$$\mathbf{A}^2 = \begin{bmatrix} 8 & 12 & 12 \\ 9 & 8 & 12 \\ 11 & 12 & 12 \end{bmatrix}.$$

(d)

$$\left(\mathbf{B}^2\right)_{1,1} = b_{1,1}b_{1,1} + b_{1,2}b_{2,1} + b_{1,3}b_{3,1}$$
$$= (-1)(-1) + (3)(5) + (0)(4) = 16;$$
$$\left(\mathbf{B}^2\right)_{1,2} = b_{1,1}b_{1,2} + b_{1,2}b_{2,2} + b_{1,3}b_{3,2}$$
$$= (-1)(3) + (3)(2) + (0)(5) = 3;$$
$$\left(\mathbf{B}^2\right)_{1,3} = b_{1,1}b_{1,3} + b_{1,2}b_{2,3} + b_{1,3}b_{3,3}$$
$$= (-1)(0) + (3)(1) + (0)(1) = 3.$$

Similarly, we find other entries of $\mathbf{B}^2$.

$$\mathbf{B}^2 = \begin{bmatrix} 16 & 3 & 3 \\ 9 & 24 & 3 \\ 25 & 27 & 6 \end{bmatrix}.$$

5. (a) $(\mathbf{AB})_{1,1} = (2)(3) + (-4)(7) + (1)(2) = -20$, etc.
Computing all elements of this product, we get

$$\mathbf{AB} = \begin{bmatrix} -20 & 3 & -15 \\ 23 & 3 & -3 \\ 10 & -6 & -6 \end{bmatrix}.$$

(b) $(\mathbf{AC})_{1,1} = (2)(0) + (-4)(1) + (1)(-3) = -7$, etc.
Computing all elements of this product, we get

$$\mathbf{AC} = \begin{bmatrix} -7 & -11 & -15 \\ -7 & 1 & 25 \\ -8 & 8 & 20 \end{bmatrix}.$$

(c) Using the distributive law and parts (a), (b), we get

$$\mathbf{A}(\mathbf{B} + \mathbf{C}) = \mathbf{AB} + \mathbf{AC}$$
$$= \begin{bmatrix} -20 & 3 & -15 \\ 23 & 3 & -3 \\ 10 & -6 & -6 \end{bmatrix} + \begin{bmatrix} -7 & -11 & -15 \\ -7 & 1 & 25 \\ -8 & 8 & 20 \end{bmatrix}$$
$$= \begin{bmatrix} -27 & -8 & -30 \\ 16 & 4 & 22 \\ 2 & 2 & 14 \end{bmatrix}.$$

7. Computing

$$(\mathbf{AB})_{1,1} = (2)(1) + (-1)(3) = -1,$$
$$(\mathbf{BA})_{1,1} = (1)(2) + (2)(-3) = -4,$$

we conclude that $\mathbf{AB} \neq \mathbf{BA}$ because (at least) two of their corresponding entries are different.

9. Let us look for a matrix $\mathbf{A}$ whose entry $a_{1,1} = 1$. We want that

$$\mathbf{A}^2 = \begin{bmatrix} 1 & a_{1,2} \\ a_{2,1} & a_{22} \end{bmatrix} \begin{bmatrix} 1 & a_{1,2} \\ a_{2,1} & a_{22} \end{bmatrix}$$
$$= \begin{bmatrix} 1 + a_{1,2}a_{2,1} & a_{1,2}(1 + a_{2,2}) \\ a_{2,1}(1 + a_{2,2}) & a_{2,1}a_{1,2} + a_{2,2}^2 \end{bmatrix} = \mathbf{0}$$

This matrix equation can be written as

$$1 + a_{1,2}a_{2,1} = 0$$
$$a_{1,2}(1 + a_{2,2}) = 0$$
$$a_{2,1}(1 + a_{2,2}) = 0$$
$$a_{2,1}a_{1,2} + a_{2,2}^2 = 0.$$

Since $a_{1,2}$ and $a_{2,1}$ cannot be zero due to the first equation, from the second and third equations we conclude that

$$1 + a_{2,2} = 0 \qquad \Rightarrow \qquad a_{2,2} = -1.$$

Thus, the system reduces to just one equation,

$$1 + a_{1,2}a_{2,1} = 0 \qquad \Rightarrow \qquad a_{1,2}a_{2,1} = -1,$$

and has infinitely many solutions. Choosing, for example,

$$a_{1,2} = -1 \qquad \Rightarrow \qquad a_{2,1} = 1,$$

we get the matrix

$$\mathbf{A} = \begin{bmatrix} 1 & -1 \\ 1 & -1 \end{bmatrix}.$$

11. First, we note that the distributive law says

$$\mathbf{AC} = \mathbf{BC} \quad \Leftrightarrow \quad \mathbf{AC} - \mathbf{BC} = \mathbf{0} \quad \Leftrightarrow \quad (\mathbf{A} - \mathbf{B})\mathbf{C} = \mathbf{0}.$$

Thus, if **C** is a solution to Problem 9, as suggested in the *hint*, then any matrices **A** and **B** with $\mathbf{A} - \mathbf{B} = \mathbf{C}$ wil give

$$(\mathbf{A} - \mathbf{B})\mathbf{C} = \mathbf{C}^2 = \mathbf{0}.$$

For example, one can choose

$$\mathbf{A} = 2\mathbf{C}, \ \mathbf{B} = \mathbf{C} \quad \text{or} \quad \mathbf{A} = \mathbf{C} + \mathbf{I}, \ \mathbf{B} = \mathbf{I},$$

where **I** is the 2-by-2 multiplicative identity matrix.

13. (a) **True:** because

$$(\mathbf{AB})_{ij} = \sum_{k=1}^{n} a_{ik}b_{kj} > 0$$

since all terms in the sum are positive numbers.

(b) **False:** For example, in the 2-by-2 case, let $\mathbf{A} = \mathbf{B}$ be the solution to Problem 9 (the matrix **C** in Problem 11). Then

$$\mathbf{AB} = \mathbf{0}$$

while neither of matrices **A**, **B** is **0**. Another example,

$$\begin{bmatrix} 1 & 0 \\ 0 & 0 \end{bmatrix} \begin{bmatrix} 0 & 0 \\ 0 & 1 \end{bmatrix} = \mathbf{0}.$$

(c) **True:** because, for $i > j$,

$$(\mathbf{AB})_{ij} = \sum_{k=1}^{n} a_{ik}b_{kj} = \sum_{k=1}^{i-1} a_{ik}b_{kj} + \sum_{k=i}^{n} a_{ik}b_{kj}$$

$$= \sum_{k=1}^{i-1} (0)b_{kj} + \sum_{k=i}^{n} a_{ik}(0) = 0.$$

(d) **False:** For example,

$$\begin{bmatrix} 1 & 0 \\ 0 & 2 \end{bmatrix} \begin{bmatrix} 0 & 0 \\ 1 & 0 \end{bmatrix} = \begin{bmatrix} 0 & 0 \\ 2 & 0 \end{bmatrix},$$

$$\begin{bmatrix} 0 & 0 \\ 1 & 0 \end{bmatrix} \begin{bmatrix} 1 & 0 \\ 0 & 2 \end{bmatrix} = \begin{bmatrix} 0 & 0 \\ 1 & 0 \end{bmatrix}$$

although one of these two matrices is diagonal.

15. (a) We set

$$\mathbf{A} = \begin{bmatrix} 0 & 1 & 3 \\ -1 & 7 & -9 \\ 6 & -4 & 0 \end{bmatrix}, \quad \mathbf{x} = \begin{bmatrix} x_1 \\ x_2 \\ x_3 \end{bmatrix}, \quad \mathbf{b} = \begin{bmatrix} 0 \\ -9 \\ 2 \end{bmatrix}.$$

Then the given system is equivalent to $\mathbf{Ax} = \mathbf{b}$. Indeed,

$$(\mathbf{Ax})_{1,1} = (0, 1, 3) \cdot (x_1, x_2, x_3) = x_2 + 3x_3 = 0 = \mathbf{b}_{1,1};$$
$$(\mathbf{Ax})_{2,1} = (-1, 7, -9) \cdot (x_1, x_2, x_3)$$
$$= -x_1 + 7x_2 - 9x_3 = -9 = \mathbf{b}_{2,1};$$
$$(\mathbf{Ax})_{3,1} = (6, -4, 0) \cdot (x_1, x_2, x_3) = 6x_1 - 4x_2 = 2 = \mathbf{b}_{3,1}.$$

(b) First, we rewrite the system as

$$\begin{array}{rrrcr} -2x & +3y & & = & -7 \\ -x & +2y & +z & = & 6 \\ -x & -y & +3z & = & -3 \end{array}$$

and denote

$$\mathbf{A} = \begin{bmatrix} -2 & 3 & 0 \\ -1 & 2 & 1 \\ -1 & -1 & 3 \end{bmatrix}, \quad \mathbf{t} = \begin{bmatrix} x \\ y \\ z \end{bmatrix}, \quad \mathbf{b} = \begin{bmatrix} -7 \\ 6 \\ -3 \end{bmatrix}.$$

With these notations, the system can be written in matrix form as $\mathbf{At} = \mathbf{b}$. Indeed,

$$(\mathbf{At})_{1,1} = (-2, 3, 0) \cdot (x, y, z) = -2x + 3y = -7 = \mathbf{b}_{1,1};$$
$$(\mathbf{At})_{2,1} = (-1, 2, 1) \cdot (x, y, z) = -2 + 2y + z = 6 = \mathbf{b}_{2,1};$$
$$(\mathbf{At})_{3,1} = (-1, -1, 3) \cdot (x, y, z) = -x - y + 3z = -3 = \mathbf{b}_{3,1}.$$

17. (a) We want to perform $\rho_2 \leftrightarrow \rho_3$. This corresponds to the elementary row operation matrix

$$\mathbf{B} = \begin{bmatrix} 1 & 0 & 0 \\ 0 & 0 & 1 \\ 0 & 1 & 0 \end{bmatrix}.$$

(b) The elementary row operation that produces the required matrix in question is $\rho_2 - (1/2)\rho_1 \rightarrow \rho_2$ with the corresponding matrix

$$\mathbf{B} = \begin{bmatrix} 1 & 0 & 0 \\ -1/2 & 1 & 0 \\ 0 & 0 & 1 \end{bmatrix}.$$

(c) The elementary row operation that produces the required matrix is $2\rho_2 \rightarrow \rho_2$ with the corresponding matrix

$$\mathbf{B} = \begin{bmatrix} 1 & 0 & 0 \\ 0 & 2 & 0 \\ 0 & 0 & 1 \end{bmatrix}.$$

19. (a) To simplify and shorten further computations, we first multiply $\mathbf{A}$ (from the left) by an elementary matrix $\mathbf{E}_1$. This just re-order the rows of $\mathbf{A}$.

$$\mathbf{E}_1 \mathbf{A} = \begin{bmatrix} 0 & 1 & 0 \\ 0 & 0 & 1 \\ 1 & 0 & 0 \end{bmatrix} \mathbf{A} = \begin{bmatrix} 2 & 1 & -1 \\ 6 & -3 & 2 \\ 0 & 0 & -1 \end{bmatrix} = \mathbf{A}^{(1)}.$$

Next, we perform $\rho_2 - 3\rho_1 \rightarrow \rho_2$ by multiplying $\mathbf{A}^{(1)}$ (from the left) by $\mathbf{E}_2$.

$$\mathbf{E}_2 \mathbf{A}^{(1)} = \begin{bmatrix} 1 & 0 & 0 \\ -3 & 1 & 0 \\ 0 & 0 & 1 \end{bmatrix} \mathbf{A}^{(1)} = \begin{bmatrix} 2 & 1 & -1 \\ 0 & -6 & 5 \\ 0 & 0 & -1 \end{bmatrix} = \mathbf{A}^{(2)}.$$

The matrix $\mathbf{A}^{(2)}$ is in a row echelon form which is obtained from $\mathbf{A}$ by left multiplication by the matrix $\mathbf{E}_1$ and then by $\mathbf{E}_2$. Namely,

$$\mathbf{A}^{(2)} = \begin{bmatrix} 1 & 0 & 0 \\ -3 & 1 & 0 \\ 0 & 0 & 1 \end{bmatrix} \begin{bmatrix} 0 & 1 & 0 \\ 0 & 0 & 1 \\ 1 & 0 & 0 \end{bmatrix} \mathbf{A}.$$

(b) First we do the elementary row operations $\rho_2 - \rho_1 \to \rho_2$ and $\rho_3 - 3\rho_1 \to \rho_3$ that correspond the left multiplication of $\mathbf{B}$ by the matrix $\mathbf{E}_1$ shown below.

$$\mathbf{E}_1\mathbf{B} = \begin{bmatrix} 1 & 0 & 0 \\ -1 & 1 & 0 \\ -3 & 0 & 1 \end{bmatrix} \mathbf{B} = \begin{bmatrix} 2 & 0 & -3 & 5 \\ 0 & 1 & 2 & -1 \\ 0 & -3 & 11 & -13 \end{bmatrix} = \mathbf{B}^{(1)}.$$

The next transformation is $\rho_3 + 3\rho_2 \to \rho_3$ performed by the matrix $\mathbf{E}_2$.

$$\mathbf{E}_2\mathbf{B}^{(1)} = \begin{bmatrix} 1 & 0 & 0 \\ 0 & 1 & 0 \\ 0 & 3 & 1 \end{bmatrix} \mathbf{B}^{(1)} = \begin{bmatrix} 2 & 0 & -3 & 5 \\ 0 & 1 & 2 & -1 \\ 0 & 0 & 17 & -16 \end{bmatrix} = \mathbf{B}^{(2)}.$$

The matrix $\mathbf{B}^{(2)}$ is in a row echelon form which is obtained from $\mathbf{B}$ by left multiplication – first by $\mathbf{E}_1$ and then by $\mathbf{E}_2$. Namely,

$$\mathbf{B}^{(2)} = \begin{bmatrix} 1 & 0 & 0 \\ 0 & 1 & 0 \\ 0 & 3 & 1 \end{bmatrix} \begin{bmatrix} 1 & 0 & 0 \\ -1 & 1 & 0 \\ -3 & 0 & 1 \end{bmatrix} \mathbf{B}.$$

(c) First we do the elementary row operation $\rho_2 - \rho_1 \to \rho_2$ that corresponds the left multiplication of $\mathbf{C}$ by the matrix $\mathbf{E}_1$ shown below.

$$\mathbf{E}_1\mathbf{C} = \begin{bmatrix} 1 & 0 & 0 \\ -1 & 1 & 0 \\ 0 & 0 & 1 \end{bmatrix} \mathbf{C}$$

$$= \begin{bmatrix} 1 & 2 & 1 & 1 & -1 & 0 \\ 0 & 0 & -1 & -2 & 2 & 0 \\ 0 & 0 & 1 & 2 & -2 & 1 \end{bmatrix} = \mathbf{C}^{(1)}.$$

The next transformation is $\rho_3 + \rho_2 \rightarrow \rho_3$ performed by the matrix $\mathbf{E}_2$.

$$\mathbf{E}_2\mathbf{C}^{(1)} = \begin{bmatrix} 1 & 0 & 0 \\ 0 & 1 & 0 \\ 0 & 1 & 1 \end{bmatrix} \mathbf{C}^{(1)}$$

$$= \begin{bmatrix} 1 & 2 & 1 & 1 & -1 & 0 \\ 0 & 0 & -1 & -2 & 2 & 0 \\ 0 & 0 & 0 & 0 & 1 & 1 \end{bmatrix} = \mathbf{C}^{(2)}.$$

The matrix $\mathbf{C}^{(2)}$ is in a row echelon form which is obtained from $\mathbf{C}$ by left multiplication – first by $\mathbf{E}_1$ and then by $\mathbf{E}_2$. Namely,

$$\mathbf{C}^{(2)} = \begin{bmatrix} 1 & 0 & 0 \\ 0 & 1 & 0 \\ 0 & 1 & 1 \end{bmatrix} \begin{bmatrix} 1 & 0 & 0 \\ -1 & 1 & 0 \\ 0 & 0 & 1 \end{bmatrix} \mathbf{C}.$$

21. In solving this problem, we will use two fundamental probability laws: with $P(\star)$ meaning the probability of the event "$\star$",

(E) If $A_1, \ldots, A_k$ are *mutually exclusive* events, then

$$P\left(\bigcup_{s=1}^{k} A_s\right) = \sum_{s=1}^{k} P(A_s).$$

(I) If $B_1, \ldots, B_m$ are *mutually independent* events, then

$$P\left(\bigcap_{s=1}^{m} B_s\right) = \prod_{s=1}^{m} P(B_s).$$

(a) Using the above stated laws, with the obvious notations for probabilities (say, $P(\text{Tue}_g)$ means the probability that Tuesday is good and P_{bg} means the probability that the next day is good if the previous day was bad) we get

$$P(\text{Tue}_g) = P(\text{Mon}_g) P_{gg} + P(\text{Mon}_b) P_{bg}$$
$$= 0.75 \cdot 0.75 + 0.25 \cdot 0.60 = 0.7125,$$
$$P(\text{Tue}_b) = P(\text{Mon}_g) P_{gb} + P(\text{Mon}_b) P_{bb}$$
$$= 0.75 \cdot 0.25 + 0.25 \cdot 0.40 = 0.2875.$$

(Another way to compute $P\left(\text{Tue}_b\right)$ is $P\left(\text{Tue}_b\right) = 1 - P\left(\text{Tue}_g\right)$.)

(b) Following the result of part (a), we can conclude that

$$\mathbf{x}_{n+1} = \begin{bmatrix} P_{gg} & P_{bg} \\ P_{gb} & P_{bb} \end{bmatrix} \mathbf{x}_n = \begin{bmatrix} 0.75 & 0.6 \\ 0.25 & 0.4 \end{bmatrix} \mathbf{x}_n.$$

(c) Since an mth day is either good or bad, we have $g_m + b_m = 1$, for any m. Regardless of the rating of the previous day, good or bad (in other words, you consider the first or the second column of the matrix $\mathbf{A}$), one of the two possibilities for the next day must happen: it is either good or bad. Since this two events are complementary, their probabilities sum to 1.

(d) All the answers obtained in part (c) remain the same: all the arguments given there can be applied with the obvious change of "two" to "four".

23. (a) The product $\mathbf{A}(\mathbf{B}\mathbf{v})$ reqires n dot products for computing the column vector $\mathbf{B}\mathbf{v}$ and the additional n dot products for multiplying $\mathbf{A}$ by it. Combined, $n+n = 2n$ dot products are required. In the product $(\mathbf{A}\mathbf{B})\mathbf{v}$. The matrix product $\mathbf{A}\mathbf{B}$ assumes n^2 dot products, and we have to add n more dot products to compute the final result. Combined, we come up with $n^2 + n$ dot products. obviously, the first way is more efficient since $2n < n^2 + n$ for any $n \geq 2$.

(b) For the first case, we need n dot products while for the second case their number is $n + n = 2n$. Therefore, the first formula is more efficient.

2.2   SOME USEFUL APPLICATIONS OF MATRIX OPERATORS

1. (a) We apply the solution in Example 1 with $\theta = 30°$ to get

$$\mathbf{A} = \begin{bmatrix} \cos 30° & -\sin 30° \\ \sin 30° & \cos 30° \end{bmatrix} = \begin{bmatrix} \sqrt{3}/2 & -1/2 \\ 1/2 & \sqrt{3}/2 \end{bmatrix}.$$

(b) A normal vector to the line $2x + y = 0$ is $\begin{bmatrix} 2 \\ 1 \end{bmatrix}$. Dividing this vector

by its length, $\sqrt{5}$, we obtain the corresponding unit normal vector

$$\mathbf{n} = \begin{bmatrix} n_1 \\ n_2 \end{bmatrix} = \begin{bmatrix} 2/\sqrt{5} \\ 1/\sqrt{5} \end{bmatrix}$$

and apply the formula obtained in Example 3 to get

$$\mathbf{A} = \begin{bmatrix} 1 - 2n_1^2 & -2n_1 n_2 \\ -2n_1 n_2 & 1 - 2n_2^2 \end{bmatrix} = \begin{bmatrix} -3/5 & -4/5 \\ -4/5 & 3/5 \end{bmatrix}.$$

(c) Here we have the direction vector

$$\mathbf{d} = \begin{bmatrix} d_1 \\ d_2 \end{bmatrix} = \begin{bmatrix} 1/\sqrt{2} \\ 1/\sqrt{2} \end{bmatrix}.$$

Applying the formula from Example 2, we get

$$\mathbf{A} = \begin{bmatrix} d_1^2 & d_1 d_2 \\ d_1 d_2 & d_2^2 \end{bmatrix} = \begin{bmatrix} 1/2 & 1/2 \\ 1/2 & 1/2 \end{bmatrix}.$$

3. The vector form of an equation of the line in the xy-plane is

$$\mathbf{x} = \mathbf{v}t + \mathbf{b} = \begin{bmatrix} v_1 \\ v_2 \end{bmatrix} t + \begin{bmatrix} b_1 \\ b_2 \end{bmatrix},$$

where $\mathbf{v} \neq \mathbf{0}$ is a direction vector, (b_1, b_2) is a point on the line, $-\infty < t < \infty$ is a real number. Applying the transformation $\mathbf{A}$ to this equation and using the distributive and associative laws, we get

$$\mathbf{Ax} = \mathbf{A}\,(\mathbf{v}t + \mathbf{b}) = \mathbf{A}(\mathbf{v}t) + \mathbf{Ab} = (\mathbf{Av})t + \mathbf{Ab} = \mathbf{w}t + \mathbf{c}.$$

If $\mathbf{w} \neq \mathbf{0}$, this equation defines the line with a direction vector $\mathbf{w}$ and passing through the point $\mathbf{c}$; if $\mathbf{w} = \mathbf{0}$, then the line $\mathbf{x}$ is mapped to a single point, $\mathbf{c}$.

Since $\mathbf{v} \neq \mathbf{0}$, we a looking for a condition on $\mathbf{A}$ that guarantees that the equation $\mathbf{Av} = \mathbf{0}$ has only *trivial* solution. This equation is equivalent to a homogeneous system of two equations with two unknowns with the coefficient matrix $\mathbf{A}$. In order that this system has only trivial solution, the matrix $\mathbf{A}$, transformed to a row-echelon form, must have the same rank as the number of unknowns, i.e., 2. First of all, necessarily, either a or c (or both) is not zero. We can assume that $a \neq 0$ (switching the rows if necessary). Thus we have

$$\begin{bmatrix} a & b \\ c & d \end{bmatrix} \rho_2 - (c/a)\rho_1 \rightarrow \rho_2 \begin{bmatrix} a & b \\ 0 & d - (bc/a) \end{bmatrix}$$
$$= \begin{bmatrix} a & b \\ 0 & (ad - bc)/a \end{bmatrix}$$

whose rank is 2 if and only if $ad - bc \neq 0$.

If $ad - bc = 0$, *some* lines can be transformed to a single point. For example, the matrix $\mathbf{A}$ with $a_{1,1} = 1$ and all other zero entries, shrinks the entire y-axis to the origin because we can take

$$\mathbf{Av} = \begin{bmatrix} 1 & 0 \\ 0 & 0 \end{bmatrix} \begin{bmatrix} 0 \\ 1 \end{bmatrix} = \mathbf{0}$$

and, for any point $(0, b_2)$ on the y-axis,

$$\mathbf{Ab} = \begin{bmatrix} 1 & 0 \\ 0 & 0 \end{bmatrix} \begin{bmatrix} 0 \\ b_2 \end{bmatrix} = \mathbf{0}.$$

5. (a) A unit normal vector to the line $x + 2y = 0$ is

$$\mathbf{n} = \frac{1}{\sqrt{1^2 + 2^2}} \begin{bmatrix} 1 \\ 2 \end{bmatrix} = \begin{bmatrix} 1/\sqrt{5} \\ 2/\sqrt{5} \end{bmatrix} = \begin{bmatrix} n_1 \\ n_2 \end{bmatrix}.$$

Therefore,

$$\mathbf{M}_{\text{ref}} = \begin{bmatrix} 1 - 2\left(1/\sqrt{5}\right)^2 & -2\left(1/\sqrt{5}\right)\left(2/\sqrt{5}\right) \\ -2\left(1/\sqrt{5}\right)\left(2/\sqrt{5}\right) & 1 - 2\left(2/\sqrt{5}\right)^2 \end{bmatrix}$$
$$= \begin{bmatrix} 3/5 & -4/5 \\ -4/5 & -3/5 \end{bmatrix}.$$

A unit direction vector to the line $y = -x$ is

$$\mathbf{d} = \frac{1}{\sqrt{1^2 + (-1)^2}} \begin{bmatrix} 1 \\ -1 \end{bmatrix} = \begin{bmatrix} 1/\sqrt{2} \\ -1/\sqrt{2} \end{bmatrix} = \begin{bmatrix} d_1 \\ d_2 \end{bmatrix}.$$

Therefore,

$$\mathbf{M}_{\text{proj}} = \begin{bmatrix} \left(1/\sqrt{2}\right)^2 & \left(1/\sqrt{2}\right)\left(-1/\sqrt{2}\right) \\ \left(1/\sqrt{2}\right)\left(-1/\sqrt{2}\right) & \left(-1/\sqrt{2}\right)^2 \end{bmatrix}$$

$$= \begin{bmatrix} 1/2 & -1/2 \\ -1/2 & 1/2 \end{bmatrix}.$$

Thus, the matrix that performs these two operations in the indicated order is

$$\mathbf{A} = \mathbf{M}_{\text{proj}}\mathbf{M}_{\text{ref}} = \begin{bmatrix} 1/2 & -1/2 \\ -1/2 & 1/2 \end{bmatrix} \begin{bmatrix} 3/5 & -4/5 \\ -4/5 & -3/5 \end{bmatrix}$$

$$= \frac{1}{10} \begin{bmatrix} 7 & -1 \\ -7 & 1 \end{bmatrix}.$$

(b) The counterclockwise rotation matrix by the angle $\theta = 120°$ is given by

$$\mathbf{M}_{\text{rot}} = \begin{bmatrix} \cos 120° & -\sin 120° \\ \sin 120° & \cos 120° \end{bmatrix} = \begin{bmatrix} -1/2 & -\sqrt{3}/2 \\ \sqrt{3}/2 & -1/2 \end{bmatrix}.$$

A unit normal vector to the line $y - 2x = 0$ is

$$\mathbf{n} = \frac{1}{\sqrt{(-2)^2 + 1^2}} \begin{bmatrix} -2 \\ 1 \end{bmatrix} = \begin{bmatrix} -2/\sqrt{5} \\ 1/\sqrt{5} \end{bmatrix}.$$

Therefore,

$$\mathbf{M}_{\text{ref}} = \begin{bmatrix} 1 - 2\left(-2/\sqrt{5}\right)^2 & -2\left(-2/\sqrt{5}\right)\left(1/\sqrt{5}\right) \\ -2\left(-2/\sqrt{5}\right)\left(1/\sqrt{5}\right) & 1 - 2\left(1/\sqrt{5}\right)^2 \end{bmatrix}$$

$$= \begin{bmatrix} -3/5 & 4/5 \\ 4/5 & 3/5 \end{bmatrix}.$$

Multiplying, we get the answer

$$A = M_{ref}M_{rot} = \begin{bmatrix} -3/5 & 4/5 \\ 4/5 & 3/5 \end{bmatrix} \begin{bmatrix} -1/2 & -\sqrt{3}/2 \\ \sqrt{3}/2 & -1/2 \end{bmatrix}$$
$$= \frac{1}{10} \begin{bmatrix} 3+4\sqrt{3} & -4+3\sqrt{3} \\ -4+3\sqrt{3} & -3-4\sqrt{3} \end{bmatrix}.$$

7. The matrix that transforms any vector $\mathbf{x}$ to the vector $(1/2)\mathbf{x}$ is $(1/2)\mathbf{I}$. The rotation by $90°$ matrix is

$$M_{rot} = \begin{bmatrix} \cos 90° & -\sin 90° \\ \sin 90° & \cos 90° \end{bmatrix} = \begin{bmatrix} 0 & -1 \\ 1 & 0 \end{bmatrix}.$$

These two transforms combined give the matrix

$$A = M_{rot}\left(\frac{1}{2}\mathbf{I}\right) = \frac{1}{2}(M_{rot}\mathbf{I}) = \frac{1}{2}M_{rot} = \begin{bmatrix} 0 & -1/2 \\ 1/2 & 0 \end{bmatrix}.$$

9. For the rotation matrix M_{rot} from Example 4, we have

$$M_{rot}^2 = \left(\frac{1}{\sqrt{2}} \begin{bmatrix} 1 & -1 \\ 1 & 1 \end{bmatrix}\right)^2 = \frac{1}{2} \begin{bmatrix} 1 & -1 \\ 1 & 1 \end{bmatrix}^2$$
$$= \frac{1}{2} \begin{bmatrix} 0 & -2 \\ 2 & 0 \end{bmatrix} = \begin{bmatrix} 0 & -1 \\ 1 & 0 \end{bmatrix},$$

we have

$$M_{rot}^4 = (M_{rot}^2)^2 = \begin{bmatrix} 0 & -1 \\ 1 & 0 \end{bmatrix}^2 = \begin{bmatrix} -1 & 0 \\ 0 & -1 \end{bmatrix} = -\mathbf{I}.$$

To explain the result of these computations, we just note that four counterclockwise rotations of a vector by $45°$ are equivalent to one rotation by $4 \times 45° = 180°$, whose rotation matrix is

$$\begin{bmatrix} \cos 180° & -\sin 180° \\ \sin 180° & \cos 180° \end{bmatrix} = \begin{bmatrix} -1 & 0 \\ 0 & -1 \end{bmatrix} = -\mathbf{I}.$$

11. We follow the arguments of Example 2 to conclude that

$$\mathbf{M}_{\text{proj}} = \mathbf{v}\mathbf{v}^T,$$

where $\mathbf{v}$ is the unit direction vector of the line on which the given vector should be orthogonally projected.

(a) For z-axis, $\mathbf{v} = [0\ 0\ 1]^T$. Thus,

$$\mathbf{M}_{\text{proj}} = \begin{bmatrix} 0 \\ 0 \\ 1 \end{bmatrix} [0\ 0\ 1] = \begin{bmatrix} 0 & 0 & 0 \\ 0 & 0 & 0 \\ 0 & 0 & 1 \end{bmatrix}.$$

(b) For the line $x = y = z$, $\mathbf{v} = (1/\sqrt{3}) [1\ 1\ 1]^T$. Therefore,

$$\mathbf{M}_{\text{proj}} = \left(\frac{1}{\sqrt{3}}\right)^2 \begin{bmatrix} 1 \\ 1 \\ 1 \end{bmatrix} [1\ 1\ 1] = \begin{bmatrix} 1/3 & 1/3 & 1/3 \\ 1/3 & 1/3 & 1/3 \\ 1/3 & 1/3 & 1/3 \end{bmatrix}.$$

13. (a) Basically, the first equation in (3) says that $\mathbf{M}_{\text{ref}} = \mathbf{I}_2 - 2\mathbf{n}\mathbf{n}^T$. All arguments leading to this equation remain valid in higher dimensions $n \geq 3$.

(b) Applying the reflection operator to a vector twice corresponds to the left multiplication of this vector by the matrix

$$\mathbf{M}_{\text{ref}}^2 = \mathbf{M}_{\text{ref}} (\mathbf{M}_{\text{ref}}) = (\mathbf{I} - 2\mathbf{n}\mathbf{n}^T) (\mathbf{I} - 2\mathbf{n}\mathbf{n}^T)$$
$$= \mathbf{I} - 4\mathbf{n}\mathbf{n}^T + 4 (\mathbf{n}\mathbf{n}^T) (\mathbf{n}\mathbf{n}^T).$$

Since, by the associative law for matrix multiplication,

$$(\mathbf{n}\mathbf{n}^T) (\mathbf{n}\mathbf{n}^T) = \mathbf{n} (\mathbf{n}^T\mathbf{n}) \mathbf{n}^T = \mathbf{n} |\mathbf{n}|^2 \mathbf{n}^T = \mathbf{n}\mathbf{n}^T$$

($\mathbf{n}$ has the unit length), we conclude that $\mathbf{M}_{\text{ref}}^2 = \mathbf{I}$.

The matrix that produces the residual vector of $\mathbf{v}$ via the left multiplication is

$$\mathbf{M}_{\text{res}} = \mathbf{I} - \mathbf{M}_{\text{ref}} = 2\mathbf{n}\mathbf{n}^T.$$

Therefore,

$$\mathbf{v}_{res} = \mathbf{M}_{res}\mathbf{v} = \left(2\mathbf{n}\mathbf{n}^T\right)\mathbf{v} = 2\mathbf{n}\left(\mathbf{n}^T\mathbf{v}\right) = \left(2\mathbf{n}^T\mathbf{v}\right)\mathbf{n},$$

which is a constant multiple of $\mathbf{n}$.

2.3   THE INVERSE AND THE TRANSPOSE

1. We have

$$\left(\begin{bmatrix} 1 & 0 & -5 \\ 0 & 1 & 0 \\ 0 & 0 & 1 \end{bmatrix}\right)^{-1} = \begin{bmatrix} 1 & 0 & -(-5) \\ 0 & 1 & 0 \\ 0 & 0 & 1 \end{bmatrix} = \begin{bmatrix} 1 & 0 & 5 \\ 0 & 1 & 0 \\ 0 & 0 & 1 \end{bmatrix}.$$

(See Example 2.)

3. Following Example 2, we conclude that

$$\left(\begin{bmatrix} 1 & 0 & 0 \\ 0 & 5 & 0 \\ 0 & 0 & 1 \end{bmatrix}\right)^{-1} = \begin{bmatrix} 1 & 0 & 0 \\ 0 & 1/(5) & 0 \\ 0 & 0 & 1 \end{bmatrix} = \begin{bmatrix} 1 & 0 & 0 \\ 0 & 1/5 & 0 \\ 0 & 0 & 1 \end{bmatrix}.$$

5. We use Example 2 to conclude that

$$\left(\begin{bmatrix} 1 & 0 & 0 & 0 \\ 0 & 0 & 1 & 0 \\ 0 & 1 & 0 & 0 \\ 0 & 0 & 0 & 1 \end{bmatrix}\right)^{-1} = \begin{bmatrix} 1 & 0 & 0 & 0 \\ 0 & 0 & 1 & 0 \\ 0 & 1 & 0 & 0 \\ 0 & 0 & 0 & 1 \end{bmatrix}.$$

7. We apply the Gauss-Jordan algorithm to the matrix $[\mathbf{A} : \mathbf{I}]$. (See Example 3.)

$$\begin{bmatrix} 2 & 1 & \vdots & 1 & 0 \\ -1 & 4 & \vdots & 0 & 1 \end{bmatrix} \quad 2\rho_2 + \rho_1 \rightarrow \rho_2 \quad \begin{bmatrix} 2 & 1 & \vdots & 1 & 0 \\ 0 & 9 & \vdots & 1 & 2 \end{bmatrix}$$

$$(1/2)\rho_1 \to \rho_1 \quad \begin{bmatrix} 1 & 1/2 & \vdots & 1/2 & 0 \\ 0 & 1 & \vdots & 1/9 & 2/9 \end{bmatrix}$$
$$(1/9)\rho_2 \to \rho_2$$

$$\rho_1 - (1/2)\rho_2 \to \rho_1 \quad \begin{bmatrix} 1 & 0 & \vdots & 4/9 & -1/9 \\ 0 & 1 & \vdots & 1/9 & 2/9 \end{bmatrix}.$$

9. Apply the Gauss-Jordan algorithm to the matrix $[\mathbf{A} : \mathbf{I}]$, we get

$$\begin{bmatrix} 1 & 1 & 1 & \vdots & 1 & 0 & 0 \\ 1 & 2 & 1 & \vdots & 0 & 1 & 0 \\ 2 & 3 & 2 & \vdots & 0 & 0 & 1 \end{bmatrix}$$

$$\rho_2 - \rho_1 \to \rho_2 \quad \begin{bmatrix} 1 & 1 & 1 & \vdots & 1 & 0 & 0 \\ 0 & 1 & 0 & \vdots & -1 & 1 & 0 \\ 0 & 1 & 0 & \vdots & -2 & 0 & 1 \end{bmatrix}$$
$$\rho_3 - 2\rho_1 \to \rho_3$$

$$\rho_3 - \rho_2 \to \rho_3 \quad \begin{bmatrix} 1 & 1 & 1 & \vdots & 1 & 0 & 0 \\ 0 & 1 & 0 & \vdots & -1 & 1 & 0 \\ 0 & 0 & 0 & \vdots & -1 & -1 & 1 \end{bmatrix}.$$

By Theorem 2, the given matrix is not invertible since, in the row echelon form, its rank (2) is not the full rank (3).

11. We re-arrange the rows of the augemented matrix as $\rho_2, \rho_4, \rho_3,$ and ρ_1.

$$\begin{bmatrix} 3 & 0 & 0 & 1 & \vdots & 1 & 0 & 0 & 0 \\ 2 & 0 & 0 & 0 & \vdots & 0 & 1 & 0 & 0 \\ 0 & 0 & 2 & 0 & \vdots & 0 & 0 & 1 & 0 \\ 0 & 1 & 2 & 0 & \vdots & 0 & 0 & 0 & 1 \end{bmatrix}$$

$$\to \begin{bmatrix} 2 & 0 & 0 & 0 & \vdots & 0 & 1 & 0 & 0 \\ 0 & 1 & 2 & 0 & \vdots & 0 & 0 & 0 & 1 \\ 0 & 0 & 2 & 0 & \vdots & 0 & 0 & 1 & 0 \\ 3 & 0 & 0 & 1 & \vdots & 1 & 0 & 0 & 0 \end{bmatrix}.$$

Working now with the new augmented matrix, we perform some elementary row operations to get

$$
\begin{array}{c}
(1/2)\rho_1 \to \rho_1 \\
(1/2)\rho_3 \to \rho_3
\end{array}
\left[
\begin{array}{cccc:cccc}
1 & 0 & 0 & 0 & 0 & 1/2 & 0 & 0 \\
0 & 1 & 2 & 0 & 0 & 0 & 0 & 1 \\
0 & 0 & 1 & 0 & 0 & 0 & 1/2 & 0 \\
3 & 0 & 0 & 1 & 1 & 0 & 0 & 0
\end{array}
\right]
$$

$$
\begin{array}{c}
\rho_2 - 2\rho_3 \to \rho_2 \\
\rho_4 - 3\rho_1 \to \rho_4
\end{array}
\left[
\begin{array}{cccc:cccc}
1 & 0 & 0 & 0 & 0 & 1/2 & 0 & 0 \\
0 & 1 & 0 & 0 & 0 & 0 & -1 & 1 \\
0 & 0 & 1 & 0 & 0 & 0 & 1/2 & 0 \\
0 & 0 & 0 & 1 & 1 & -3/2 & 0 & 0
\end{array}
\right].
$$

Therefore,

$$
\left[
\begin{array}{cccc}
3 & 0 & 0 & 1 \\
2 & 0 & 0 & 0 \\
0 & 0 & 2 & 0 \\
0 & 1 & 2 & 0
\end{array}
\right]^{-1}
=
\left[
\begin{array}{cccc}
0 & 1/2 & 0 & 0 \\
0 & 0 & -1 & 1 \\
0 & 0 & 1/2 & 0 \\
1 & -3/2 & 0 & 0
\end{array}
\right].
$$

13. To find $\mathbf{A}^{-1}$, we proceed as follows.

$$
\left[
\begin{array}{ccc:ccc}
1 & 0 & 3 & 1 & 0 & 0 \\
0 & 2 & 4 & 0 & 1 & 0 \\
4 & -3 & 8 & 0 & 0 & 1
\end{array}
\right]
$$

$$
\begin{array}{c}
\rho_3 - 4\rho_1 \to \rho_3 \\
(1/2)\rho_2 \to \rho_2
\end{array}
\left[
\begin{array}{ccc:ccc}
1 & 0 & 3 & 1 & 0 & 0 \\
0 & 1 & 2 & 0 & 1/2 & 0 \\
0 & -3 & -4 & -4 & 0 & 1
\end{array}
\right]
$$

$$
(1/2)\rho_3 + (3/2)\rho_2 \to \rho_3
\left[
\begin{array}{ccc:ccc}
1 & 0 & 3 & 1 & 0 & 0 \\
0 & 1 & 2 & 0 & 1/2 & 0 \\
0 & 0 & 1 & -2 & 3/4 & 1/2
\end{array}
\right]
$$

$$\begin{array}{l} \rho_1 - 3\rho_3 \to \rho_1 \\ \rho_2 - 2\rho_3 \to \rho_2 \end{array} \qquad \left[\begin{array}{ccc:ccc} 1 & 0 & 0 & 7 & -9/4 & -3/2 \\ 0 & 1 & 0 & 4 & -1 & -1 \\ 0 & 0 & 1 & -2 & 3/4 & 1/2 \end{array} \right]$$

Therefore,

$$\mathbf{A}^{-1} = \left[\begin{array}{ccc} 7 & -9/4 & -3/2 \\ 4 & -1 & -1 \\ -2 & 3/4 & 1/2 \end{array} \right].$$

(a)

$$\mathbf{x} = \mathbf{A}^{-1}[1\ 0\ 3]^T = \left[\begin{array}{ccc} 7 & -9/4 & -3/2 \\ 4 & -1 & -1 \\ -2 & 3/4 & 1/2 \end{array} \right] \left[\begin{array}{c} 1 \\ 0 \\ 3 \end{array} \right] = \left[\begin{array}{c} 5/2 \\ 1 \\ -1/2 \end{array} \right].$$

(b)

$$\mathbf{x} = \mathbf{A}^{-1}[-3\ 7\ 1]^T = \left[\begin{array}{ccc} 7 & -9/4 & -3/2 \\ 4 & -1 & -1 \\ -2 & 3/4 & 1/2 \end{array} \right] \left[\begin{array}{c} -3 \\ 7 \\ 1 \end{array} \right]$$

$$= \left[\begin{array}{c} -153/4 \\ -20 \\ 47/4 \end{array} \right].$$

(c)

$$\mathbf{x} = \mathbf{A}^{-1}[0\ -4\ 1]^T = \left[\begin{array}{ccc} 7 & -9/4 & -3/2 \\ 4 & -1 & -1 \\ -2 & 3/4 & 1/2 \end{array} \right] \left[\begin{array}{c} 0 \\ -4 \\ 1 \end{array} \right]$$

$$= \left[\begin{array}{c} 15/2 \\ 3 \\ -5/2 \end{array} \right].$$

15. First, we compute $\mathbf{A}^{-1}$.

$$
\left[\begin{array}{ccc:ccc}
1 & 1 & 1 & 1 & 0 & 0 \\
1 & 2 & 3 & 0 & 1 & 0 \\
0 & 1 & 1 & 0 & 0 & 1
\end{array}\right]
$$

$$
\begin{array}{c}
\rho_2 - \rho_1 \to \rho_2 \\
\rho_1 - \rho_3 \to \rho_1
\end{array}
\left[\begin{array}{ccc:ccc}
1 & 0 & 0 & 1 & 0 & -1 \\
0 & 1 & 2 & -1 & 1 & 0 \\
0 & 1 & 1 & 0 & 0 & 1
\end{array}\right]
$$

$$
-\rho_3 + \rho_2 \to \rho_3
\left[\begin{array}{ccc:ccc}
1 & 0 & 0 & 1 & 0 & -1 \\
0 & 1 & 2 & -1 & 1 & 0 \\
0 & 0 & 1 & -1 & 1 & -1
\end{array}\right]
$$

$$
\rho_2 - 2\rho_3 \to \rho_2
\left[\begin{array}{ccc:ccc}
1 & 0 & 0 & 1 & 0 & -1 \\
0 & 1 & 0 & 1 & -1 & 2 \\
0 & 0 & 1 & -1 & 1 & -1
\end{array}\right]
$$

Therefore,

$$
\mathbf{A}^{-1} = \left[\begin{array}{ccc}
1 & 0 & -1 \\
1 & -1 & 2 \\
-1 & 1 & -1
\end{array}\right]
$$

and

$$
\mathbf{X} = \mathbf{A}^{-1} \left[\begin{array}{ccc}
1 & 0 & 3 \\
0 & 2 & 4 \\
4 & -3 & 8
\end{array}\right] = \left[\begin{array}{ccc}
1 & 0 & -1 \\
1 & -1 & 2 \\
-1 & 1 & -1
\end{array}\right] \left[\begin{array}{ccc}
1 & 0 & 3 \\
0 & 2 & 4 \\
4 & -3 & 8
\end{array}\right]
$$

$$
= \left[\begin{array}{ccc}
-3 & 3 & -5 \\
9 & -8 & 15 \\
-5 & 5 & -7
\end{array}\right].
$$

17. (a) False: say, $\mathbf{I}$ is invertible $(\mathbf{I}^{-1} = \mathbf{I})$, $(-\mathbf{I})$ is invertible $((-\mathbf{I})^{-1} = -\mathbf{I})$, but $\mathbf{I} + (-\mathbf{I}) = \mathbf{0}$ is not invertible.

(b) True: by Theorem 3 (see its generalization (12)), $\mathbf{A}^3 := \mathbf{AAA}$ is invertible and, moreover,

$$\left(\mathbf{A}^3\right)^{-1} = \left(\mathbf{A}^{-1}\right)^3.$$

(c) True: by Theorem 3,

$$\left(\mathbf{A}^2\right)^T = (\mathbf{AA})^T = \mathbf{A}^T\mathbf{A}^T = \left(\mathbf{A}^T\right)^2 = \mathbf{A}^2$$

since $\mathbf{A}$ is symmetric.

(d) True: by Theorem 3, we have

$$\mathbf{A}^T\left(\mathbf{A}^{-1}\right)^T = \left(\mathbf{A}^{-1}\mathbf{A}\right)^T = \mathbf{I}^T = \mathbf{I}.$$

Therefore, $\mathbf{A}^T$ is invertible and, moreover, $\left(\mathbf{A}^T\right)^{-1} = \left(\mathbf{A}^{-1}\right)^T$.

19. If $ad - bc \neq 0$, then the matrix $\mathbf{A}$ is invertible because

$$\left(\frac{1}{ad-bc}\begin{bmatrix} d & -b \\ -c & a \end{bmatrix}\right)\begin{bmatrix} a & b \\ c & d \end{bmatrix}$$

$$= \frac{1}{ad-bc}\left(\begin{bmatrix} d & -b \\ -c & a \end{bmatrix}\begin{bmatrix} a & b \\ c & d \end{bmatrix}\right)$$

$$= \frac{1}{ad-bc}\begin{bmatrix} ad-bc & db-bd \\ -ca+ac & -cb+ad \end{bmatrix} = \mathbf{I}.$$

Therefore, the matrix

$$\frac{1}{ad-bc}\begin{bmatrix} d & -b \\ -c & a \end{bmatrix}$$

is the left inverse (and so the inverse) of $\mathbf{A}$.

On the other hand, if $ad - bc = 0$, then the rank of $\mathbf{A}$ is not the full rank. Indeed, if both, a and c, are zero, then $\mathbf{A}$ has a zero first column, and so its rank is less than 2. If at least one of a, c is not zero (say, $a \neq 0$), then transforming $\mathbf{A}$ to a row echelon form yields

$$\begin{bmatrix} a & b \\ c & d \end{bmatrix} \quad a\rho_2 - c\rho_1 \rightarrow \rho_2 \quad \begin{bmatrix} a & b \\ 0 & ad-bc \end{bmatrix} = \begin{bmatrix} a & b \\ 0 & 0 \end{bmatrix},$$

whose rank is $1 < 2$. By Theorem 2, $\mathbf{A}$ is singular.

21. Applying Theorem 3 twice and using the associative property of the matrix product, we get

$$(\mathbf{ABC})^T = (\mathbf{A(BC)})^T = (\mathbf{BC})^T \mathbf{A}^T = \mathbf{C}^T \mathbf{B}^T \mathbf{A}^T.$$

23. From Example 1 we know that

$$\mathbf{M}_{\text{rot}}(\theta)^{-1} = \mathbf{M}_{\text{rot}}(-\theta) = \begin{bmatrix} \cos(-\theta) & -\sin(-\theta) \\ \sin(-\theta) & \cos(-\theta) \end{bmatrix}$$
$$= \begin{bmatrix} \cos\theta & \sin\theta \\ -\sin\theta & \cos\theta \end{bmatrix}$$

which is the same as

$$\mathbf{M}_{\text{rot}}(\theta)^T = \begin{bmatrix} \cos\theta & -\sin\theta \\ \sin\theta & \cos\theta \end{bmatrix}^T = \begin{bmatrix} \cos\theta & \sin\theta \\ -\sin\theta & \cos\theta \end{bmatrix}.$$

25. We multiply the right-hand side of the Sherman-Morrison-Woodbury formula by $\mathbf{A} + \mathbf{BCD}$. If the product simplifies to $\mathbf{I}$, then the factors are inverses to each other. Simplifying the product, we will frequently use the distributive and associative properties of matrix multiplication.

$$(\mathbf{A} + \mathbf{BCD})\,\mathbf{A}^{-1}\left[\mathbf{I} - \mathbf{B}\left(\mathbf{C}^{-1} + \mathbf{DA}^{-1}\mathbf{B}\right)^{-1}\mathbf{DA}^{-1}\right]$$
$$= \mathbf{I} - \mathbf{B}\left(\mathbf{C}^{-1} + \mathbf{DA}^{-1}\mathbf{B}\right)^{-1}\mathbf{DA}^{-1}$$
$$\quad + \mathbf{BCDA}^{-1}\left[\mathbf{I} - \mathbf{B}\left(\mathbf{C}^{-1} + \mathbf{DA}^{-1}\mathbf{B}\right)^{-1}\mathbf{DA}^{-1}\right]$$
$$= \mathbf{I} + \mathbf{BCDA}^{-1} - \mathbf{B}\left[\left(\mathbf{C}^{-1} + \mathbf{DA}^{-1}\mathbf{B}\right)^{-1}\mathbf{DA}^{-1}\right.$$
$$\quad + \left.\mathbf{CDA}^{-1}\mathbf{B}\left(\mathbf{C}^{-1} + \mathbf{DA}^{-1}\mathbf{B}\right)^{-1}\mathbf{DA}^{-1}\right]$$
$$= \mathbf{I} + \mathbf{BCDA}^{-1} - \mathbf{B}\left(\mathbf{I} + \mathbf{CDA}^{-1}\mathbf{B}\right)\left(\mathbf{C}^{-1} + \mathbf{DA}^{-1}\mathbf{B}\right)^{-1}\mathbf{DA}^{-1}$$
$$= \mathbf{I} + \mathbf{BCDA}^{-1} - \mathbf{B}\left(\mathbf{CC}^{-1} + \mathbf{CDA}^{-1}\mathbf{B}\right)\left(\mathbf{C}^{-1} + \mathbf{DA}^{-1}\mathbf{B}\right)^{-1}\mathbf{DA}^{-1}$$
$$= \mathbf{I} + \mathbf{BCDA}^{-1} - \mathbf{BC}\left(\mathbf{C}^{-1} + \mathbf{DA}^{-1}\mathbf{B}\right)\left(\mathbf{C}^{-1} + \mathbf{DA}^{-1}\mathbf{B}\right)^{-1}\mathbf{DA}^{-1}$$
$$= \mathbf{I} + \mathbf{BCDA}^{-1} - \mathbf{BCIDA}^{-1} = \mathbf{I}.$$

With $\mathbf{A} = \mathbf{I}$ (so that $\mathbf{A}^{-1} = \mathbf{I}$), the formula simplifies to

$$(\mathbf{I} + \mathbf{BCD})^{-1} = \mathbf{I} - \mathbf{B}\left(\mathbf{C}^{-1} + \mathbf{DB}\right)^{-1}\mathbf{D}$$

and, since for the given $\mathbf{B}$, $\mathbf{C}$, and $\mathbf{D}$, we have

$$\mathbf{C}^{-1} = 10, \quad \mathbf{DB} = [1\ 1\ 1\ 1\ 1][0\ 0\ 0\ 0\ 1]^T = [1],$$

$$\left(\mathbf{C}^{-1} + \mathbf{DB}\right)^{-1} = [1/11]$$

$$\Rightarrow \quad (\mathbf{I} + \mathbf{BCD})^{-1} = \mathbf{I} - [0\ 0\ 0\ 0\ 1]^T[1/11][1\ 1\ 1\ 1\ 1]$$

$$= \mathbf{I} - \begin{bmatrix} 0 & 0 & 0 & 0 & 0 \\ 0 & 0 & 0 & 0 & 0 \\ 0 & 0 & 0 & 0 & 0 \\ 0 & 0 & 0 & 0 & 0 \\ 1/11 & 1/11 & 1/11 & 1/11 & 1/11 \end{bmatrix}$$

$$= \begin{bmatrix} 1 & 0 & 0 & 0 & 0 \\ 0 & 1 & 0 & 0 & 0 \\ 0 & 0 & 1 & 0 & 0 \\ 0 & 0 & 0 & 1 & 0 \\ -1/11 & -1/11 & -1/11 & -1/11 & 10/11 \end{bmatrix}$$

so that

$$\begin{bmatrix} 1 & 0 & 0 & 0 & 0 \\ 0 & 1 & 0 & 0 & 0 \\ 0 & 0 & 1 & 0 & 0 \\ 0 & 0 & 0 & 1 & 0 \\ 0.1 & 0.1 & 0.1 & 0.1 & 1.1 \end{bmatrix}^{-1} = \begin{bmatrix} 1 & 0 & 0 & 0 & 0 \\ 0 & 1 & 0 & 0 & 0 \\ 0 & 0 & 1 & 0 & 0 \\ 0 & 0 & 0 & 1 & 0 \\ -\dfrac{1}{11} & -\dfrac{1}{11} & -\dfrac{1}{11} & -\dfrac{1}{11} & \dfrac{10}{11} \end{bmatrix}.$$

2.4 DETERMINANTS

1. We expand the determinant using the first row to have

$$\det\begin{bmatrix} -1 & -1 \\ -1 & -1 \end{bmatrix} = (-1)\mathrm{cof}(1,1) + (-1)\mathrm{cof}(1,2)$$

$$= (-1)\cdot(-1)^{1+1}\det[-1] + (-1)\cdot(-1)^{1+2}\det[-1]$$

$$= (-1)(1)(-1) + (-1)(-1)(-1) = 0.$$

3. We expand the determinant using the first row.

$$\det \begin{bmatrix} 1 & 2 & 2 \\ 2 & 1 & 1 \\ 1 & 1 & 3 \end{bmatrix}$$

$$= (1)(-1)^{1+1}\det \begin{bmatrix} 1 & 1 \\ 1 & 3 \end{bmatrix} + (2)(-1)^{1+2}\det \begin{bmatrix} 2 & 1 \\ 1 & 3 \end{bmatrix}$$

$$+ (2)(-1)^{1+3}\det \begin{bmatrix} 2 & 1 \\ 1 & 1 \end{bmatrix} = (2) - (10) + (2) = -6.$$

5. We expand the determinant using the second row.

$$\det \begin{bmatrix} 2 & 1 & 4+i \\ 3 & 3 & 0 \\ -1 & 4 & 2-3i \end{bmatrix} = (3)(-1)^{2+1}\det \begin{bmatrix} 1 & 4+i \\ 4 & 2-3i \end{bmatrix}$$

$$+ (3)(-1)^{2+2}\det \begin{bmatrix} 2 & 4+i \\ -1 & 2-3i \end{bmatrix}$$

$$= -(3)(-14-7i) + (3)(8-5i)$$

$$= 66 + 6i.$$

7. We use the expansions along the top rows and use the formula for the determinant of 2-by-2 matrix (see (7) or Exercise 1).

$$\begin{vmatrix} 0 & 0 & 0 & a \\ 0 & 0 & b & c \\ 0 & d & e & f \\ g & h & i & j \end{vmatrix} = (a)(-1)^{1+4}\begin{vmatrix} 0 & 0 & b \\ 0 & d & e \\ g & h & i \end{vmatrix}$$

$$= -a(b)(-1)^{1+3}\begin{vmatrix} 0 & d \\ g & h \end{vmatrix}$$

$$= -ab(0 \cdot h - d \cdot g) = abdg.$$

9. We expand the determinant of the given matrix along the first row.

$$\det \begin{bmatrix} x & 1 & 0 & 0 \\ 0 & x & 1 & 0 \\ 0 & 0 & x & 1 \\ a & b & c & d \end{bmatrix}$$

$$= (x)\det \begin{bmatrix} x & 1 & 0 \\ 0 & x & 1 \\ b & c & d \end{bmatrix} - (1)\det \begin{bmatrix} 0 & 1 & 0 \\ 0 & x & 1 \\ a & c & d \end{bmatrix}$$

$$= x \left((x) \det \begin{bmatrix} x & 1 \\ c & d \end{bmatrix} - (1) \begin{bmatrix} 0 & 1 \\ b & d \end{bmatrix} \right) + (1) \det \begin{bmatrix} 0 & 1 \\ a & d \end{bmatrix}$$

$$= x\,[x(xd - c) - (-b)] + (-a) = dx^3 - cx^2 + bx - a.$$

11. We use corollaries of Theorem 5 to simplify calculations. First,

$$\det \begin{bmatrix} x & 1 & 1 & 1 \\ 1 & x & 1 & 1 \\ 1 & 1 & x & 1 \\ 1 & 1 & 1 & x \end{bmatrix} \quad \left(\begin{array}{c} \rho_1 - \rho_2 \to \rho_1 \\ \rho_2 - \rho_3 \to \rho_2 \\ \rho_3 - \rho_4 \to \rho_3 \end{array} \right)$$

$$= \begin{vmatrix} x-1 & 1-x & 0 & 0 \\ 0 & x-1 & 1-x & 0 \\ 0 & 0 & x-1 & 1-x \\ 1 & 1 & 1 & x \end{vmatrix}.$$

Next, we factor $(x - 1)$ from the first three rows to continue the equation with

$$\ldots = (x - 1)^3 \begin{vmatrix} 1 & -1 & 0 & 0 \\ 0 & 1 & -1 & 0 \\ 0 & 0 & 1 & -1 \\ 1 & 1 & 1 & x \end{vmatrix}$$

$$= (x - 1)^3 \left(\begin{vmatrix} 1 & -1 & 0 \\ 0 & 1 & -1 \\ 1 & 1 & x \end{vmatrix} + \begin{vmatrix} 0 & -1 & 0 \\ 0 & 1 & -1 \\ 1 & 1 & x \end{vmatrix} \right)$$

$$= (x - 1)^3 \left(\begin{vmatrix} 0 & -2 & -x \\ 0 & 1 & -1 \\ 1 & 1 & x \end{vmatrix} + \begin{vmatrix} 0 & -1 & 0 \\ 0 & 1 & -1 \\ 1 & 1 & x \end{vmatrix} \right)$$

$$= (x - 1)^3 \left(\begin{vmatrix} -2 & -x \\ 1 & -1 \end{vmatrix} + \begin{vmatrix} -1 & 0 \\ 1 & -1 \end{vmatrix} \right) = (x - 1)^3 (x + 3).$$

13. Let us denote the variable entry, a_{ij}, by x. Using the cofactor expansion (11) along the ith row, we get

$$f(x) = \det(A) = x \cdot \text{cof}(i, j) = ax + b,$$

which is a linear function of x. Its graph is a non-vertical line. If this line is horizontal (i.e, $a = 0$), then it is either the x-axis (for $b = 0$)

so that $f(x) \equiv 0$, or (for $b \neq 0$) this line does not intersect the x-axis so that $f(x) \neq 0$ for any x. If $a \neq 0$, the line $y = f(x)$ has exactly one x-intercept ($f(x) = 0$ for a unique value of x).

Examples demonstrating these cases are:

$$\det \begin{bmatrix} 1 & 0 & 0 \\ 0 & 0 & x \\ 0 & 0 & 0 \end{bmatrix} = 0 \quad \text{for any } x;$$

$$\det \begin{bmatrix} 1 & 0 & 0 \\ 0 & 1 & x \\ 0 & 0 & 1 \end{bmatrix} = 1 \neq 0 \quad \text{for any } x;$$

$$\det \begin{bmatrix} 1 & 0 & 0 \\ 0 & x & 0 \\ 0 & 0 & 1 \end{bmatrix} = x = 0 \quad \text{only for } x = 0.$$

15. We have to compute $(\mathbf{a} \times \mathbf{b}) \cdot \mathbf{c}$. Using corollaries 2 and 4 from Theorem 5, we get

$$(\mathbf{a} \times \mathbf{b}) \cdot \mathbf{c} = \det \begin{bmatrix} c_1 & c_2 & c_3 \\ a_1 & a_2 & a_3 \\ b_1 & b_2 & b_3 \end{bmatrix} = \det \begin{bmatrix} 4 & 0 & -4 \\ -2 & -3 & 1 \\ 2 & 6 & -2 \end{bmatrix}$$

$$= 8 \cdot \det \begin{bmatrix} 1 & 0 & -1 \\ -2 & -3 & 1 \\ 1 & 3 & -1 \end{bmatrix} = 8 \cdot \det \begin{bmatrix} 1 & 0 & -1 \\ 0 & -3 & -1 \\ 0 & 3 & 0 \end{bmatrix}$$

$$= 8(-3) \cdot \det \begin{vmatrix} 1 & -1 \\ 0 & -1 \end{vmatrix} = 24.$$

17. In all questions, (a) through (h), we denote by $\mathbf{B}$ the matrix in question. For the rows of $\mathbf{A}$ and $\mathbf{B}$ we will use superscripts to indicate the corresponding matrix.

   (a) Since $\rho_1^{(\mathbf{B})} = \rho_1^{(\mathbf{A})} + \rho_2^{(\mathbf{A})}$, by Corollary 4, $\det(\mathbf{B}) = \det(\mathbf{A})$.

   (b) $\rho_2^{(\mathbf{B})} = 3\rho_2^{(\mathbf{A})}$. Therefore, by Corollary 2, $\det(\mathbf{B}) = 3 \cdot \det(\mathbf{A})$.

   (c) $\rho_3^{(\mathbf{B})} = -\rho_3^{(\mathbf{A})}$. Therefore, by Corollary 2, $\det(\mathbf{B}) = -\det(\mathbf{A})$.

   (d) $\mathbf{B} = -\mathbf{A}$. Therefore, $\det(\mathbf{B}) = (-1)^4 \cdot \det(\mathbf{A}) = \det(\mathbf{A})$.

   (e) Since $\mathbf{B} = 3\mathbf{A}$, we have $\det(\mathbf{B}) = (3)^4 \cdot \det(\mathbf{A}) = 81 \cdot \det(\mathbf{A})$.

   (f) Here we have $\rho_1^{(\mathbf{B})} = \rho_3^{(\mathbf{A})}$, $\rho_3^{(\mathbf{B})} = \rho_1^{(\mathbf{A})}$, $\rho_2^{(\mathbf{B})} = \rho_4^{(\mathbf{A})}$, and $\rho_4^{(\mathbf{B})} = \rho_2^{(\mathbf{A})}$. In other words, $\mathbf{B}$ is obtained from $\mathbf{A}$ by switching its rows: $1 \leftrightarrow 3$ and $2 \leftrightarrow 4$. By Theorem 4, each procedure results the opposite sign of the determinant. Therefore, $\det(\mathbf{B}) = (-1)^2 \cdot \det(\mathbf{A}) = \det(\mathbf{A})$.

(g) Since $\rho_2^{(\mathbf{B})} = 3\rho_2^{(\mathbf{A})} - \rho_1^{(\mathbf{A})}$, from Corollaries 2 and 4 we conclude that $\det(\mathbf{B}) = 3 \cdot \det(\mathbf{A})$.

(h) Since $\rho_2^{(\mathbf{B})} = \rho_2^{(\mathbf{A})} - 3\rho_1^{(\mathbf{A})}$, by Corollary 4 we have $\det(\mathbf{B}) = \det(\mathbf{A})$.

19. (a) We perform the following elementary row operations in the indicated order.

$$\begin{bmatrix} 1 & x & x^2 & x^3 \\ x & 1 & x & x^2 \\ x^2 & x & 1 & x \\ x^3 & x^2 & x & 1 \end{bmatrix}$$

$$\begin{matrix} \rho_1 - x\rho_2 \rightarrow \rho_1 \\ \rho_2 - x\rho_3 \rightarrow \rho_2 \\ \rho_3 - x\rho_4 \rightarrow \rho_3 \end{matrix} \quad \begin{bmatrix} 1-x^2 & 0 & 0 & 0 \\ x-x^3 & 1-x^2 & 0 & 0 \\ x^2-x^4 & x-x^3 & 1-x^2 & 0 \\ x^3 & x^2 & x & 1 \end{bmatrix}.$$

By Example 5, the determinant of this lower triangular matrix is the product of its diagonal elements, which equals the determinant of the original matrix.

$$\det \begin{bmatrix} 1 & x & x^2 & x^3 \\ x & 1 & x & x^2 \\ x^2 & x & 1 & x \\ x^3 & x^2 & x & 1 \end{bmatrix} = (1-x^2)(1-x^2)(1-x^2)(1)$$

$$= (1-x^2)^3.$$

The computations made for 4-by-4 matrix suggest that, for all $n \geq 1$,

$$\det \begin{bmatrix} 1 & x & x^2 & \cdots & x^{n-1} \\ x & 1 & x & \cdots & x^{n-2} \\ & & \vdots & & \\ x^{n-2} & x^{n-2} & x^{n-3} & \cdots & x \\ x^{n-1} & x^{n-2} & x^{n-3} & \cdots & 1 \end{bmatrix} = (1-x^2)^{n-1}.$$

(b) We perform the following elementary row operations to reduce this matrix to a lower triangular form.

$$\begin{bmatrix} 1 & 1 & 1 & 1 \\ 1 & 2 & 2 & 2 \\ 1 & 2 & 3 & 3 \\ 1 & 2 & 3 & 4 \end{bmatrix} \quad \rho_1 - (1/2)\rho_2 \to \rho_1 \quad \begin{bmatrix} 1/2 & 0 & 0 & 0 \\ 1 & 2 & 2 & 2 \\ 1 & 2 & 3 & 3 \\ 1 & 2 & 3 & 4 \end{bmatrix}$$

$$\rho_2 - (2/3)\rho_3 \to \rho_2 \quad \begin{bmatrix} 1/2 & 0 & 0 & 0 \\ 1/3 & 2/3 & 0 & 0 \\ 1 & 2 & 3 & 3 \\ 1 & 2 & 3 & 4 \end{bmatrix}$$

$$\rho_3 - (3/4)\rho_4 \to \rho_3 \quad \begin{bmatrix} 1/2 & 0 & 0 & 0 \\ 1/3 & 2/3 & 0 & 0 \\ 1/4 & 2/4 & 3/4 & 0 \\ 1 & 2 & 3 & 4 \end{bmatrix}.$$

By Example 5, the determinant of this lower triangular matrix is the product of its diagonal elements; that is,

$$\det \begin{bmatrix} 1/2 & 0 & 0 & 0 \\ 1/3 & 2/3 & 0 & 0 \\ 1/4 & 2/4 & 3/4 & 0 \\ 1 & 2 & 3 & 4 \end{bmatrix} = \frac{1}{2} \cdot \frac{2}{3} \cdot \frac{3}{4} \cdot 4 = 1.$$

The procedure of reducing an n-by-n matrix $\mathbf{A}$ with the indicated pattern of entries to a lower triangular form suggests that $\det(\mathbf{A}) = 1$ for any n.

21. Let $P(x) = \alpha x^4 + \beta x^3 + \gamma x^2 + \delta x + \varepsilon$ denote the determinant of the given matrix. In the expansion of $P(x)$ from Problem 20, the only term that contain x to a power larger than 2 is

$$a_{1,1}a_{2,2}a_{3,3}a_{4,4}\det(P_{1,2,3,4}) = (x+a)(x+f)(x+k)(x+p)\det(\mathbf{I})$$
$$= x^4 + (a+f+k+p)x^3 + \cdots.$$

Therefore,

$$\alpha = 1, \quad \beta = a+f+k+p.$$

The simplest way to find ε is

$$\varepsilon = P(0) = \det \begin{bmatrix} a & b & c & d \\ e & f & g & h \\ i & j & k & l \\ m & n & o & p \end{bmatrix}.$$

23. (a) Expanding $\det(V)$ along the first row yields

$$\det(V) = (1)\text{cof}(1,1) + (z)\text{cof}(1,2) + (z^2)\,\text{cof}(1,3).$$

Since

$$a_2 = \text{cof}(1,3) = (-1)^{1+3}\det \begin{bmatrix} 1 & x_1 \\ 1 & x_2 \end{bmatrix} = x_2 - x_1 \neq 0,$$

$\det(V)$ is a quadratic polynomial with the leading coefficient $(x_2 - x_1)$.

(b) Clearly, a_2 can be expressed as the value of 2-by-2 Vandermonde determinant

$$\det \begin{bmatrix} 1 & t \\ 1 & x_2 \end{bmatrix} = x_2 - t$$

at the point $t = x_1$.

(c) The values $z = x_1$, $z = x_2$ result determinants with two identical rows. By Corollary 1 from Theorem 4, these determinants equal zero. Any quadratic polynomial with these two roots has the form $a_2\,(z - x_1)\,(z - x_2)$, where a_2 is the leading coefficient.

(d) From part (a), we know that $a_2 = x_2 - x_1$. Therefore,

$$\det(V) = (x_2 - x_1)\,(z - x_1)\,(z - x_2).$$

(e) Expanding this 4-by-4 Vandermonde determinant using the first row, we get

$$\begin{vmatrix} 1 & z & z^2 & z^3 \\ 1 & x_1 & x_1^2 & x_1^3 \\ 1 & x_2 & x_2^2 & x_2^3 \\ 1 & x_3 & x_3^2 & x_3^3 \end{vmatrix} = 1 \cdot \text{cof}(1,1) + z \cdot \text{cof}(1,2) + z^2 \cdot \text{cof}(1,3)$$

$$+ z^3 \cdot \text{cof}(1,4).$$

The coefficient at z^3 can be expressed as a particular value of a 3-by-3 Vandermonde determinant. Indeed,

$$\text{cof}(1,4) = (-1)^{1+4} \begin{vmatrix} 1 & x_1 & x_1^2 \\ 1 & x_2 & x_2^2 \\ 1 & x_3 & x_3^2 \end{vmatrix} = - \begin{vmatrix} 1 & z & z^2 \\ 1 & x_2 & x_2^2 \\ 1 & x_3 & x_3^2 \end{vmatrix}$$

evaluated at $z = x_1$. Applying the formula for 3-by-3 Vandermonde determinant from part (d), we find that

$$\begin{aligned} \text{cof}(1,4) &= - (x_3 - x_2)(x_1 - x_2)(x_1 - x_3) \\ &= (x_1 - x_2)(x_1 - x_3)(x_2 - x_3) \neq 0 \end{aligned}$$

because the points x_1, x_2, and x_3 are pairwise distinct. Therefore, a 4-by-4 Vandermonde determinant is a cubic polynomial. Since this polynomial has zeros at $z = x_1, x_2$, and x_3 (these values lead to a determinant with two equal rows), we get the formula

$$\begin{vmatrix} 1 & z & z^2 & z^3 \\ 1 & x_1 & x_1^2 & x_1^3 \\ 1 & x_2 & x_2^2 & x_2^3 \\ 1 & x_3 & x_3^2 & x_3^3 \end{vmatrix} = (x_1 - x_2)(x_1 - x_3)(x_2 - x_3)$$

$$\times (z - x_1)(z - x_2)(z - x_3).$$

2.5   THREE IMPORTANT DETERMINANT RULES

1. In Example 1, Section 1.2, this matrix is reduced the upper triangular form using only the elementary row operations of type (v), which do not change the determinant. Therefore,

$$\begin{vmatrix} 1 & 2 & 2 \\ 2 & 1 & 1 \\ 1 & 1 & 3 \end{vmatrix} = \begin{vmatrix} 1 & 2 & 2 \\ 0 & -3 & -3 \\ 0 & 0 & 2 \end{vmatrix} = (1)(-3)(2) = -6.$$

3. In Example 2, Section 1.1, this matrix is reduced the upper triangular form using only the elementary row operations of type (v), which do not change the determinant. Therefore,

$$\begin{vmatrix} .202131 & .732543 & .141527 & .359867 \\ .333333 & -.112987 & .412989 & .838838 \\ -.486542 & .500000 & .989989 & -.246801 \\ .101101 & .321111 & -.444444 & .245542 \end{vmatrix}$$

$$\approx \begin{vmatrix} .202131 & .732543 & .141527 & .359867 \\ 0 & -1.32101 & .179598 & .245384 \\ 0 & 0 & 1.63832 & 1.03176 \\ 0 & 0 & 0 & .388032 \end{vmatrix}$$

$$= (.202131)(-1.32101)(1.63832)(.388032) \approx -.169748,$$

which is accurate to four decimal places.

5. First, we find that $\det(\mathbf{A}) = (0)(0) - (1)(1) = -1$. Next,

$$\mathbf{A}^{adj} = \begin{bmatrix} \mathrm{cof}(1,1) & \mathrm{cof}(2,1) \\ \mathrm{cof}(1,2) & \mathrm{cof}(2,2) \end{bmatrix} = \begin{bmatrix} 0 & -1 \\ -1 & 0 \end{bmatrix}.$$

Therefore,

$$\mathbf{A}^{-1} = \frac{1}{\det(\mathbf{A})}\mathbf{A}^{adj} = \frac{1}{-1}\begin{bmatrix} 0 & -1 \\ -1 & 0 \end{bmatrix} = \begin{bmatrix} 0 & 1 \\ 1 & 0 \end{bmatrix}.$$

7. Since, interchanging the second and the third rows in $\mathbf{A}$ results the identity matrix $\mathbf{I}$, $\det(\mathbf{A}) = -1$. Next we compute the entries of $\mathbf{A}^{adj}$. For the first row, we get

$$\mathrm{cof}(1,1) = (-1)^{1+1}\det\begin{bmatrix} 0 & 1 \\ 1 & 0 \end{bmatrix} = -1$$

$$\mathrm{cof}(2,1) = (-1)^{2+1}\det\begin{bmatrix} 0 & 0 \\ 1 & 0 \end{bmatrix} = 0$$

$$\mathrm{cof}(3,1) = (-1)^{3+1}\det\begin{bmatrix} 0 & 0 \\ 0 & 1 \end{bmatrix} = 0.$$

Similarly, we compute all other cofactors to conclude that

$$\mathbf{A}^{adj} = \begin{bmatrix} -1 & 0 & 0 \\ 0 & 0 & -1 \\ 0 & -1 & 0 \end{bmatrix}$$

so that

$$\mathbf{A}^{-1} = \frac{1}{\det(\mathbf{A})}\mathbf{A}^{\text{adj}} = \frac{1}{-1}\begin{bmatrix} -1 & 0 & 0 \\ 0 & 0 & -1 \\ 0 & -1 & 0 \end{bmatrix} = \begin{bmatrix} 1 & 0 & 0 \\ 0 & 0 & 1 \\ 0 & 1 & 0 \end{bmatrix}.$$

9. The matrix $\mathbf{A}$ is diagonal, so that

$$\det(\mathbf{A}) = (1)(\beta)(1) = \beta.$$

The entries of $\mathbf{A}^{\text{adj}}$ are as follows:

$$\begin{array}{llll} \rho_1: & \text{cof}(1,1) = \beta, & \text{cof}(2,1) = 0, & \text{cof}(3,1) = 0; \\ \rho_2: & \text{cof}(1,2) = 0, & \text{cof}(2,2) = 1, & \text{cof}(3,2) = 0; \\ \rho_3: & \text{cof}(1,3) = 0, & \text{cof}(2,3) = 0, & \text{cof}(3,1) = \beta. \end{array}$$

Therefore,

$$\mathbf{A}^{-1} = \frac{1}{\beta}\begin{bmatrix} \beta & 0 & 0 \\ 0 & 1 & 0 \\ 0 & 0 & \beta \end{bmatrix} = \begin{bmatrix} 1 & 0 & 0 \\ 0 & 1/\beta & 0 \\ 0 & 0 & 1 \end{bmatrix}.$$

11. To simplify computations, we note that

$$\mathbf{A} = \frac{1}{2}\begin{bmatrix} 3 & -2 & 1 \\ 1 & 0 & -1 \\ -3 & 2 & 1 \end{bmatrix} = \frac{1}{2}\mathbf{B}$$

so that $\mathbf{A}^{-1} = 2\mathbf{B}^{-1}$. To compute $\mathbf{B}^{-1}$, we find (by expanding along the second row) that

$$\det(\mathbf{B}) = (-1)^{2+1}(1)\begin{vmatrix} -2 & 1 \\ 2 & 1 \end{vmatrix} + (-1)^{2+3}(-1)\begin{vmatrix} 3 & -2 \\ -3 & 2 \end{vmatrix} = 4,$$

and that the entries of $\mathbf{B}^{adj}$ are

$$\text{cof}(1, 1) = \begin{vmatrix} 0 & -1 \\ 2 & 1 \end{vmatrix} = 2,$$

$$\text{cof}(2, 1) = - \begin{vmatrix} -2 & 1 \\ 2 & 1 \end{vmatrix} = 4,$$

$$\text{cof}(3, 1) = \begin{vmatrix} -2 & 1 \\ 0 & -1 \end{vmatrix} = 2;$$

$$\text{cof}(1, 2) = 2, \quad \text{cof}(2, 2) = 6, \quad \text{cof}(3, 2) = 4;$$

$$\text{cof}(1, 3) = 2, \quad \text{cof}(2, 3) = 0, \quad \text{cof}(3, 3) = 2.$$

Therefore,

$$\mathbf{B}^{-1} = \frac{1}{4} \begin{bmatrix} 2 & 4 & 2 \\ 2 & 6 & 4 \\ 2 & 0 & 2 \end{bmatrix}$$

and

$$\mathbf{A}^{-1} = \frac{1}{2} \begin{bmatrix} 2 & 4 & 2 \\ 2 & 6 & 4 \\ 2 & 0 & 2 \end{bmatrix} = \begin{bmatrix} 1 & 2 & 1 \\ 1 & 3 & 2 \\ 1 & 0 & 1 \end{bmatrix}.$$

13. We compute

$$\det(\mathbf{A}) = \det \begin{bmatrix} 2 & 2 \\ 3 & 1 \end{bmatrix} = -4;$$

$$x_1 = \frac{1}{-4} \det \begin{bmatrix} 4 & 2 \\ 2 & 1 \end{bmatrix} = 0;$$

$$x_2 = \frac{1}{-4} \det \begin{bmatrix} 2 & 4 \\ 3 & 2 \end{bmatrix} = 2.$$

15. We use the row echelon form of the coefficient matrix $\mathbf{A}$ (found in Example 2, Section 1.2) to conclude that

$$\det(\mathbf{A}) = (2)(3/2)(22) = 66.$$

Therefore,

$$x_1 = \frac{1}{66} \begin{vmatrix} -2 & 1 & 4 \\ 6 & 3 & 0 \\ -5 & 4 & 2 \end{vmatrix} = -\frac{6}{66} \begin{vmatrix} 1 & 4 \\ 4 & 2 \end{vmatrix} + \frac{3}{66} \begin{vmatrix} -2 & 4 \\ -5 & 2 \end{vmatrix} = 2;$$

$$x_2 = \frac{1}{66} \begin{vmatrix} 2 & -2 & 4 \\ 3 & 6 & 0 \\ -1 & -5 & 2 \end{vmatrix} = -\frac{3}{66} \begin{vmatrix} 2 & 4 \\ -5 & 2 \end{vmatrix} + \frac{6}{66} \begin{vmatrix} 2 & 4 \\ -2 & 2 \end{vmatrix} = 0;$$

$$x_3 = \frac{1}{66} \begin{vmatrix} 2 & 1 & -2 \\ 3 & 3 & 6 \\ -1 & 4 & -5 \end{vmatrix} = -\frac{3}{2},$$

which, of course, is the same as the answer obtained in Example 2, Section 1.2.

17. First, we compute $\det(\mathbf{A})$ (expanding it along the third row) to make sure that the Cramer's Rule applies.

$$\begin{vmatrix} 1 & 2 & 1 \\ 1 & 3 & 2 \\ 1 & 0 & 1 \end{vmatrix} = (-1)^{3+1}(1) \begin{vmatrix} 2 & 1 \\ 3 & 2 \end{vmatrix} + (-1)^{3+3}(1) \begin{vmatrix} 1 & 2 \\ 1 & 3 \end{vmatrix} = 2 \neq 0.$$

The Cramer's Rule yields

$$x_1 = \frac{1}{|\mathbf{A}|} \begin{vmatrix} 2 & 2 & 1 \\ \alpha & 3 & 2 \\ 1 & 0 & 0 \end{vmatrix} = \frac{1}{|\mathbf{A}|} \left[2\mathrm{cof}(1,1) + \alpha\mathrm{cof}(2,1) \right];$$

$$x_2 = \frac{1}{|\mathbf{A}|} \begin{vmatrix} 1 & 2 & 1 \\ 1 & \alpha & 2 \\ 1 & 0 & 1 \end{vmatrix} = \frac{1}{|\mathbf{A}|} \left[2\mathrm{cof}(1,2) + \alpha\mathrm{cof}(2,2) \right];$$

$$x_3 = \frac{1}{|\mathbf{A}|} \begin{vmatrix} 1 & 2 & 2 \\ 1 & 3 & \alpha \\ 1 & 0 & 0 \end{vmatrix} = \frac{1}{|\mathbf{A}|} \left[2\mathrm{cof}(1,3) + \alpha\mathrm{cof}(2,3) \right].$$

Therefore,

$$\frac{dx_1}{d\alpha} = \frac{\mathrm{cof}(2,1)}{|\mathbf{A}|}; \quad \frac{dx_2}{d\alpha} = \frac{\mathrm{cof}(2,2)}{|\mathbf{A}|}; \quad \frac{dx_3}{d\alpha} = \frac{\mathrm{cof}(2,3)}{|\mathbf{A}|}.$$

19. First, we compute $\det(\mathbf{A})$ and check for what values of α the Cramer's Rule can be used. Using the third row expansion, we get

$$\begin{vmatrix} 2 & 2 & 1 \\ \alpha & 3 & 2 \\ 1 & 0 & 1 \end{vmatrix} = \begin{vmatrix} 2 & 1 \\ 3 & 2 \end{vmatrix} + \begin{vmatrix} 2 & 2 \\ \alpha & 3 \end{vmatrix} = 7 - 2\alpha.$$

Therefore, the Cramer's Rule applies for any $\alpha \neq 7/2$ when $\det(\mathbf{A}) \neq 0$.

(a) Using the Cramer's Rule, we get

$$x_1 = \frac{1}{7 - 2\alpha} \begin{vmatrix} 2 & 2 & 1 \\ 3 & 3 & 2 \\ 4 & 0 & 1 \end{vmatrix} = \frac{4}{7 - 2\alpha}.$$

Therefore,

$$\frac{dx_1}{d\alpha} = 4\left[(7 - 2\alpha)^{-1}\right]' = \frac{8}{(7 - 2\alpha)^2}.$$

(b) We compute

$$x_2 = \frac{1}{7 - 2\alpha} \begin{vmatrix} 2 & 2 & 1 \\ \alpha & 3 & 2 \\ 1 & 4 & 1 \end{vmatrix} = -\frac{9 - 2\alpha}{7 - 2\alpha}.$$

The simplest way to find the derivative of x_2 is to notice that

$$x_2 = -\frac{(7 - 2\alpha) + (2)}{7 - 2\alpha} = -1 - \frac{2}{7 - 2\alpha} = -1 - \frac{1}{2}x_1$$

so that

$$\frac{dx_2}{d\alpha} = -\frac{1}{2} \cdot \frac{dx_1}{d\alpha} = -\frac{4}{(7 - 2\alpha)^2}.$$

21. Let's denote the given matrices by $\mathbf{A}$ and $\mathbf{B}$, resp., and let

$$
\mathbf{C}_4 = \begin{bmatrix} 0 & 0 & 0 & 1 \\ 0 & 0 & 1 & 0 \\ 0 & 1 & 0 & 0 \\ 1 & 0 & 0 & 0 \end{bmatrix}.
$$

The left-multiplication of any 4-by-4 matrix by $\mathbf{C}_4$ corresponds to two elementary row operations: $\rho_1 \leftrightarrow \rho_4$ and $\rho_2 \leftrightarrow \rho_3$. The right-multiplication by $\mathbf{C}_4$ performs the similar operations with the columns. Keeping this in mind, one can easily check that $\mathbf{A} = \mathbf{C}_4 \mathbf{B} \mathbf{C}_4$. Therefore,

$$
\begin{aligned}
\det(\mathbf{A}) &= \det(\mathbf{C}_4 \mathbf{B} \mathbf{C}_4) = \det(\mathbf{C}_4) \det(\mathbf{B}) \det(\mathbf{C}_4) \\
&= \det(\mathbf{C}_4)^2 \det(\mathbf{B}).
\end{aligned}
$$

Since $\det(\mathbf{C}_4) = 1$, we conclude that $\det(\mathbf{A}) = \det(\mathbf{B})$.

A generalization of this problem for n-by-n case can be stated as follows: For matrices $\mathbf{A}$ and $\mathbf{B}$ of the form

$$
\mathbf{A} = \begin{bmatrix}
a_{1,1} & a_{1,2} & \cdots & a_{1,n-1} & a_{1,n} \\
a_{2,1} & a_{2,2} & \cdots & a_{2,n-1} & a_{2,n} \\
\cdots & \cdots & \vdots & \cdots & \cdots \\
a_{n-1,1} & a_{n-1,2} & \cdots & a_{n-1,n-1} & a_{n-1,n} \\
a_{n,1} & a_{n,2} & \cdots & a_{n,n-1} & a_{n,n}
\end{bmatrix},
$$

$$
\mathbf{B} = \begin{bmatrix}
a_{n,n} & a_{n,n-1} & \cdots & a_{n,2} & a_{n,1} \\
a_{n-1,n} & a_{n-1,n-1} & \cdots & a_{n-1,2} & a_{n-1,1} \\
\cdots & \cdots & \vdots & \cdots & \cdots \\
a_{2,n} & a_{2,n-1} & \cdots & a_{2,2} & a_{2,1} \\
a_{1,n} & a_{1,n-1} & \cdots & a_{1,2} & a_{1,1}
\end{bmatrix},
$$

express $\det(\mathbf{A})$ in terms of $\det(\mathbf{B})$.

To solve this problem, we indicated the simple connection between these two matrices: to get $\mathbf{A}$ from $\mathbf{B}$, rewrite the rows and the columns of $\mathbf{B}$ backward. The matrix operations, that correspond to these two procedures, can be performed by the matrix $\mathbf{C}_n$ of the shape of $\mathbf{C}_4$: the left-multiplication of $\mathbf{B}$ by $\mathbf{C}_n$ writes its rows backward, while the right-multiplication writes backward the columns. Therefore, $\mathbf{A} = \mathbf{C}_n\mathbf{B}\mathbf{C}_n$. Since $\det(\mathbf{C}_n) = (-1)^n$, we get

$$\det(\mathbf{A}) = \det(\mathbf{C}_n\mathbf{B}\mathbf{C}_n) = \det(\mathbf{C}_n)^2\det(\mathbf{B})$$
$$= (-1)^{2n}\det(\mathbf{B}) = \det(\mathbf{B}).$$

23. We have

$$D_1 = \det[a] = a;$$

$$D_2 = \det\begin{bmatrix} a & b \\ b & a \end{bmatrix} = a^2 - b^2;$$

$$D_3 = \det\begin{bmatrix} a & b & 0 \\ b & a & b \\ 0 & b & a \end{bmatrix} = a\begin{vmatrix} a & b \\ b & a \end{vmatrix} - b\begin{vmatrix} b & b \\ 0 & a \end{vmatrix} = a^3 - 2ab^2.$$

25. (a) For $n \geq 3$, we use the first row expansion of F_n and then the first column expansion of the second determinant in the result to get

$$F_n = a \cdot \det\begin{bmatrix} a & -b & 0 & 0 & \cdots \\ b & a & -b & 0 & \cdots \\ 0 & b & a & -b & \cdots \\ \vdots & \vdots & \vdots & \vdots & \vdots \end{bmatrix}$$

$$+ b \cdot \det\begin{bmatrix} b & -b & 0 & 0 & \cdots \\ 0 & a & -b & 0 & \cdots \\ 0 & b & a & -b & \cdots \\ \vdots & \vdots & \vdots & \vdots & \vdots \end{bmatrix}$$

$$= aF_{n-1} + b^2\det\begin{bmatrix} a & -b & 0 & \cdots \\ b & a & -b & \cdots \\ \vdots & \vdots & \vdots & \vdots \end{bmatrix} = aF_{n-1} + b^2F_{n-2}.$$

(b) For $a = b = 1$, the recurrence relation in (a) becomes $F_n = F_{n-1} + F_{n-2}$, $n \geq 3$. For $n = 1$ and $n = 2$ we evaluate F_n directly.

Thus, the first six members of the sequence $\{F_n\}$ are

$$F_1 = \det[1] = 1; \qquad F_2 = \det \begin{bmatrix} 1 & -1 \\ 1 & 1 \end{bmatrix} = 2;$$
$$F_3 = F_2 + F_1 = 3; \quad F_4 = F_3 + F_2 = 5;$$
$$F_5 = F_4 + F_3 = 8; \quad F_6 = F_5 + F_4 = 13.$$

Defining $F_0 := 1$, we get the Fibonacci sequence

$$1, \ 1, \ 2, \ 3, \ 5, \ 8, \ 13, \ \ldots$$

since the recurrence relation obtained in part (a) is the same as the relation that defines the Fibonacci sequence.

27. (a) True, because $\det\left(A^T A\right) = \det\left(A^T\right)\det(A) = \det(A)^2$, and it is 1 if $A^T A = I$. Solving for $\det(A)$ yields $\det(A) = \pm 1$.

(b) Generally, it is False, because this determinant should be multiplied by $(-1)^{i+j}$ to get the cofactor.

(c) True, because $\left|S^{-1}AS\right| = \left|S^{-1}\right| \cdot |A| \cdot |S| = (1/|S|) \cdot |A| \cdot |S| = |A|$.

(d) True, because if $\det(A) = 0$, then (in the row echelon form) the rank(A) is less than the full rank and so this homogeneous system has infinitely many solutions. (See Theorem 3 and Corollary 1 in Section 1.4.)

29. (a) Using the first column expansion, we get

$$|A| = (1) \begin{vmatrix} 4 & 5 \\ 0 & 6 \end{vmatrix} = 24.$$

Next, we compute cofactors. For the first row entries of A^{adj},

$$\mathrm{cof}(1,1) = \begin{vmatrix} 4 & 5 \\ 0 & 6 \end{vmatrix} = 24, \quad \mathrm{cof}(2,1) = -\begin{vmatrix} 2 & 3 \\ 0 & 6 \end{vmatrix} = -12,$$
$$\mathrm{cof}(3,1) = \begin{vmatrix} 2 & 3 \\ 4 & 5 \end{vmatrix} = -2.$$

Similarly, we find

$$
\begin{array}{lll}
\cof(1,2) = 0, & \cof(2,2) = 6, & \cof(3,2) = -5; \\
\cof(1,3) = 0, & \cof(2,3) = 0, & \cof(3,3) = 4.
\end{array}
$$

Therefore,

$$
\mathbf{A}^{\mathrm{adj}} = \begin{bmatrix} 24 & -12 & -2 \\ 0 & 6 & -5 \\ 0 & 0 & 4 \end{bmatrix};
$$

$$
\mathbf{A}^{-1} = \frac{1}{|\mathbf{A}|}\mathbf{A}^{\mathrm{adj}} = \frac{1}{24}\begin{bmatrix} 24 & -12 & -2 \\ 0 & 6 & -5 \\ 0 & 0 & 4 \end{bmatrix}
$$

$$
= \begin{bmatrix} 1 & -1/2 & -1/12 \\ 0 & 1/4 & -5/24 \\ 0 & 0 & 1/6 \end{bmatrix}.
$$

(b) Using elementary row operations, we find that

$$
|\mathbf{A}| = \begin{vmatrix} 1 & 2 & 3 \\ 2 & 3 & 4 \\ 3 & 4 & 1 \end{vmatrix} = \begin{vmatrix} 1 & 2 & 3 \\ 0 & -1 & -2 \\ 0 & -2 & -8 \end{vmatrix} = 2\begin{vmatrix} 1 & 2 \\ 1 & 4 \end{vmatrix} = 4.
$$

Next, we find the entries of $\mathbf{A}^{\mathrm{adj}}$. For the first row of this matrix,

$$
\cof(1,1) = \begin{vmatrix} 3 & 4 \\ 4 & 1 \end{vmatrix} = -13, \quad \cof(2,1) = -\begin{vmatrix} 2 & 3 \\ 4 & 1 \end{vmatrix} = 10,
$$

$$
\cof(3,1) = \begin{vmatrix} 2 & 3 \\ 3 & 4 \end{vmatrix} = -1.
$$

Similarly, we find

$$
\begin{array}{lll}
\cof(1,2) = 10, & \cof(2,2) = -8, & \cof(3,2) = 2; \\
\cof(1,3) = -1, & \cof(2,3) = 2, & \cof(3,3) = -1.
\end{array}
$$

Therefore,

$$\mathbf{A}^{\text{adj}} = \begin{bmatrix} -13 & 10 & -1 \\ 10 & -8 & 2 \\ -1 & 2 & -1 \end{bmatrix};$$

$$\mathbf{A}^{-1} = \frac{\mathbf{A}^{\text{adj}}}{|\mathbf{A}|} = \frac{1}{4} \begin{bmatrix} -13 & 10 & -1 \\ 10 & -8 & 2 \\ -1 & 2 & -1 \end{bmatrix}$$

$$= \begin{bmatrix} -13/4 & 5/2 & -1/4 \\ 5/2 & -2 & 1/2 \\ -1/4 & 1/2 & -1/4 \end{bmatrix}.$$

31. Using the product rule (2) for determinants, statements (ix) (in the first question) and (1) (in the second question) we get

$$\left|\mathbf{BAB}^{-1}\right| = |\mathbf{B}| \cdot |\mathbf{A}| \cdot \left|\mathbf{B}^{-1}\right| = |\mathbf{B}| \cdot |\mathbf{A}| \cdot \frac{1}{|\mathbf{B}|} = |\mathbf{A}| ;$$

$$\left|\mathbf{BAB}^{T}\right| = |\mathbf{B}| \cdot |\mathbf{A}| \cdot \left|\mathbf{B}^{T}\right| = |\mathbf{B}| \cdot |\mathbf{A}| \cdot |\mathbf{B}| = |\mathbf{B}|^{2} |\mathbf{A}| .$$

PART I

REVIEW PROBLEMS FOR PART I

1. We use the Gauss elimination to get the coefficient matrix in row echelon form.

$$\begin{bmatrix} -3 & 2 & 1 & \vdots & 4 \\ 1 & -1 & 2 & \vdots & 1 \\ -2 & 1 & 3 & \vdots & \gamma \end{bmatrix} \quad \rho_1 \leftrightarrow \rho_2 \quad \begin{bmatrix} 1 & -1 & 2 & \vdots & 1 \\ -3 & 2 & 1 & \vdots & 4 \\ -2 & 1 & 3 & \vdots & \gamma \end{bmatrix}$$

$$\begin{matrix} \rho_2 + 3\rho_1 \rightarrow \rho_2 \\ \rho_3 + 2\rho_1 \rightarrow \rho_3 \end{matrix} \quad \begin{bmatrix} 1 & -1 & 2 & \vdots & 1 \\ 0 & -1 & 7 & \vdots & 7 \\ 0 & -1 & 7 & \vdots & \gamma + 2 \end{bmatrix}$$

$$\rho_3 - \rho_2 \rightarrow \rho_3 \quad \begin{bmatrix} 1 & -1 & 2 & \vdots & 1 \\ 0 & -1 & 7 & \vdots & 7 \\ 0 & 0 & 0 & \vdots & \gamma - 5 \end{bmatrix} .$$

The system is consistent if and only if the rank of the coefficient matrix (which is 2) equals the rank of the augmented matrix, which is 2 only for $\gamma - 5 = 0$, or $\gamma = 5$. In this case, the system has infinitely many

Solutions Manual to Accompany Fundamentals of Matrix Analysis with Applications,
First Edition. Edward Barry Saff and Arthur David Snider.
© 2016 John Wiley & Sons, Inc. Published 2016 by John Wiley & Sons, Inc.

solutions because the rank of the coefficient matrix is less than the full rank (i.e., 3).

3. (a)

$$\begin{bmatrix} 2 & 3 & -1 \\ 4 & 2 & 3 \end{bmatrix} \begin{bmatrix} 2 \\ 4 \\ 6 \end{bmatrix} = \begin{bmatrix} (2)(2)+(3)(4)+(-1)(6) \\ (4)(2)+(2)(4)+(3)(6) \end{bmatrix} = \begin{bmatrix} 10 \\ 34 \end{bmatrix}.$$

(b)

$$\begin{bmatrix} 2 & 3 & -1 \end{bmatrix} \begin{bmatrix} 2 & 0 & 1 \\ 4 & 2 & -1 \\ 6 & 2 & 3 \end{bmatrix}$$
$$= \begin{bmatrix} 4+12-6 & 0+6-2 & 2-3-3 \end{bmatrix}$$
$$= \begin{bmatrix} 0 & 4 & -4 \end{bmatrix}.$$

(c)

$$\begin{bmatrix} 2 \\ 3 \\ -1 \end{bmatrix} \begin{bmatrix} 2 & 4 & 6 \end{bmatrix} = \begin{bmatrix} (2)(2) & (2)(4) & (2)(6) \\ (3)(2) & (3)(4) & (3)(6) \\ (-1)(2) & (-1)(4) & (-1)(6) \end{bmatrix}$$
$$= \begin{bmatrix} 4 & 8 & 12 \\ 6 & 12 & 18 \\ -2 & -4 & -6 \end{bmatrix}.$$

(d)

$$\begin{bmatrix} 2 & 3 & -1 \\ 4 & 2 & 3 \end{bmatrix} \begin{bmatrix} 1 & 0 & 0 \\ 0 & 1 & 0 \\ 0 & 0 & 1 \end{bmatrix} = \begin{bmatrix} 2 & 3 & -1 \\ 4 & 2 & 3 \end{bmatrix} I_3$$
$$= \begin{bmatrix} 2 & 3 & -1 \\ 4 & 2 & 3 \end{bmatrix}.$$

(e) One can go with direct computations. Another (simpler) way to go is to notice that the given matrix **A** performs the following elementary row operations: $\rho_1 \leftrightarrow \rho_2$ and $\rho_3 \leftrightarrow \rho_4$. Therefore, $A^2 = AA$ changes the rows back to the original. Thus, $A^2 = I$ and

$$A^3 = \left(A^2\right) A = IA = A = \begin{bmatrix} 0 & 1 & 0 & 0 \\ 1 & 0 & 0 & 0 \\ 0 & 0 & 0 & 1 \\ 0 & 0 & 1 & 0 \end{bmatrix}.$$

(f) We represent the given matrix A as

$$A = \begin{bmatrix} -1 & 0 & 0 \\ 0 & 1 & 0 \\ 0 & 0 & -1 \end{bmatrix} + \begin{bmatrix} 0 & 0 & 1 \\ 0 & 0 & 0 \\ 0 & 0 & 0 \end{bmatrix} = B + C.$$

Concerning matrices B and C, we remark the following:

$$B^2 = I \quad \Rightarrow \quad B^{2k} = \left(B^2\right)^k = I,$$
$$B^{2k+1} = \left(B^2\right)^k B = B, \; k \geq 1;$$
$$C^2 = 0 \quad \Rightarrow \quad C^n = \left(C^2\right) C^{n-2} = 0, \; n \geq 2.$$

Since the matrices B and C commute, i.e.,

$$BC = CB \left(= \begin{bmatrix} 0 & 0 & -1 \\ 0 & 0 & 0 \\ 0 & 0 & 0 \end{bmatrix} \right),$$

similarly to Problem 10 in Exercises 2.1 we can conclude that the binomial formula

$$(B + C)^{25} = \sum_{n=0}^{25} C_{25}^n B^{25-n} C^n$$

holds. In this sum, only the first two terms (for $n = 0, 1$) survive, and we have

$$\mathbf{A}^{25} = C_{25}^0 \mathbf{B}^{25} \mathbf{C}^0 + C_{25}^1 \mathbf{B}^{24} \mathbf{C}^1 = \mathbf{B} + 25\mathbf{C}$$

$$= \begin{bmatrix} -1 & 0 & 0 \\ 0 & 1 & 0 \\ 0 & 0 & -1 \end{bmatrix} + 25 \begin{bmatrix} 0 & 0 & 1 \\ 0 & 0 & 0 \\ 0 & 0 & 0 \end{bmatrix}$$

$$= \begin{bmatrix} -1 & 0 & 25 \\ 0 & 1 & 0 \\ 0 & 0 & -1 \end{bmatrix}.$$

5. We can take

$$\mathbf{A} = \mathbf{B} = \begin{bmatrix} 0 & 1 \\ 0 & 0 \end{bmatrix} \quad \Rightarrow \quad \mathbf{A}^2 + \mathbf{B}^2 = 2\mathbf{A}^2 = 2 \cdot \mathbf{0} = \mathbf{0}.$$

7. (a) Using the notations in Theorem 1, Section 1.1, the matrix $\mathbf{E}$, that performs the corresponding elementary row operation with a matrix $\mathbf{A}$ via the left multiplication is as follows. To get $\mathbf{E}$, in the identity matrix $\mathbf{I}$,

(i) to add c times the ith equation to the jth equation, $i \neq j$, replace in the row ρ_j the zero ith entry by c;

(ii) to multiply the jth equation by $c \neq 0$, do this with ρ_j, i.e., replace the diagonal element 1 by c;

(iii) to reorder the equations, make the equivalent reordering of the rows of $\mathbf{I}$. For example, to switch the first and the second equations, do $\rho_1 \leftrightarrow \rho_2$.

(b) Following the concept of "undo" for elementary row operations, we describe the inverses $\mathbf{E}^{-1}$ of matrices $\mathbf{E}$ in part (a) as follows: In $\mathbf{I}$,

(i) replace in ρ_j the zero ith entry by $-c$;

(ii) replace the diagonal element 1 by $1/c$;

(iii) $\mathbf{E}^{-1} = \mathbf{E}$. For instance, $\mathbf{E}^{-1}$ for $\mathbf{E}$ in example in part (a) (iii), $\mathbf{E}^{-1}$ switches the first and the second row in $\mathbf{I}$.

(c) The right multiplication of $\mathbf{A}$ by $\mathbf{E}$ performs the operations with the *columns* of $\mathbf{A}$ similar to those for rows when $\mathbf{A}$ is multiplied by $\mathbf{E}$ from the left. For example,

(i) The left multiplication of a 2-by-2 matrix $\mathbf{A}$ by

$$\mathbf{E} = \begin{bmatrix} 1 & 0 \\ c & 1 \end{bmatrix}$$

replaces the second row by the sum of the second row and the first row multiplied by c, while the right multiplication of $\mathbf{A}$ by $\mathbf{E}$ replaces the first column by the sum of the first column and c times the second column.

(ii) The left multiplication of a 2-by-2 matrix $\mathbf{A}$ by

$$\mathbf{E} = \begin{bmatrix} 1 & 0 \\ 0 & c \end{bmatrix}$$

multiplies the second row of $\mathbf{A}$ by c, while the right multiplication multiplies the second column by c.

(iii) Switching the rows of a 2-by-2 matrix $\mathbf{A}$ corresponds to its left multiplication by

$$\mathbf{E} = \begin{bmatrix} 0 & 1 \\ 1 & 0 \end{bmatrix},$$

while the product $\mathbf{AE}$ switches the columns of $\mathbf{A}$.

(d) The matrix $\mathbf{E}$ corresponding to the elementary row operation of switching two rows of $\mathbf{A}$ obviously satisfies $\mathbf{E}^{-1} = \mathbf{E}$, since $\mathbf{E}^2 = \mathbf{E}(\mathbf{E}) = \mathbf{I}$. Therefore, by Theorem 3 in Section 2.3,

$$(\mathbf{A}')^{-1} = (\mathbf{EA})^{-1} = \mathbf{A}^{-1}\mathbf{E}^{-1} = \mathbf{A}^{-1}\mathbf{E}.$$

According to the part (c), this right multiplication of $\mathbf{A}^{-1}$ by $\mathbf{E}$ just switches the corresponding columns of $\mathbf{A}^{-1}$.

(e) The elementary row operation corresponding to multiplying the second row of $\mathbf{A}$ by 7 can be written in the matrix form as

$$\mathbf{A}' = \begin{bmatrix} 1 & 0 & 0 & \cdots & 0 \\ 0 & 7 & 0 & \cdots & 0 \\ 0 & 0 & 1 & \cdots & 0 \\ & & \vdots & & \\ 0 & 0 & 0 & \cdots & 1 \end{bmatrix} \mathbf{A} = \mathbf{EA}.$$

Therefore, by Theorem 3 in Section 2.3,

$$
(\mathbf{A}')^{-1} = \mathbf{A}^{-1}\mathbf{E}^{-1} = \mathbf{A}^{-1}
\begin{bmatrix}
1 & 0 & 0 & \cdots & 0 \\
0 & 7 & 0 & \cdots & 0 \\
0 & 0 & 1 & \cdots & 0 \\
& & \vdots & & \\
0 & 0 & 0 & \cdots & 1
\end{bmatrix}^{-1}
$$

$$
= \mathbf{A}^{-1}
\begin{bmatrix}
1 & 0 & 0 & \cdots & 0 \\
0 & 1/7 & 0 & \cdots & 0 \\
0 & 0 & 1 & \cdots & 0 \\
& & \vdots & & \\
0 & 0 & 0 & \cdots & 1
\end{bmatrix},
$$

where we have used the "undo" concept to simplify the computation of $\mathbf{E}^{-1}$. According to the part (c), this right multiplication of $\mathbf{A}^{-1}$ by $\mathbf{E}^{-1}$ just multiplies the second column of $\mathbf{A}^{-1}$ by $1/7$.

(f) The elementary row operation corresponding to adding the first row of $\mathbf{A}$ to its second row can be written in the matrix form as

$$
\mathbf{A}' =
\begin{bmatrix}
1 & 0 & 0 & \cdots & 0 \\
1 & 1 & 0 & \cdots & 0 \\
0 & 0 & 1 & \cdots & 0 \\
& & \vdots & & \\
0 & 0 & 0 & \cdots & 1
\end{bmatrix} \mathbf{A} = \mathbf{E}\mathbf{A} \quad \Rightarrow
$$

$$
(\mathbf{A}')^{-1} = \mathbf{A}^{-1}
\begin{bmatrix}
1 & 0 & 0 & \cdots & 0 \\
1 & 1 & 0 & \cdots & 0 \\
0 & 0 & 1 & \cdots & 0 \\
& & \vdots & & \\
0 & 0 & 0 & \cdots & 1
\end{bmatrix}^{-1}
$$

$$
= \mathbf{A}^{-1}
\begin{bmatrix}
1 & 0 & 0 & \cdots & 0 \\
-1 & 1 & 0 & \cdots & 0 \\
0 & 0 & 1 & \cdots & 0 \\
& & \vdots & & \\
0 & 0 & 0 & \cdots & 1
\end{bmatrix}
$$

where we have used the "undo" concept to simplify the computation of $\mathbf{E}^{-1}$. According to the part (c), this right multiplication of

$\mathbf{A}^{-1}$ by $\mathbf{E}^{-1}$ just subtracts the second column of $\mathbf{A}^{-1}$ from its first column.

9. Let $\mathbf{M}$ be an r-by-s matrix, $\mathbf{A} = [a_k]$ is the 1-by-r row vector with $a_i = 1$ and all other zero entries, and $\mathbf{B} = [b_k]$ is the s-by-1 column vector with $b_j = 1$ and all other zero entries. Then

$$\mathbf{AMB} = \begin{bmatrix} 0 & \cdots & 1 & 0 & \cdots \end{bmatrix} \begin{bmatrix} m_{11} & \cdots & m_{1j} & \cdots & m_{1s} \\ & & \vdots & & \\ m_{i1} & \cdots & m_{ij} & \cdots & m_{is} \\ & & \vdots & & \\ m_{r1} & \cdots & m_{rj} & \cdots & m_{rs} \end{bmatrix} \begin{bmatrix} 0 \\ \vdots \\ 1 \\ 0 \\ \vdots \end{bmatrix}$$

$$= \begin{bmatrix} m_{i1} & \cdots & m_{ij} & \cdots & m_{is} \end{bmatrix} \begin{bmatrix} 0 & \cdots & 1 & 0 & \cdots \end{bmatrix}^T = m_{ij}.$$

11. (a) True. Moreover, $\left(\mathbf{A}^2\right)^{-1} = (\mathbf{AA})^{-1} = \left(\mathbf{A}^{-1}\right)\left(\mathbf{A}^{-1}\right) = \left(\mathbf{A}^{-1}\right)^2.$
    (b) True. If

$$\mathbf{x} = \begin{bmatrix} x_1 \\ \vdots \\ x_n \end{bmatrix},$$

then

$$\mathbf{xy} = \begin{bmatrix} x_1\mathbf{y} \\ \vdots \\ x_n\mathbf{y} \end{bmatrix}.$$

If $\mathbf{x} = \mathbf{0}$, then $\mathbf{xy} = \mathbf{0}$, which is a singular matrix. If $\mathbf{x} \neq \mathbf{0}$, then it has at least one nonzero entry, say x_k. In this case, in the product $\mathbf{xy}$, the rows satisfy

$$\rho_i = \frac{x_i}{x_k}\rho_k,$$

i.e., the (all) rows of $\mathbf{xy}$ are multiples of one of them; that is, of x_k. Even one of these relations imply that $\det(\mathbf{xy}) = 0$ (see properties (i) and (iv) in Section 2.5) so that the matrix is singular due to the property (viii) in the same section.

(c) True. (See Theorem 2 in Section 2.3.)

(d) False. For a counterexample, see Problem 13(b) in Exercises 2.1.

(e) Can be False. For example, the row vector $x = [1\ 0\ 0]$ cannot be expressed as a linear combination of the given three vectors. Indeed, in trying to find such representation we have to solve

$$c_1[1\ 3\ -1] + c_2[4\ 6\ 1] + c_3[1\ -3\ 4] = [1\ 0\ 0],$$

which, in the matrix form, is equivalent to

$$\begin{bmatrix} 1 & 4 & 1 \\ 3 & 6 & -3 \\ -1 & 1 & 4 \end{bmatrix} \begin{bmatrix} c_1 \\ c_2 \\ c_3 \end{bmatrix} = \begin{bmatrix} 1 \\ 0 \\ 0 \end{bmatrix}.$$

Since the $\det(A)$ of the coefficient matrix A is

$$\det(A) = (1)\begin{vmatrix} 6 & -3 \\ 1 & 4 \end{vmatrix} - (4)\begin{vmatrix} 3 & -3 \\ -1 & 4 \end{vmatrix} + (1)\begin{vmatrix} 3 & 6 \\ -1 & 1 \end{vmatrix} = 0,$$

the rank of A is less than the full rank. (Combine (viii) in Section 2.5 and Theorem 2 in Section 2.3.) On the other hand, because

$$\begin{vmatrix} 1 & 4 & 1 \\ 3 & 6 & 0 \\ -1 & 1 & 0 \end{vmatrix} = (1)\begin{vmatrix} 3 & 6 \\ -1 & 1 \end{vmatrix} = 9 \neq 0,$$

the augmented matrix has the full rank. By Theorem 3 in Section 1.4, the system has no solution.

(f) False. For example, for $n = 2$,

$$\mathbf{A} = \begin{bmatrix} 1 & 0 \\ 1 & 0 \end{bmatrix};$$

$$\mathbf{A}\mathbf{A}^T = \begin{bmatrix} 1 & 0 \\ 1 & 0 \end{bmatrix} \begin{bmatrix} 1 & 1 \\ 0 & 0 \end{bmatrix} = \begin{bmatrix} 1 & 1 \\ 1 & 1 \end{bmatrix};$$

$$\mathbf{A}^T\mathbf{A} = \begin{bmatrix} 1 & 1 \\ 0 & 0 \end{bmatrix} \begin{bmatrix} 1 & 0 \\ 1 & 0 \end{bmatrix} = \begin{bmatrix} 2 & 0 \\ 0 & 0 \end{bmatrix}.$$

13. Since, for any $\mathbf{v}$, $\mathbf{v}^T\mathbf{K}\mathbf{v}$ is a number, we have

$$\mathbf{v}^T\mathbf{K}\mathbf{v} = \left(\mathbf{v}^T\mathbf{K}\mathbf{v}\right)^T = \mathbf{v}^T\mathbf{K}^T\left(\mathbf{v}^T\right) = \mathbf{v}^T(-\mathbf{K})\mathbf{v} = -\mathbf{v}^T\mathbf{K}\mathbf{v}$$

so that $\mathbf{v}^T\mathbf{K}\mathbf{v} = 0$.

Multiplying the homogeneous equation $(\mathbf{I} + \mathbf{K})\,\mathbf{x} = \mathbf{x} + \mathbf{K}\mathbf{x} = \mathbf{0}$ from the left by $\mathbf{x}^T$, we get

$$\mathbf{x}^T\mathbf{x} + \mathbf{x}^T\mathbf{K}\mathbf{x} = \mathbf{x}^T\mathbf{x} = x_1^2 + x_2^2 + x_3^2 = 0,$$

meaning that $x_1 = x_2 = x_3 = 0$, where x_j's are the components of $\mathbf{x}$. Thus, this homogeneous equation has only the trivial solution and, by Theorem 2 in Section 2.3 , the matrix $\mathbf{I} + \mathbf{K}$ is invertible.

15. (a) First, expanding the determinant using the second row, we get

$$\det(\mathbf{A}) = (-1) \begin{vmatrix} -3 & 1 \\ -2 & 3 \end{vmatrix} - (2) \begin{vmatrix} -3 & 2 \\ -2 & 1 \end{vmatrix} = 7 - 2 = 5.$$

By Cramer's Rule,

$$x_1 = \frac{1}{5} \begin{vmatrix} 4 & 2 & 1 \\ 1 & -1 & 2 \\ \gamma & 1 & 3 \end{vmatrix}$$

$$= \frac{1}{5} \left((4) \begin{vmatrix} -1 & 2 \\ 1 & 3 \end{vmatrix} - (1) \begin{vmatrix} 2 & 1 \\ 1 & 3 \end{vmatrix} + (\gamma) \begin{vmatrix} 2 & 1 \\ -1 & 2 \end{vmatrix} \right)$$

$$= \frac{5\gamma - 25}{5}.$$

Similarly, we find that

$$x_2 = \frac{6\gamma - 23}{5}, \quad x_3 = \frac{3\gamma - 9}{5}.$$

Therefore,

$$\frac{dx_1}{d\gamma} = 1, \quad \frac{dx_2}{d\gamma} = \frac{6}{5}, \quad \frac{dx_3}{d\gamma} = \frac{3}{5}.$$

(b) Expanding the determinant of the coefficient matrix $\mathbf{A}$ using, say, the first column, we get

$$\det(\mathbf{A}) = (-3)\begin{vmatrix} -1 & 2 \\ 1 & 3 \end{vmatrix} - (\gamma)\begin{vmatrix} 2 & 1 \\ 1 & 3 \end{vmatrix} + (-2)\begin{vmatrix} 2 & 1 \\ -1 & 2 \end{vmatrix}$$

$$= 5 - 5\gamma.$$

Applying the Cramer's Rule, we solve the system when $\det(\mathbf{A}) \neq 0 \Leftrightarrow \gamma \neq 1.$

$$x_1 = \frac{1}{5 - 5\gamma}\begin{vmatrix} 4 & 2 & 1 \\ 1 & -1 & 2 \\ 0 & 1 & 3 \end{vmatrix}$$

$$= \frac{1}{5 - 5\gamma}\left((4)\begin{vmatrix} -1 & 2 \\ 1 & 3 \end{vmatrix} - (1)\begin{vmatrix} 2 & 1 \\ 1 & 3 \end{vmatrix} \right) = \frac{25}{5\gamma - 5}.$$

Similarly, we compute

$$x_2 = \frac{12\gamma + 23}{5\gamma - 5}, \quad x_3 = \frac{4\gamma - 9}{5 - 5\gamma}.$$

Therefore,

$$\frac{dx_1}{d\gamma} = -\frac{125}{(5\gamma - 5)^2}, \quad \frac{dx_2}{d\gamma} = -\frac{175}{(5\gamma - 5)^2},$$

$$\frac{dx_3}{d\gamma} = -\frac{25}{(5\gamma - 5)^2}.$$

3

VECTOR SPACES

3.1 GENERAL SPACES, SUBSPACES, AND SPANS

1. Collection (a) is the set of all solutions of

$$\begin{bmatrix} 1 & 1 & 1 \end{bmatrix} \mathbf{x} = 0,$$

so it suffices to use Theorem 1 of Section 3.1 to conclude it is a subspace. Furthermore, vector addition and scalar multiplication preserve the form of vectors in (c) making it a subspace as well.

On the other hand, (b) and (d) are both not subspaces. Indeed, consider vectors $[1 -1\ 0]$, $[1\ 0 -1]$ from (b). Their sum is $[2 -1 -1]$, not in (b). Vectors $[1\ 1\ 0]$, $[0\ 0\ 1]$ both belong to (d), yet their sum $[1\ 1\ 1]$ doesn't.

3. The set (c) is not a subspace because the function $f(x) \equiv 1$ belongs to (c) but $2f(x) \equiv 2$ doesn't.

To verify that a set is a subspace of $C[-1, 1]$ we must check that for any pair of its elements f, g and any scalars α, β, the linear combination $\alpha f + \beta g$ also belongs to the set. From linearity of integration and differentiation:

Solutions Manual to Accompany Fundamentals of Matrix Analysis with Applications,
First Edition. Edward Barry Saff and Arthur David Snider.
© 2016 John Wiley & Sons, Inc. Published 2016 by John Wiley & Sons, Inc.

(a) $\int_{-1}^{1} \alpha f(x) + \beta g(x)\, dx = \alpha \int_{-1}^{1} f(x)\, dx + \beta \int_{-1}^{1} g(x)\, dx = 0.$

(b) $(\alpha f(x) + \beta g(x))''' = \alpha f'''(x) + \beta g'''(x)$ is a sum of two functions, continuous on $[-1, 1]$, and is therefore continuous itself.

(d) $(\alpha f + \beta g)'(0) = \alpha f'(0) + \beta g'(0) = 0.$

From the above, (a), (b) and (d) are subspaces.

5. (a) Consider a polynomial $p(x)$ from $\mathbf{P}_2$. Then its Taylor series is precisely $p(x) = p(1) + p'(1)(x - 1) + p''(x)(x - 1)^2/2$ since all the higher derivatives are zero. This expression is the linear combination of polynomials 1, $x - 1$, $(x - 1)^2$ with coefficients $p(1)$, $p'(1)$, $p''(1)/2$ respectively.

(b) It is clear that $\mathbf{P}_2$ is spanned by the polynomials 1, x, x^2. Let us identify every polynomial $a + bx + cx^2$ with the vector $a\mathbf{i} + b\mathbf{j} + c\mathbf{k}$ of $\mathbf{R}^3$. It is obviously a 1-to-1 correspondence. If it were that the whole of $\mathbf{P}_2$ is spanned by a pair of polynomials, corresponding vectors of $\mathbf{R}^3$ would span the whole space. But as known from geometry, no two vectors span $\mathbf{R}^3$.

7. As noted in Section 3.1, a subset of a vector space is a subspace if and only if it is closed under vector addition and scalar multiplication. It is easy to see that for any two symmetric/upper triangular/skew-symmetric matrices, their sum is also an element of the corresponding class of matrices (because all of the above properties are preserved under addition). The same is true for a scalar multiple of a matrix from any of these subsets of $\mathbf{R}^{3,3}$, and therefore (a), (b), (e) are subspaces of $\mathbf{R}^{3,3}$.

To verify that (c) and (f) are subspaces as well, we identify matrices $\mathbf{R}^{3,3}$ with appropriate column vectors of $\mathbf{R}^9_{col}$. Then the sets (c) and (f) are all the vectors $\mathbf{x}$ of $\mathbf{R}^9_{col}$ such that

$$\left(\begin{bmatrix} 1 & 0 & 0 \\ 0 & 1 & 0 \\ 0 & 0 & 1 \end{bmatrix}_{\mathbf{R}^9_{col}} \right)^T \mathbf{x} = 0 \text{ for } (c), \qquad \left(\begin{bmatrix} 1 & 2 & 3 \\ 5 & 6 & 7 \\ 8 & 9 & 10 \end{bmatrix}_{\mathbf{R}^9_{col}} \right)^T \mathbf{x} = 0 \text{ for } (f),$$

where the parentheses contain the element of $\mathbf{R}^9_{col}$ corresponding to the matrix. Thus in $\mathbf{R}^9_{col}$ the sets (c) and (f) are all the vectors whose scalar product with a fixed vector is zero. It remains to observe that by Theorem 1 from Section 3.1, sets of solutions of linear homogeneous algebraic systems are vector spaces (and equality to zero of scalar product with a fixed vector is indeed a homogeneous system, see Exercise 1 above).

Finally, the set (d) is not a subspace since it does not contain the zero matrix.

9. (a) True: the first vector multiplied by $1/2$ is $[1,\ 0]$, and the second multiplied by $1/\pi$ is $[0,\ 1]$, so it suffices to note that vectors $[1,\ 0]$ and $[0,\ 1]$ span $\mathbf{R}^2_{\text{row}}$.

   (b) False: take $\mathbf{w}_1$, $\mathbf{w}_2$ to be the basis vectors from (a). Let furthermore $\mathbf{v}_1 = \mathbf{0} = 0\mathbf{w}_1 + 0\mathbf{w}_2$ and $\mathbf{v}_2 = \mathbf{w}_2$. Then it is immediate that $\mathbf{w}_i$'s span the whole $\mathbf{R}^2_{\text{row}}$ while $\mathbf{v}_i$'s do not.

   (c) True: take $\{1,\ i\}$ to be the spanning vectors. Any complex number can be represented as $\alpha 1 + \beta i$ with real α, β. All the properties of a vector space are trivially satisfied because of definition of operations with complex numbers.

   (d) True: by definition, any subspace contains the zero vector $\mathbf{0}$.

11. (a) Because both vector operations (addiiton and scalar multiplication) are defined elementwise, the sequence consisting entirely of zeros $(0,0,\ldots)$ is the zero vector.

   In (b)–(d) we verify that for any two elements of the given subset, their linear combination also belongs to the subset.

   (b) Yes: for a pair of sequences $\mathbf{s}$, $\mathbf{s}'$ converging to c and c' respectively, and a pair of scalars α, β the sequence $\alpha\mathbf{s} + \beta\mathbf{s}'$ is a converging sequence with the limit $\alpha c + \beta c'$. The given subset is therefore closed under addition and scalar multiplication and is a subspace.

   (c) Yes: Let us write $(\mathbf{s})_i$ for the i-th element of sequence $\mathbf{s}$. In view of the triangle inequality for absolute value, for any two sequences $\mathbf{s}$, $\mathbf{s}'$ bounded by constants M, M' respectively, and any scalars α, β it holds: $|(\alpha\mathbf{s} + \beta\mathbf{s}')_i| \leq |\alpha s_i| + |\beta s'_i| \leq |\alpha|M + |\beta|M'$ for all $i = 0,\ 1,\ldots$, and therefore $\alpha\mathbf{s} + \beta\mathbf{s}'$ is also a bounded sequence.

   (d) Yes: in the above notation, $\sum_{i=0}^{\infty}(as+bs')_i = a\sum_{i=0}^{\infty}s_i + b\sum_{i=0}^{\infty}s'_i$, in other words, a linear combination of convergent series is a convergent series itself.

13. First of all note that it is only necessary to verify that the given collection is a subspace of the space of twice differentiable functions,

the latter being a vector space by the argument similar to that in Exercise 3.(b).

Let now f, g be functions satisfying the given equation for $x > 0$, and let α, β be any two scalars. Substituting $\alpha f + \beta g$ into the differential equation gives

$$x^2(\alpha f + \beta g)'' + e^x(\alpha f + \beta g)' - (\alpha f + \beta g)$$
$$= \alpha(x^2 f'' + e^x f' - f) + \beta(x^2 g'' + e^x g' - g) = 0.$$

The given set is therefore closed under vector addition and scalar multiplication, and is a subspace.

15. Just as above, we will use that to verify whether a subset of a vector space is a subspace it suffices to check whether it is closed under addition and scalar multiplication. Let $\mathbf{v} = \mathbf{s} + \mathbf{t}$ and $\mathbf{v}' = \mathbf{s}' + \mathbf{t}'$ be vectors of the given form, i.e., $\mathbf{s}$, $\mathbf{s}' \in S$, $\mathbf{t}$, $\mathbf{t}' \in T$. Then for any scalars α, β, we have

$$\alpha\mathbf{v} + \beta\mathbf{v}' = (\alpha\mathbf{s} + \beta\mathbf{s}') + (\alpha\mathbf{t} + \beta\mathbf{t}'),$$

where the inside of the first pair of parentheses belongs to S, and the inside of the second belongs to T because of S, T being subspaces. Hence the whole last display has the required form, and the given subset is indeed a subspace of V.

17. One can classify subspaces of $\mathbf{R}^2_{\text{col}}$ by the number of linearly independent vectors inside them that span the whole subspace (it is called the *dimension* of the subspace, see Section 3.2). It is an elementary geometric fact that a plane can contain at most two linearly independent vectors. A subspace containing two such vectors is therefore the whole plane. A subspace with only one linearly-independent vector is the line spanned by it (and passing through the origin). This leaves us with the last case of a subspace containing only the origin.

By a similar argument, for $\mathbf{R}^3_{\text{col}}$ one may have a subspace consisting of the whole space, a plane passing through the origin, a line passing through the origin, the origin itself, depending on the number of linearly-independent vectors that span the given subspace.

19. No, they don't. The two sets of vectors span the same subspace if and only if vector $[2, \ 1, \ 1]^T$ lies in the span of vectors $[1, \ 1, \ 1]^T$ and $[1, \ 0, \ -1]^T$. To check if this is indeed the case, let

$$\alpha \begin{bmatrix} 1 \\ 1 \\ 1 \end{bmatrix} + \beta \begin{bmatrix} 1 \\ 0 \\ -1 \end{bmatrix} = \begin{bmatrix} 2 \\ 1 \\ 1 \end{bmatrix}.$$

This system is not consistent: the second row gives $\alpha = 1$, from the third one $\beta = 0$, which does not satisfy the first row.

21. Assume that x lies in the span of $\{e^x, \sin x\}$, that is, $x = c_1 e^x + c_2 \sin x$ holds for all $-\infty < x < \infty$. Substituting $x = 0$ and $x = \pi$ in the last equation gives a system for constants c_1, c_2:

$$\begin{bmatrix} 1 & 0 & \vdots & 0 \\ e^\pi & 0 & \vdots & \pi \end{bmatrix}.$$

This system is obviously not consistent, and we arrive at a contradiction.

3.2   LINEAR DEPENDENCE

1. Vectors (c) and (d) clearly lie in the span of $\{\mathbf{a}_1, \mathbf{a}_2\}$ as the former equals $\mathbf{a}_1$, and the latter is the trivial combination $0\mathbf{a}_1 + 0\mathbf{a}_2$. We will only consider (a) and (b). Reducing $\begin{bmatrix} 1 & 2 & \vdots & 1 & 3 \\ -1 & 5 & \vdots & 6 & 4 \\ 2 & 6 & \vdots & 7 & 8 \end{bmatrix}$ to the row echelon form gives $\begin{bmatrix} 1 & 2 & \vdots & 1 & 3 \\ 0 & 7 & \vdots & 7 & 7 \\ 0 & 0 & \vdots & 3 & 0 \end{bmatrix}$, so the first system is not consistent while the second one is. Thus (b) lies in the span of $\{\mathbf{a}_1, \mathbf{a}_2\}$ while (a) doesn't.

3. It is immediate that $\mathbf{b}_2 = -\mathbf{a}_2$. For the remaining vectors we have $\begin{bmatrix} 1 & 1 & 1 & \vdots & 1 & 1 \\ 0 & 1 & 0 & \vdots & -1 & 0 \\ 2 & 1 & 2 & \vdots & 2 & 1 \\ 0 & 1 & 1 & \vdots & 1 & 0 \end{bmatrix}$, and in row-echelon form:

$$\begin{bmatrix} 1 & 1 & 1 & \vdots & 1 & 1 \\ 0 & 1 & 0 & \vdots & -1 & 0 \\ 0 & 0 & 0 & \vdots & 1 & -1 \\ 0 & 0 & 1 & \vdots & 2 & 0 \end{bmatrix}$$, where both systems are not consistent, so

we conclude that $\mathbf{b}_2$ is the only vector spanned by $\{\mathbf{a}_1, \mathbf{a}_2, \mathbf{a}_3\}$.

5. Just as in Example 2 of Section 3.2, we row reduce the systems that have the given vectors as columns to check their consistency. Firstly,

since $\begin{bmatrix} 1 & 0 & 2 & \vdots & 2 & 0 \\ 0 & 1 & -3 & \vdots & -2 & -1 \\ 0 & 1 & -3 & \vdots & -2 & -1 \\ 1 & 3 & -7 & \vdots & -4 & -3 \end{bmatrix}$ after reduction is

$\begin{bmatrix} 1 & 0 & 2 & \vdots & 2 & 0 \\ 0 & 1 & -3 & \vdots & -2 & -1 \\ 0 & 0 & 0 & \vdots & 0 & 0 \\ 0 & 3 & 0 & \vdots & 0 & 0 \end{bmatrix}$, both systems are consistent, and the

second collection of vectors lies in the span of the first collection. On

the other hand, $\begin{bmatrix} 2 & 0 & \vdots & 1 & 0 & 2 \\ -2 & -1 & \vdots & 0 & 1 & -3 \\ -2 & -1 & \vdots & 0 & 1 & -3 \\ -4 & -3 & \vdots & 1 & 3 & -7 \end{bmatrix}$ has the row echelon form

$\begin{bmatrix} 2 & 0 & \vdots & 1 & 0 & 2 \\ 0 & -1 & \vdots & 1 & 1 & -1 \\ 0 & 0 & \vdots & 0 & 0 & 0 \\ 0 & 0 & \vdots & 0 & 0 & 0 \end{bmatrix}$, which gives the inverse inclusion. Summa-

rizing, spans of both the collections of vectors coincide.

7. The given vectors are clearly not multiples of each other and therefore linearly independent.

In the solutions of Problems 7–15 we argue similarly to Example 3 of Section 3.2 reducing the matrix of given vectors to row-echelon form. The nonzero rows are linearly independent and span a subspace which contains all the other rows as well.

9. Matrix $\begin{bmatrix} -2 & 1 & -1 & 1 \\ 4 & 0 & 6 & 0 \\ 5 & 4 & -2 & 1 \end{bmatrix}$ has the row echelon form

$\begin{bmatrix} -2 & 1 & -1 & 1 \\ 0 & 1 & 2 & 1 \\ 0 & 0 & -17.5 & -3 \end{bmatrix}$, thus its rows are linearly independent.

11. Matrix $\begin{bmatrix} 2+3i & 1+i & 4+i \\ 6-2i & 2-i & 2-4i \\ -6+13i & -1+5i & 8+11i \end{bmatrix}$ has the row echelon form

$\begin{bmatrix} 1 & 0 & -3 \\ 0 & 1 & 10 \\ 0 & 0 & 0 \end{bmatrix}$, so its rows are linearly dependent. Coefficients for a non-trivial linear combination equal to zero are, for example, $[3 \ -10 \ 1]^T$.

13. The given matrix is already reduced to row echelon form, and its rows are clearly linearly independent because of positions of zeros.

15. Similarly to Example 6 of Section 3.2, we regard the given matrices as vectors in $\mathbf{R}^4_{row}$ and apply the above argument:

$\begin{bmatrix} 2 & -3 & 3 & 1 \\ -8 & 0 & -9 & -4 \\ 8 & -2 & 10 & 6 \end{bmatrix}$ has the row echelon form

$\begin{bmatrix} 2 & -3 & 3 & 1 \\ 0 & -4 & 1 & 0 \\ 0 & 0 & 0.5 & 2 \end{bmatrix}$, so the matrices are linearly independent.

17. The first row is obviously linearly independent of the other two. In order to verify the linear independence sequentially (as described in the Remark after Theorem 2, Section 3.2), we only need to check that rows 2 and 3 are linearly independent. Because the first coordinate is zero in both rows, it suffices to check whether rows of the minor $\mathbf{M} := \begin{bmatrix} b & 1 \\ 1 & c \end{bmatrix}$ are linearly independent. By Theorem 3 this can be stated as $\det(\mathbf{M}) \neq 0$. Summarizing, the necessary and sufficient condition is $bc \neq 1$.

19. (a) True: zero vector is always linearly dependent of the rest of the set being the trivial linear combination with all coefficients equal zero.

(b) False: consider vectors $[1 \ 0]$, $[0 \ 1]$, $[1 \ 1]$ in $\mathbf{R}^2_{row}$. They are obviously linearly dependent, but none of them is a multiple of another.

(c) False: take the example of (b). The first two vectors are linearly independent (in fact, they constitute a basis of R_{row}^2, see Section 3.3 for a discussion of bases).

(d) True: it is the statement of Theorem 3.

(e) True: note that any row operation on a matrix can be reversed to obtain the original matrix. Given that row operations can be expressed as left multiplication by appropriate matrices (see Section 2.1), it follows that the reversed sequence of row operations corresponds to the inverse matrix.

21. Use the test of linear dependence from Theorem 2 of Section 3.2. Let for some scalars c_1, c_2, c_3, c_4,

$$c_1 v_1 + c_2 v_2 + c_3 v_3 + c_4 v_4 = 0.$$

Here necessarily $c_4 = 0$, since otherwise $v_4 = -c_1/c_4\, v_1 - c_2/c_4\, v_2 - c_3/c_4\, v_3$, which contradicts the assumption that v_4 does not lie in the span of $\{v_1,\ v_2,\ v_3\}$. This reduces the display above to a linear combination of vectors v_1, v_2, v_3, which are assumed to be linearly independent, so $c_1 = c_2 = c_3 = 0$. Thus all the four scalars must equal zero, and the whole set $\{v_1,\ v_2,\ v_3,\ v_4\}$ is linearly independent.

23. If it were that $a\mathbf{AD} + b\mathbf{BD} + c\mathbf{CD} = 0$ for scalars a, b, c, multiplying by $\mathbf{D}^{-1}$ on the right would give $a\mathbf{A} + b\mathbf{B} + c\mathbf{C} = 0$, whence by linear independence of $\mathbf{A}, \mathbf{B}, \mathbf{C}$ follows $a = b = c = 0$.

25. For any n-by-1 column vector $\mathbf{x}$, $\mathbf{Ax}$ is a linear combination of columns of A:

$$\mathbf{Ax} = \begin{bmatrix} \mathbf{a}_1 & \mathbf{a}_2 & \cdots & \mathbf{a}_n \end{bmatrix} \begin{bmatrix} x_1 \\ x_2 \\ \vdots \\ x_n \end{bmatrix} = x_1 \mathbf{a}_1 + \ldots + x_n \mathbf{a}_n.$$

Furthermore, as $\mathbf{A}^2\mathbf{x} = \mathbf{A}(\mathbf{Ax})$, and $\mathbf{Ax}$ is an n-by-1 column vector, it is also a linear combination of columns of $\mathbf{A}$. Similarly, $\mathbf{A}^3\mathbf{x} = \mathbf{A}(\mathbf{A}^2\mathbf{x})$, etc.

27. We will argue similarly in both the (a) and (b) case. Let a linear combination of the given vectors be zero. Expressing it in terms of v_1, v_2, v_3, obtain a homogeneous linear system for coefficients at w_i. It then suffices to verify that the obtained system only has the trivial solution to conclude that $\mathbf{w}_i$ are linearly independent.

(a) Let $a\mathbf{w}_1 + b\mathbf{w}_2 = \mathbf{0}$. Hence $a(2\mathbf{v}_1 - 3\mathbf{v}_2) + b(4\mathbf{v}_1 - 3\mathbf{v}_2) = \mathbf{0}$, and because $\mathbf{v}_i$ are linearly independent, collecting coefficients at $\mathbf{v}_1$ and $\mathbf{v}_2$ gives

$$\begin{cases} 2a + 4b = 0 \\ -3a - 3b = 0 \end{cases}$$

The unique solution is $a = b = 0$.

(b) Let $a\mathbf{w}_1 + b\mathbf{w}_2 + c\mathbf{w}_3 = \mathbf{0}$, then $a + 4b = 0$, $-3a + 2c = 0$, $-3b - 3c = 0$, which gives $a = b = c = 0$, so $\mathbf{w}_i$ are linearly independent.

29. From the test of linear independence, vectors $\mathbf{v}_i$ in $\mathbf{R}^3_{\text{col}}$ are linearly dependent if and only if the system

$$\begin{bmatrix} \mathbf{v}_1 & \mathbf{v}_2 & \mathbf{v}_3 \end{bmatrix} \begin{bmatrix} \mathbf{x} \end{bmatrix} = \mathbf{0}$$

has a non-trivial solution. Because of Theorem 3 of Section 3.2, this is equivalent to the transposed homogeneous system having a non-trivial solution. Writing the latter in the form

$$\begin{bmatrix} \mathbf{v}_1^T \\ \mathbf{v}_2^T \\ \mathbf{v}_3^T \end{bmatrix} \begin{bmatrix} \mathbf{n} \end{bmatrix} = \begin{bmatrix} \mathbf{v}_1 \cdot \mathbf{n} \\ \mathbf{v}_2 \cdot \mathbf{n} \\ \mathbf{v}_3 \cdot \mathbf{n} \end{bmatrix} = \mathbf{0},$$

we conclude that it is equivalent to existence of a non-zero vector $\mathbf{n}$, orthogonal to every $\mathbf{v}_i$, which is the condition of $\mathbf{v}_i$ being coplanar.

3.3   BASES, DIMENSION, AND RANK

1. For convenience we will work with column vectors and transpose the solution. The vectors in $\mathbf{R}^6_{\text{col}}$ whose first three and last three entries sum to zero are precisely the solutions of the homogeneous system

$$\begin{bmatrix} 1 & 1 & 1 & 0 & 0 & 0 & \vdots & 0 \\ 0 & 0 & 0 & 1 & 1 & 1 & \vdots & 0 \end{bmatrix}.$$

The matrix above is already in row echelon form. According to the discussion in Section 3.3, every column without bold entries introduces a free parameter. Back substituting, obtain solutions of the above system:

$$\mathbf{x} = t_1 \begin{bmatrix} -1 \\ 1 \\ 0 \\ 0 \\ 0 \\ 0 \end{bmatrix} + t_2 \begin{bmatrix} -1 \\ 0 \\ 1 \\ 0 \\ 0 \\ 0 \end{bmatrix} + t_3 \begin{bmatrix} 0 \\ 0 \\ 0 \\ -1 \\ 1 \\ 0 \end{bmatrix} + t_4 \begin{bmatrix} 0 \\ 0 \\ 0 \\ -1 \\ 0 \\ 1 \end{bmatrix},$$

which shows basis vectors of the given subspace of $\mathbf{R}^6_{\text{row}}$ are for example $[-1\ 1\ 0\ 0\ 0\ 0]$, $[-1\ 0\ 1\ 0\ 0\ 0]$, $[0\ 0\ 0\ -1\ 1\ 0]$, $[0\ 0\ 0\ -1\ 0\ 1]$.

3. Vectors $\mathbf{v}_1 = [1\ -1\ 0\ 0\ 0\ 0]$, $\mathbf{v}_2 = [0\ 0\ 1\ -1\ 0\ 0]$, $\mathbf{v}_3 = [0\ 0\ 0\ 0\ 1\ -1]$ are linearly independent due to the positions of zeros, and span the given subspace of $\mathbf{R}^6_{\text{row}}$, as can be seen from

$$[a\ -a\ b\ -b\ c\ -c] = a\mathbf{v}_1 + b\mathbf{v}_2 + c\mathbf{v}_3.$$

5. As shown in Section 1.2, the matrix $\begin{bmatrix} 2 & 1 & 4 \\ 3 & 3 & 0 \\ -1 & 4 & 2 \end{bmatrix}$ has the row echelon form $\begin{bmatrix} 2 & 1 & 4 \\ 0 & 3/2 & -6 \\ 0 & 0 & 22 \end{bmatrix}$. All the rows and columns of the original matrix are therefore independent, and together form bases of the row and column space respectively. The null space of the given matrix consists solely of the zero vector.

7. According to Section 1.3, after reduction to row echelon form (which involves changing the order of rows), the matrix $\begin{bmatrix} 0 & 2 & 1 & 1 & 0 \\ 2 & 4 & 4 & 2 & 2 \\ 3 & 6 & 6 & 0 & 0 \\ 0 & -2 & -1 & -2 & 2 \\ 0 & 2 & 1 & 2 & 4 \end{bmatrix}$

$$\begin{bmatrix} \mathbf{3} & 6 & 6 & 0 & 0 \\ 0 & \mathbf{2} & 1 & 1 & 0 \\ 0 & 0 & 0 & \mathbf{2} & 2 \\ 0 & 0 & 0 & 0 & \mathbf{3} \\ 0 & 0 & 0 & 0 & 0 \end{bmatrix} \begin{matrix} 3 \\ 1 \\ 2 \\ 4 \\ 5 \end{matrix}$$

takes the form, where for every row of the result-
ing matrix we have marked its number in the original one. Again, bold
entries flag the basis rows and vectors. Accordingly, rows 1–4 are a
basis of the row space; columns 1–2 and 4–5 are a basis of the column
space. This shows the null space is one-dimensional, so it suffices to
observe $[-1 \ \ -0.5 \ \ 1 \ \ 0 \ \ 0]^T$ is annihilated by our matrix.

9. According to Example 2 of Section 1.3, reducing to row echelon form

the matrix $\begin{bmatrix} 1 & 0 & 1 \\ 1 & 1 & 1 \\ 0 & 1 & 1 \\ 3 & 3 & 6 \end{bmatrix}$ gives $\begin{bmatrix} 1 & 0 & 1 \\ 0 & 1 & 0 \\ 0 & 0 & 1 \\ 0 & 0 & 0 \end{bmatrix}$.

Hence rows 1–3 are a basis of the row space, columns 1–3 are a basis
of the column space, and the null space contains only the vector **0**.

11. Having reduced the given matrix $\begin{bmatrix} 2 & 2 & 2 & 2 & 2 \\ 2 & 2 & 4 & 6 & 4 \\ 0 & 0 & 2 & 3 & 1 \\ 1 & 4 & 3 & 2 & 3 \end{bmatrix}$ to row echelon

form, we obtain $\begin{bmatrix} \mathbf{1} & 1 & 1 & 1 & 1 \\ 0 & \mathbf{3} & 2 & 1 & 2 \\ 0 & 0 & \mathbf{2} & 4 & 2 \\ 0 & 0 & 0 & \mathbf{-1} & -1 \end{bmatrix} \begin{matrix} 1 \\ 4 \\ 3 \\ 2 \end{matrix}$.

Thus rows 1–4 are a basis of the row space, columns 1–4 are a basis
of the column space. Back-solving the above homogeneous system
shows its solutions are $\mathbf{x} = t[0 \ -1 \ 1 \ -1 \ 1]^T$, which gives the basis
element of the null space.

13. No. Let $\mathbf{a}_1, \ldots, \mathbf{a}_k$ be all the rows of **A**, such that for each $\mathbf{a}_i$ some of
its terms form a row $\mathbf{b}_i$ of **B** (so that k is the number of rows in **B**). If
for some scalars c_i,

$$c_1 \mathbf{a}_1 + \ldots + c_k \mathbf{a}_k = \mathbf{0},$$

then it also holds

$$c_1 \mathbf{b}_1 + \ldots + c_k \mathbf{b}_k = \mathbf{0},$$

which implies rank of the set $\mathbf{a}_1, \ldots, \mathbf{a}_k$ is at least that of $\mathbf{b}_1, \ldots, \mathbf{b}_k$. Now observe that for any collection of vectors, adding new ones to it does not decrease the overall rank, so (row) rank of $\mathbf{A}$ is at least that of $\mathbf{B}$.

15. The dimension of space of upper triangular matrices in $\mathbf{R}^{5,5}$ is 15: a basis of this space consists of the matrices having 0 on all except for a single position among those marked with "X", where they have 1 instead:

$$\begin{bmatrix} X & X & X & X & X \\ 0 & X & X & X & X \\ 0 & 0 & X & X & X \\ 0 & 0 & 0 & X & X \\ 0 & 0 & 0 & 0 & X \end{bmatrix}.$$

Such matrices are obviously linearly independent and span the whole space.

17. Dimension of the given space is 10: each basis element has a single 1 on a position marked with "X", and -1 on the corresponding position marked with "$-$":

$$\begin{bmatrix} 0 & X & X & X & X \\ - & 0 & X & X & X \\ - & - & 0 & X & X \\ - & - & - & 0 & X \\ - & - & - & - & 0 \end{bmatrix}.$$

19. Dimension of this space is 20. To verify this, let us think about matrices in $\mathbf{R}^{5,5}$ as elements of $\mathbf{R}^{25}_{\text{col}}$, and write $[a_1 \ a_2 \ \ldots \ a_{25}]^T$ for a vector corresponding to some matrix $\mathbf{A}$. The assumption of column sums of $\mathbf{A}$ being zero then becomes

$$\begin{cases} a_1 & + & a_2 & + & a_3 & + & a_4 & + & a_5 & = 0 \\ \vdots & & \vdots & & \vdots & & \vdots & & \vdots & \\ a_{21} & + & a_{22} & + & a_{23} & + & a_{24} & + & a_{25} & = 0 \end{cases}.$$

It is easy to see that the matrix of this homogeneous system of 5 equations and 25 unknowns is already in row echelon form, and it contains exactly 5 flagged (as we call them in Section 3.3) columns which form a basis of the column space. The terms corresponding to such columns are bold in the above display. Accordingly, Theorem 6 of Section 3.3 implies that the dimension of the null space is $25 - 5 = 20$.

It remains to find 20 linearly independent matrices in the given space. We construct them in the following way: each matrix of the basis will have 1 on a single position of those marked with "X", and -1 in the same column in the fifth row (marked with "$-$"), see the following display.

$$\begin{bmatrix} X & X & X & X & X \\ X & X & X & X & X \\ X & X & X & X & X \\ X & X & X & X & X \\ - & - & - & - & - \end{bmatrix}.$$

Such matrices are linearly independent because of positions of 1's, and there are 20 of them.

21. One: a single row $[1 \ 1 \ 1 \ 1 \ 1]$ spans the whole row space (all other rows being the same).

23. To give an answer to all the three questions we need to recall that the rank of a matrix equals the dimension of its range, the latter being the set of all vectors **b** for which the system **Ax=b** has solutions.

   The range of the rotation matrix of Section 2.2 has dimension 2, as image of a rotation of the plane is again the whole plane, thus the matrix has rank 2. Range of the orthogonal projection matrix is one-dimensional, spanned by the corresponding unit vector. Rank of the projection matrix is therefore 1. Finally, image of the plane under mirror-reflection is the whole plane, thus the corresponding matrix has rank 2.

25. The three vectors $[1 \ 2 \ 3]$, $[0 \ 1 \ 0]$, $[0 \ 0 \ 1]$ are clearly linearly independent (we could just as well take any two canonical base vectors and $[1 \ 2 \ 3]$).

27. One can take for example $[0 \ 0 \ 1 \ 0]$ and $[0 \ 0 \ 0 \ 1]$ as well as the two given vectors, since the resulting matrix

$$\begin{bmatrix} 1 & 2 & 3 & 4 \\ 0 & 1 & 2 & 2 \\ 0 & 0 & 1 & 0 \\ 0 & 0 & 0 & 1 \end{bmatrix}$$

is in row echelon form and clearly has rank 4.

29. It is easy to notice that the first row of the given matrix is the sum of the second and third, thus the matrix has rank 2. The first and the third columns are linearly independent and must therefore span the column space. Following the hint from Exercise 28, we take matrix $\mathbf{B}$ to be $\begin{bmatrix} 1 & 1 & 0 \\ 1 & 0 & 1 \end{bmatrix}^T$, and solve the system $\mathbf{Bc} = \mathbf{a}$ with respect to vector $\mathbf{c}$ for $\mathbf{a}$ running through columns of $\mathbf{A}$, then assemble all the resulting columns $\mathbf{c}$ into matrix $\mathbf{C}$. The result is

$$\mathbf{A} = \mathbf{BC} = \begin{bmatrix} 1 & 1 \\ 1 & 0 \\ 0 & 1 \end{bmatrix} \begin{bmatrix} 1 & 2 & 0 & -1 & 1 & 0 \\ 0 & 0 & 1 & 2 & -2 & 0 \end{bmatrix}.$$

31. Observe that for all α, the third of the three given vectors equals the sum of the first two. Since elements of a basis of $\mathbf{R}_{\mathrm{row}}^3$ must be linearly independent, the given vectors do not form one for any value of α.

33. Matrix $\mathbf{A}$ has 14 columns, thus by Theorem 6 of Section 3.3 the dimension of the null space of $\mathbf{A}$ is $14 - 10 = 4$. Similarly, $\mathbf{A}^T$ has 23 columns, and the dimension of its null space is $23 - 10 = 13$.

35. The space Span $\{[1\ 1\ 0], [0\ 1\ 1]\}$ is an example. Indeed, let us write $\mathbf{v}_1$, $\mathbf{v}_2$ for the two basis vectors of this span, and assume that for some scalars c_1, c_2,

$$c_1\mathbf{v}_1 + c_2\mathbf{v}_2 = [1\ 0\ 0].$$

This equation means that of the three quantities, c_1, c_2 and $c_1 + c_2$, two are equal to zero, whence $c_1 = c_2 = 0$, a contradiction. The remaining vectors $[0\ 1\ 0]$ and $[0\ 0\ 1]$ can be dealt with in the same manner.

37. We will need the following simple observation: adding to any matrix $\mathbf{M}$ linear combinations of its rows (columns) does not increase its rank. Thus rank of any matrix comprised of linear combinations of rows (columns) of $\mathbf{M}$ is at most rank($\mathbf{M}$).

Notice also that from the definition of matrix multiplication, each row of the matrix $\mathbf{AB}$ is a linear combination of rows of $\mathbf{B}$ (with scalar coefficients taken from the corresponding row of $\mathbf{A}$). This allows to conclude that rank($\mathbf{AB}$) $\leq$ rank($\mathbf{B}$). Similarly, each column of $\mathbf{AB}$ is a linear combination of columns of $\mathbf{A}$, which proves rank($\mathbf{AB}$) $\leq$ min\{rank($\mathbf{A}$), rank($\mathbf{B}$)\}.

39. (a) Consider an m-by-n matrix constructed from a geometric sequence as described:

$$\mathbf{G} = \begin{bmatrix} a & a^2 & a^2 & \cdots & a^n \\ a^{n+1} & a^{n+2} & a^{n+3} & \cdots & a^{2n} \\ \vdots & \vdots & \vdots & \ddots & \vdots \\ a^{(m-2)n+1} & a^{(m-2)n+2} & a^{(m-2)n+3} & \cdots & a^{(m-1)n} \\ a^{(m-1)n+1} & a^{(m-1)n+2} & a^{(m-1)n+3} & \cdots & a^{mn} \end{bmatrix}.$$

The k-th row is obviously a multiple of the first one with the factor $a^{(k-1)n}$. Hence rank of this matrix equals 1.

(b) The general form of an m-by-n matrix constructed from an arithmetic sequence is

$$\mathbf{A} = \begin{bmatrix} a & a+d & \cdots \\ a+nd & a+(n+1)d & \cdots \\ \vdots & \vdots & \ddots \\ a+(m-2)nd & a+((m-2)n+1)d & \cdots \\ a+(m-1)nd & a+((m-1)n+1)d & \cdots \end{bmatrix}$$

$$\begin{matrix} a+(n-1)d \\ a+(2n-1)d \\ \vdots \\ a+((m-1)n-1)d \\ a+(mn-1)d \end{matrix}.$$

If we introduce notation for the vectors $\mathbf{a} = [a\ \ a+d\ \ a+2d\ \ \cdots\ \ a+(n-1)d]$ and $\mathbf{d} = [nd\ \ nd\ \ nd\ \ \cdots\ \ nd]$, the k-th row of $\mathbf{A}$ can be represented as $\mathbf{a}+(k-1)\mathbf{d}$. This shows rank$(\mathbf{A}) = 2$.

(c) As above, we first describe the general form of an m-by-n matrix constructed from the Fibonacci sequence, writing f_k for the k-th element of Fibonacci sequence.

$$\mathbf{F} = \begin{bmatrix} f_1 & f_2 & \cdots & f_n \\ f_{n+1} & f_{n+2} & \cdots & f_{2n} \\ \vdots & \vdots & \ddots & \vdots \\ f_{(m-1)n+1} & f_{(m-1)n+2} & \cdots & f_{mn} \end{bmatrix}.$$

It is immediate from the definition of Fibonacci numbers that in the above matrix j-th column is the sum of $j - 2$-nd and $j - 1$-st for $j = 3, 4, \ldots, n$. In particular, each column is a linear combination of the first two. On the other hand, since $f_1 = f_2 = 1$, but all the other f_k are distinct, it follows that the first two columns are linearly independent, and $\text{rank}(\mathbf{F}) = 2$.

4

ORTHOGONALITY

4.1 ORTHOGONAL VECTORS AND THE GRAM-SCHMIDT ALGORITHM

1. Let $\mathbf{v_1} = [3/5\ 4/5]$ and $\mathbf{v_2} = [-4/5\ 3/5]$. We compute

$$\mathbf{v_1} \cdot \mathbf{v_2} = (3/5)(-4/5) + (4/5)(3/5) = -12/25 + 12/25 = 0$$
$$\mathbf{v_1} \cdot \mathbf{v_1} = (3/5)(3/5) + (4/5)(4/5) = (9 + 16)/25 = 1,$$
$$\mathbf{v_2} \cdot \mathbf{v_2} = (-4/5)(-4/5) + (3/5)(3/5) = (16 + 9)/25 = 1.$$

Hence, the set $\{\mathbf{v_1}, \mathbf{v_2}\}$ forms an orthonormal basis of $\mathbb{R}^2_{\text{row}}$. Let $\mathbf{v} = [5\ 10]$, and set $\mathbf{v} = c_1\mathbf{v_1} + c_2\mathbf{v_2}$. To compute the coefficients c_1 and c_2, by Theorem 1, Section 4.1, it suffices to compute $\mathbf{v} \cdot \mathbf{v_1}$ and $\mathbf{v} \cdot \mathbf{v_2}$:

$$c_1 = \mathbf{v} \cdot \mathbf{v_1} = [5\ 10] \cdot [3/5\ 4/5] = 5(3/5) + 10(4/5) = 3 + 8 = 11$$
$$c_2 = \mathbf{v} \cdot \mathbf{v_2} = [5\ 10] \cdot [-4/5\ 3/5] = 5(-4/5) + 10(3/5) = -4 + 6 = 2.$$

Therefore, we have,

$$[5\ 10] = 11\,[3/5\ 4/5] + 2\,[-4/5\ 3/5].$$

Solutions Manual to Accompany Fundamentals of Matrix Analysis with Applications,
First Edition. Edward Barry Saff and Arthur David Snider.
© 2016 John Wiley & Sons, Inc. Published 2016 by John Wiley & Sons, Inc.

3. Let $v_1 = [1/3 \; 2/3 \; 2/3]$ and $v_2 = [-2/3 \; -1/3 \; 2/3]$, and $v_3 = [-2/3 \; 2/3 \; -1/3]$. We compute

$$
\begin{aligned}
v_1 \cdot v_1 &= (1/3)(1/3) + (2/3)(2/3) + (2/3)(2/3) \\
&= (1 + 4 + 4)/9 = 1, \\
v_2 \cdot v_2 &= (-2/3)(-2/3) + (-1/3)(-1/3) + (2/3)(2/3) \\
&= (4 + 1 + 4)/9 = 1, \\
v_3 \cdot v_3 &= (-2/3)(-2/3) + (2/3)(2/3) + (-1/3)(-1/3) \\
&= (4 + 4 + 1)/9 = 1, \\
v_1 \cdot v_2 &= (1/3)(-2/3) + (2/3)(-1/3) + (2/3)(2/3) \\
&= (-2 - 2 + 4)/9 = 0, \\
v_1 \cdot v_3 &= (1/3)(-2/3) + (2/3)(2/3) + (2/3)(-1/3) \\
&= (-2 + 4 - 2)/9 = 0, \\
v_2 \cdot v_3 &= (-2/3)(-2/3) + (-1/3)(2/3) + (2/3)(-1/3) \\
&= (4 - 2 - 2)/9 = 0.
\end{aligned}
$$

Hence, the set $\{v_1, v_2, v_3\}$ forms an orthonormal basis of $\mathbb{R}^3_{\text{row}}$. Let $v = [3 \; 6 \; 9]$, and set $v = c_1 v_1 + c_2 v_2 + c_3 v_3$. To compute the coefficients c_1, c_2 and c_3, by Theorem 1, Section 4.1, it suffices to compute $v \cdot v_1$, $v \cdot v_2$ and $v \cdot v_3$:

$$
\begin{aligned}
c_1 = v \cdot v_1 &= [3 \; 6 \; 9] \cdot [1/3 \; 2/3 \; 2/3] = 3(1/3) + 6(2/3) \\
&+ 9(2/3) = 1 + 4 + 6 = 11, \\
c_2 = v \cdot v_2 &= [3 \; 6 \; 9] \cdot [-2/3 \; -1/3 \; 2/3] = 3(-2/3) + 6(-1/3) \\
&+ 9(2/3) = -2 - 2 + 6 = 2, \\
c_2 = v \cdot v_3 &= [3 \; 6 \; 9] \cdot [-2/3 \; 2/3 \; -1/3] = 3(-2/3) + 6(2/3) \\
&+ 9(-1/3) = -2 + 4 - 3 = -1.
\end{aligned}
$$

Therefore, we have

$$
\begin{aligned}
[3 \; 6 \; 9] = 11\,[1/3 \; 2/3 \; 2/3] &+ 2\,[-2/3 \; -1/3 \; 2/3] \\
&+ (-1)\,[-2/3 \; 2/3 \; -1/3].
\end{aligned}
$$

5. Let $\mathbf{v_1} = [-8\,6]$ and $\mathbf{v_2} = [-1\,7]$. First, we renormalize $\mathbf{v_1}$ to a unit vector, and let

$$\mathbf{w_1} = \frac{1}{\|\mathbf{v_1}\|}\mathbf{v_1} = \frac{1}{\sqrt{(-8)^2 + 6^2}}[-8\,6] = \frac{1}{10}[-8\,6] = [-4/5\,3/5].$$

We compute next

$$\begin{aligned}
\mathbf{w_2} &= \mathbf{v_2} - (\mathbf{v_2} \cdot \mathbf{w_1}) \\
&= [-1\,7] - ([-1\,7] \cdot [-4/5\,3/5])[-4/5\,3/5] \\
&= [-1\,7] - 5[-4/5\,3/5] \\
&= [-1\,7] - [-4\,3] \\
&= [3\,4].
\end{aligned}$$

We renormalize $\mathbf{w_2}$ to a unit vector, and let

$$\mathbf{w_2}^{\text{unit}} = \frac{1}{\|\mathbf{w_2}\|}\mathbf{w_2} = \frac{1}{\sqrt{3^2 + 4^2}}[3\,4] = \frac{1}{5}[3\,4] = [3/5\,4/5].$$

Therefore, the set $\{[-4/5\,3/5], [3/5\,4/5]\}$ of orthonormal vectors spans the same space as $\{\mathbf{v_1}, \mathbf{v_2}\}$.

7. Let $\mathbf{v_1} = [1\,-1\,1\,-1]$ and $\mathbf{v_2} = [3\,-1\,3\,-1]$. First, we renormalize $\mathbf{v_1}$ to a unit vector, and let

$$\begin{aligned}
\mathbf{w_1} &= \frac{1}{\|\mathbf{v_1}\|}\mathbf{v_1} = \frac{1}{\sqrt{1^2 + (-1)^2 + 1^2 + (-1)^2}}[1\,-1\,1\,-1] \\
&= \frac{1}{2}[1\,-1\,1\,-1] = [1/2\,-1/2\,1/2\,-1/2].
\end{aligned}$$

We compute next

$$\begin{aligned}
\mathbf{w_2} &= \mathbf{v_2} - (\mathbf{v_2} \cdot \mathbf{w_1}) \\
&= [3\,-1\,3\,-1] - ([3\,-1\,3\,-1] \cdot [1/2\,-1/2\,1/2\,-1/2]) \\
&\quad [1/2\,-1/2\,1/2\,-1/2] \\
&= [3\,-1\,3\,-1] - ((3)(1/2) + (-1)(-1/2) + (3)(1/2) \\
&\quad + (-1)(-1/2))[1/2\,-1/2\,1/2\,-1/2]
\end{aligned}$$

$$= [3 \; -1 \, 3 \; -1] - 4[1/2 \; -1/2 \, 1/2 \; -1/2]$$
$$= [3 \; -1 \, 3 \; -1] - [2 \; -2 \, 2 \; -2]$$
$$= [1 \, 1 \, 1 \, 1].$$

We renormalize $\mathbf{w_2}$ to a unit vector, and let

$$\mathbf{w_2}^{\text{unit}} = \frac{1}{\|\mathbf{w_2}\|}\mathbf{w_2} = \frac{1}{\sqrt{1^2 + 1^1 + 1^2 + 1^2}}[1\,1\,1\,1] = \frac{1}{2}[1\,1\,1\,1]$$
$$= [1/2 \; 1/2 \; 1/2 \; 1/2].$$

Therefore, the set $\{[1/2 \; -1/2 \; 1/2 \; -1/2], [1/2 \; 1/2 \; 1/2 \; 1/2]\}$ of orthonormal vectors spans the same space as $\{\mathbf{v_1}, \mathbf{v_2}\}$.

9. Let $\mathbf{v_1} = [2 \, 2 \; -2 \; -2]$, $\mathbf{v_2} = [3 \, 1 \; -1 \; -3]$ and $\mathbf{v_3} = [2 \, 0 \; -2 \; -4]$. We take $\mathbf{w_1} = \mathbf{v_1}$. Since $\mathbf{w_1} \cdot \mathbf{w_1} = \|\mathbf{w_1}\|^2 = 2^2 + 2^2 + (-2)^2 + (-2)^2 = 16$, we get

$$\mathbf{w_1}^{\text{unit}} = \frac{1}{\sqrt{16}}\mathbf{w_1} = [1/2 \, 1/2 \; -1/2 \; -1/2].$$

Let

$$\mathbf{w_2} = \mathbf{v_2} - (\mathbf{w_2} \cdot \mathbf{w_1}^{\text{unit}})\mathbf{w_1}^{\text{unit}}$$
$$= [3 \, 1 \; -1 \; -3] - ([3 \, 1 \; -1 \; -3] \cdot [1/2 \, 1/2 \; -1/2 \; -1/2])$$
$$[1/2 \, 1/2 \; -1/2 \; -1/2]$$
$$= [3 \, 1 \; -1 \; -3] - 4[1/2 \, 1/2 \; -1/2 \; -1/2]$$
$$= [1 \; -1 \, 1 \; -1],$$

and we have

$$\mathbf{w_2}^{\text{unit}} = \frac{1}{\|\mathbf{w_2}\|}\mathbf{w_2} = \frac{1}{\sqrt{1^2 + (-1)^2 + 1^2 + (-1)^2}}[1\,1\,1\;-1]$$
$$= [1/2 \; -1/2 \, 1/2 \; -1/2].$$

Finally, let

$$\mathbf{w_3} = \mathbf{v_3} - (\mathbf{w_3} \cdot \mathbf{w_1}^{\text{unit}})\mathbf{w_1}^{\text{unit}} - (\mathbf{w_3} \cdot \mathbf{w_2}^{\text{unit}})\mathbf{w_2}^{\text{unit}}$$
$$= [2 \, 0 \; -2 \; -4] - ([2 \, 0 \; -2 \; -4] \cdot [1/2 \, 1/2 \; -1/2 \; -1/2])$$
$$[1/2 \, 1/2 \; -1/2 \; -1/2]$$
$$- ([2 \, 0 \; -2 \; -4] \cdot [1/2 \; -1/2 \, 1/2 \; -1/2])$$
$$[1/2 \; -1/2 \, 1/2 \; -1/2]$$

$$= [2\,0\,-2\,-4] - 4[1/2\,1/2\,-1/2\,-1/2]$$
$$-\,2[1/2\,-1/2\,1/2\,-1/2]$$
$$= [-1\,-1\,-1\,-1],$$

and we have

$$\mathbf{w_3}^{\text{unit}} = \frac{1}{\|\mathbf{w_3}\|}\mathbf{w_3} = \frac{1}{\sqrt{(-1)^2+(-1)^2+(-1)^2+(-1)^2}}[1\,1\,1\,-1]$$
$$= [-1/2\,-1/2\,-1/2\,-1/2].$$

Therefore, the set

$$\{[1/2\,1/2\,-1/2\,-1/2],\,[1/2\,-1/2\,1/2\,-1/2],$$
$$[-1/2\,-1/2\,-1/2\,-1/2]\}$$

of orthonormal vectors spans the same space as $\{\mathbf{v_1}, \mathbf{v_2}, \mathbf{v_2}, \mathbf{v_3}\}$.

11. (a) Pick any vector which is not a scalar multiple of $[5\ 7]$ (for example, one can pick $[1\ 0]$). (b) Pick any two vectors $\mathbf{u}$ and $\mathbf{v}$ such that $\{[1\ 2\ 3], \mathbf{u}, \mathbf{v}\}$ are linearly independent. For example, iGram can apply the Gram-Schmidt algorithm on $\{[1\ 2\ 3], [1\ 0\ 0], [0\ 1\ 0]\}$.

13. It suffices to append one more vector to the vectors $\{\mathbf{v_1}, \mathbf{v_2}, \mathbf{v_2}, \mathbf{v_3}\}$ in Problem 9, Exercises 4.1. Let $\mathbf{v_4} = [1\ 0\ 0\ 0]$. We continue the Gram-Schmidt algorithm in the aforementioned problem with the newly appended vector.

Let

$$\mathbf{w_4} = \mathbf{v_4} - (\mathbf{w_4}\cdot\mathbf{w_1}^{\text{unit}})\mathbf{w_1}^{\text{unit}} - (\mathbf{w_4}\cdot\mathbf{w_2}^{\text{unit}})\mathbf{w_2}^{\text{unit}}(\mathbf{w_4}\cdot\mathbf{w_3}^{\text{unit}})\mathbf{w_2}^{\text{unit}}$$
$$= [1\,0\,0\,0] - ([1\,0\,0\,0]\cdot[1/2\,1/2\,-1/2\,-1/2])$$
$$[1/2\,1/2\,-1/2\,-1/2]$$
$$-\,([1\,0\,0\,0]\cdot[1/2\,-1/2\,1/2\,-1/2])[1/2\,-1/2\,1/2\,-1/2]$$
$$-\,([1\,0\,0\,0]\cdot[-1/2\,-1/2\,-1/2\,-1/2])$$
$$[-1/2\,-1/2\,-1/2\,-1/2]$$
$$= [1\,0\,0\,0] - (1/2)[1/2\,1/2\,-1/2\,-1/2]$$
$$-\,(1/2)[1/2\,-1/2\,1/2\,-1/2] + (1/2)$$
$$[-1/2\,-1/2\,-1/2\,-1/2]$$
$$= [1/4\,-1/4\,1/4\,1/4],$$

and we have

$$\mathbf{w_4}^{\text{unit}} = \frac{1}{\|\mathbf{w_4}\|}\mathbf{w_4}$$

$$= \frac{1}{\sqrt{(1/4)^2 + (-1/4)^2 + (1/4)^2 + (1/4)^2}}[1/4 - 1/4\,1/4\,1/4]$$

$$= 2[1/4 - 1/4\,1/4\,1/4]$$

$$= [1/2 - 1/2\,1/2\,1/2].$$

Hence, we found the additional orthonormal vector $[1/2 - 1/2 -1/2\,1/2]$.

15. As in Problem 13, Exercises 4.1, we append the vectors $\mathbf{v_3} = [1\,0\,0\,0]$ and $\mathbf{v_4} = [0\,1\,0\,0]$, and continue the Gram-Schmidt algorithm, to find the additional two orthogonal vectors.

Let

$$\mathbf{w_3} = \mathbf{v_3} - (\mathbf{w_3} \cdot \mathbf{w_1}^{\text{unit}})\mathbf{w_1}^{\text{unit}} - (\mathbf{w_3} \cdot \mathbf{w_2}^{\text{unit}})\mathbf{w_2}^{\text{unit}}$$

$$= [1\,0\,0\,0] - ([1\,0\,0\,0] \cdot [1/2 - 1/2\,1/2 - 1/2])$$
$$[1/2 - 1/2\,1/2 - 1/2]$$
$$- ([1\,0\,0\,0] \cdot [1/2\,1/2\,1/2\,1/2])[1/2\,1/2\,1/2\,1/2]$$

$$= [1\,0\,0\,0] - (1/2)[1/2 - 1/2\,1/2 - 1/2]$$
$$- (1/2)[1/2\,1/2\,1/2\,1/2]$$

$$= [1/2\,0 - 1/2\,0],$$

and we have

$$\mathbf{w_3}^{\text{unit}} = \frac{1}{\|\mathbf{w_3}\|}\mathbf{w_3} = \frac{1}{\sqrt{(1/2)^2 + (0)^2 + (-1/2)^2 + (0)^2}}[1\,1\,1 - 1]$$

$$= \left[-1/\sqrt{2}\,0 - 1/\sqrt{2}\,0\right].$$

Now let

$$\mathbf{w_4} = \mathbf{v_4} - (\mathbf{w_4} \cdot \mathbf{w_1}^{\text{unit}})\mathbf{w_1}^{\text{unit}} - (\mathbf{w_4} \cdot \mathbf{w_2}^{\text{unit}})\mathbf{w_2}^{\text{unit}}(\mathbf{w_4} \cdot \mathbf{w_3}^{\text{unit}})\mathbf{w_2}^{\text{unit}}$$

$$= [0\,1\,0\,0] - ([0\,1\,0\,0] \cdot [1/2 - 1/2\,1/2 - 1/2])$$
$$[1/2 - 1/2\,1/2 - 1/2]$$

$$- ([0\ 1\ 0\ 0] \cdot [1/2\ 1/2\ 1/2\ 1/2])\,[1/2\ 1/2\ 1/2\ 1/2]$$
$$- ([0\ 1\ 0\ 0] \cdot \left[-1/\sqrt{2}\ 0\ -1/\sqrt{2}\ 0\right])\left[-1/\sqrt{2}\ 0\ -1/\sqrt{2}\ 0\right]$$
$$= [0\ 1\ 0\ 0] + (1/2)\,[1/2\ -1/2\ 1/2\ -1/2]$$
$$- (1/2)\,[1/2\ 1/2\ 1/2\ 1/2]$$
$$= [0\ 1/2\ 0\ -1/2].$$

We compute next

$$\mathbf{w_4}^{\text{unit}} = \frac{1}{\|\mathbf{w_4}\|}\mathbf{w_4}$$

$$= \frac{1}{\sqrt{(0)^2 + (1/2)^2 + (0)^2 + (-1/2)^2}}[0\ 1/2\ 0\ -1/2]$$
$$= \sqrt{2}[1/4\ -1/4\ 1/4\ 1/4]$$
$$= [0\ 1/\sqrt{2}\ 0\ -1/\sqrt{2}].$$

Hence, we found the two additional orthonormal vectors

$$\left\{\left[\frac{\sqrt{2}}{2}\ 0\ -\frac{\sqrt{2}}{2}\ 0\right],\left[0\ \frac{\sqrt{2}}{2}\ 0\ -\frac{\sqrt{2}}{2}\right]\right\}.$$

17. We find first a basis for the space of vectors $[x\ y\ z]$ in the hyperplane $2x - y + z + 3w = 0$. Setting $x = s$, $z = t$, $w = u$, we find $y = 2s + t + 3u$. Hence, a basis is formed by the vectors $\mathbf{v_1} = [1\ 2\ 0\ 0]$, $\mathbf{v_2} = [0\ 1\ 1\ 0]$, and $\mathbf{v_3} = [0\ 3\ 0\ 1]$. We apply next the Gram-Schmidt algorithm. Let $\mathbf{w_1} = \mathbf{v_1}$, and compute

$$\mathbf{w_1}^{\text{unit}} = \frac{1}{\sqrt{5}}\mathbf{w_1} = \frac{1}{\sqrt{5}}[1\ 2\ 0\ 0].$$

Let

$$\mathbf{w_2} = \mathbf{v_2} - (\mathbf{w_2} \cdot \mathbf{w_1}^{\text{unit}})\mathbf{w_1}^{\text{unit}}$$

$$= [0\ 1\ 1\ 0] - \frac{1}{5}([0\ 1\ 1\ 0] \cdot [1\ 2\ 0\ 0])\,[1\ 2\ 0\ 0]$$

$$= [0\ 1\ 1\ 0] - \frac{2}{5}[1\ 2\ 0\ 0]$$

$$= \frac{1}{5}[-2\ 1\ 5\ 0],$$

and we have

$$\mathbf{w_2}^{unit} = \frac{1}{\|\mathbf{w_2}\|}\mathbf{w_2} = \frac{5}{\sqrt{(-2)^2 + 1^2 + 5^2 + 0^2}}[-2\ 1\ 5\ 0]$$

$$= \frac{1}{\sqrt{30}}[-2\ 1\ 5\ 0].$$

Let

$$\mathbf{w_3} = \mathbf{v_3} - (\mathbf{w_3} \cdot \mathbf{w_1}^{unit})\mathbf{w_1}^{unit} - (\mathbf{w_3} \cdot \mathbf{w_2}^{unit})\mathbf{w_2}^{unit}$$

$$= [0\ 3\ 0\ 1] - \frac{1}{5}([0\ 3\ 0\ 1] \cdot [1\ 2\ 0\ 0])[1\ 2\ 0\ 0]$$

$$- \frac{1}{30}([0\ 3\ 0\ 1] \cdot [-2\ 1\ 5\ 0])[-2\ 1\ 5\ 0]$$

$$= [0\ 3\ 0\ 1] - \frac{6}{5}[1\ 2\ 0\ 0] - \frac{1}{10}[-2\ 1\ 5\ 0]$$

$$= [-1\ 1/2\ -1/2\ 1],$$

and we have

$$\mathbf{w_3}^{unit} = \frac{1}{\|\mathbf{w_3}\|}\mathbf{w_3} = \frac{1}{\sqrt{(-1)^2 + (1/2)^2 + (-1/2)^2 + (1)^2}}$$

$$[-1\ 1/2\ -1/2\ 1] = \frac{1}{\sqrt{10}}[-2\ 1\ -1\ 2].$$

Hence, an orthonormal basis for the given hyperplane is

$$\left\{ \left[\frac{1}{\sqrt{5}}\ \frac{2}{\sqrt{5}}\ 0\ 0\right], \left[-\frac{2}{\sqrt{30}}\ \frac{1}{\sqrt{30}}\ \frac{5}{\sqrt{30}}\ 0\right], \right.$$

$$\left. \left[-\frac{2}{\sqrt{10}}\ \frac{1}{\sqrt{10}}\ -\frac{1}{\sqrt{10}}\ \frac{2}{\sqrt{10}}\right] \right\}.$$

19. By Gaussian elimination (subtracting twice the second row from the first row of the given matrix **A**), the matrix equation

$$\begin{bmatrix} 2 & 3 & 1 & 0 \\ 1 & 1 & 1 & 1 \end{bmatrix} \begin{bmatrix} u_1 \\ u_2 \\ u_3 \\ u_4 \end{bmatrix} = \begin{bmatrix} 0 \\ 0 \\ 0 \\ 0 \end{bmatrix}$$

is equivalent to the system of scalar equations

$$u_1 + u_2 + u_3 + u_4 = 0$$
$$u_2 - u_3 - 2u_4 = 0.$$

We assign arbitrary values for u_3 (say $u_3 = s$) and $u_4 = t$. We can set now $u_2 = s + 2t$, and $u_1 = -2s - 3t$. Therefore, a basis for the null space of $\mathbf{A}$ is formed by the vectors $\mathbf{v}_1 = [-2\ 1\ 1\ 0]^T$ and $\mathbf{v}_2 = [-3\ 2\ 0\ 1]^T$. We apply next the Gram-Schmidt algorithm on the vectors $\mathbf{v}_1$ and $\mathbf{v}_2$. Let $\mathbf{w}_1 = \mathbf{v}_1$. Then

$$\mathbf{w}_1{}^{unit} = \frac{1}{\|\mathbf{w}_1\|}\mathbf{w}_1 = \frac{1}{\sqrt{(-2)^2 + 1^2 + 1^2 + 0^2}}\begin{bmatrix} -2 \\ 1 \\ 1 \\ 0 \end{bmatrix} = \frac{1}{\sqrt{6}}\begin{bmatrix} -2 \\ 1 \\ 1 \\ 0 \end{bmatrix}.$$

Let

$$\mathbf{w}_2 = \mathbf{v}_2 - (\mathbf{w}_2 \cdot \mathbf{w}_1{}^{unit})\mathbf{w}_1{}^{unit}$$

$$= \begin{bmatrix} -3 \\ 2 \\ 0 \\ 1 \end{bmatrix} - \frac{1}{6}\left(\begin{bmatrix} -3 \\ 2 \\ 0 \\ 1 \end{bmatrix} \cdot \begin{bmatrix} -2 \\ 1 \\ 1 \\ 0 \end{bmatrix}\right)\begin{bmatrix} -2 \\ 1 \\ 1 \\ 0 \end{bmatrix}$$

$$= \begin{bmatrix} -3 \\ 2 \\ 0 \\ 1 \end{bmatrix} - \frac{4}{3}\begin{bmatrix} -2 \\ 1 \\ 1 \\ 0 \end{bmatrix}$$

$$= \frac{1}{3}\begin{bmatrix} -1 \\ 2 \\ -4 \\ 3 \end{bmatrix}.$$

We have

$$\|\mathbf{w}_2\|^2 = \frac{1}{3^2}\left((-1)^2 + 2^2 + (-4)^2 + 3^2\right) = \frac{30}{9}$$

Hence, an orthonormal basis for the null space of the matrix $\mathbf{A}$ is

$$\left\{\left[-\frac{2}{\sqrt{6}} \ \frac{1}{\sqrt{6}} \ \frac{1}{\sqrt{6}} \ 0\right]^T, \left[-\frac{1}{\sqrt{30}} \ -\frac{2}{\sqrt{30}} \ -\frac{4}{\sqrt{30}} \ \frac{3}{\sqrt{30}}\right]^T\right\}.$$

21. Let $\mathbf{v}_1 = \begin{bmatrix} r_1 \sin \phi_1 \cos \theta_1 \\ r_1 \sin \phi_1 \sin \theta_1 \\ r_1 \cos \phi_1 \end{bmatrix}$ and $\mathbf{v}_2 = \begin{bmatrix} r_2 \sin \phi_2 \cos \theta_2 \\ r_2 \sin \phi_2 \sin \theta_2 \\ r_2 \cos \phi_2 \end{bmatrix}$ be the spherical

coordinates description of two 3-dimensional vectors.

The vectors $\mathbf{v}_1$ and $\mathbf{v}_2$ are orthogonal if and only if $\mathbf{v}_1 \cdot \mathbf{v}_2 = 0$, which is equivalent to

$$(r_1 \sin \phi_1 \cos \theta_1)(r_2 \sin \phi_2 \cos \theta_2) + (r_1 \sin \phi_1 \sin \theta_1)(r_2 \sin \phi_2 \sin \theta_2)$$
$$+ (r_1 \cos \phi_1)(r_2 \cos \phi_2) = 0.$$

However, this is the same as

$$r_1 r_2 (\sin \phi_1 \sin \phi_2 \cos \theta_1 \cos \theta_2 + \sin \phi_1 \sin \phi_2 \sin \theta_1 \sin \theta_2$$
$$+ \cos \phi_1 \cos \phi_2) = 0,$$

which is equivalent to

$$\sin \phi_1 \sin \phi_2 \cos(\theta_1 - \theta_2) + \cos \phi_1 \cos \phi_2 = 0.$$

23. Following the steps in Problem 22, Exercises 4.1, it follows that the equality holds in the Cauchy-Schwartz inequality if and only if the quadratic polynomial $f(x) = \|\|\mathbf{v}+x\mathbf{w}\|\|^2$ has a double root x_0. However, $\|\|\mathbf{v} + x_0\mathbf{w}\|\| = 0$ implies $\mathbf{v} + x_0\mathbf{w} = 0$, that is $\mathbf{v}$ and $\mathbf{w}$ are linearly dependent.

25. We compute

$$\begin{aligned}
\|\mathbf{v} + \mathbf{w}\|^2 + \|\mathbf{v} - \mathbf{w}\|^2 &= (\mathbf{v} + \mathbf{w}) \cdot (\mathbf{v} + \mathbf{w}) + (\mathbf{v} - \mathbf{w}) \cdot (\mathbf{v} - \mathbf{w}) \\
&= \mathbf{v} \cdot \mathbf{v} + 2(\mathbf{v} \cdot \mathbf{w}) + \mathbf{w} \cdot \mathbf{w} \\
&\quad + \mathbf{v} \cdot \mathbf{v} - 2(\mathbf{v} \cdot \mathbf{w}) + \mathbf{w} \cdot \mathbf{w} \\
&= 2\|\mathbf{v}\|^2 + 2\|\mathbf{w}\|^2.
\end{aligned}$$

4.2 ORTHOGONAL MATRICES

1. Let $\mathbf{Q}$ be an orthogonal matrix. Taking determinants in the formula $\mathbf{Q}\mathbf{Q}^T = \mathbf{I}$ from Theorem 2, we find that $\det(\mathbf{Q}\mathbf{Q}^T) = \det\mathbf{I} = 1$. However, $\det(\mathbf{Q}\mathbf{Q}^T) = \det\mathbf{Q}\,\det\mathbf{Q}^T = (\det\mathbf{Q})^2$. Hence, the determinant of an orthogonal matrix can only be ± 1.

3. We use the characterization of orthogonal matrices provided by Theorem 2 (ii), as the real square matrices with orthonormal columns.

   (a) The columns of matrices of the form $\begin{bmatrix} x & y \\ y & x \end{bmatrix}$ are orthogonal if and only if

   $$\begin{bmatrix} x \\ y \end{bmatrix} \cdot \begin{bmatrix} y \\ x \end{bmatrix} = 0 \text{ and } x^2 + y^2 = 1.$$

   This is equivalent to $2xy = 0$, and $x^2 + y^2 = 0$. The only possible matrices of the required form satisfying these conditions are.

   $$\begin{bmatrix} a & 0 \\ 0 & a \end{bmatrix} \text{ and } \begin{bmatrix} 0 & a \\ a & 0 \end{bmatrix},$$

   where $a = \pm 1$.

   (b) For matrices of the form $\begin{bmatrix} x & y \\ -y & x \end{bmatrix}$, notice that the columns are already orthogonal:

   $$\begin{bmatrix} x \\ y \end{bmatrix} \cdot \begin{bmatrix} -y \\ x \end{bmatrix} = x(-y) + yx = 0,$$

   and so the only condition to impose is that the columns are unit vectors. Hence, the matrices we are looking for are all matrices of the form

   $$\begin{bmatrix} a & -b \\ b & a \end{bmatrix},$$

   where $a^2 + b^2 = 1$.

5. We follow the first part of Example 3, Section 4.2.
   The (x, z) coordinates change according to

$$x_2 = x_1 \cos \theta - z_1 \sin \theta$$
$$z_2 = x_1 \sin \theta + z_1 \cos \theta,$$

while the y-coordinate remains unchanged $y_2 = y_1$. In matrix formulation

$$\begin{bmatrix} x_2 \\ y_2 \\ z_2 \end{bmatrix} = \begin{bmatrix} \cos \theta & 0 & -\sin \theta \\ 0 & 1 & 0 \\ \sin \theta & 0 & \cos \theta \end{bmatrix} \begin{bmatrix} x_1 \\ y_1 \\ z_1 \end{bmatrix}.$$

Hence the 3×3 rotation matrix we search is

$$\begin{bmatrix} \cos \theta & 0 & -\sin \theta \\ 0 & 1 & 0 \\ \sin \theta & 0 & \cos \theta \end{bmatrix}.$$

7. Notice that the matrices must have exactly one non-zero entry on every column, otherwise the columns would not be unit vectors. Moreover, the matrices must have exactly one non-zero entry on every row, otherwise the columns would not be orthogonal. The only possibilities are the following:

$$\begin{bmatrix} 1 & 0 & 0 \\ 0 & 1 & 0 \\ 0 & 0 & 1 \end{bmatrix}, \begin{bmatrix} 1 & 0 & 0 \\ 0 & 0 & 1 \\ 0 & 1 & 0 \end{bmatrix}, \begin{bmatrix} 0 & 0 & 1 \\ 1 & 0 & 0 \\ 0 & 1 & 0 \end{bmatrix}, \begin{bmatrix} 0 & 1 & 0 \\ 1 & 0 & 0 \\ 0 & 0 & 1 \end{bmatrix},$$

$$\begin{bmatrix} 0 & 1 & 0 \\ 0 & 0 & 1 \\ 1 & 0 & 0 \end{bmatrix}, \begin{bmatrix} 0 & 0 & 1 \\ 0 & 1 & 0 \\ 1 & 0 & 0 \end{bmatrix}.$$

9. We follow the strategy in Example 4.2, Section 4.2.
   Since reflections preserve length (norm), the final vector must be (plus or minus) $\sqrt{5^2 + 12^2 + 13^2}\mathbf{i} = 13\sqrt{2}\mathbf{i}$. The normal to the mirror must then be parallel to the vector connecting the tips of (say) $13\sqrt{2}\mathbf{i}$ and $5\mathbf{i} + 12\mathbf{j} + 13\mathbf{k}$, that is,

$$\mathbf{n} = \begin{bmatrix} 5 \\ 12 \\ 13 \end{bmatrix} - \begin{bmatrix} 13\sqrt{2} \\ 0 \\ 0 \end{bmatrix} = \begin{bmatrix} -13\sqrt{2} + 5 \\ 12 \\ 13 \end{bmatrix}.$$

Therefore

$$\mathbf{n}^{unit} = \frac{1}{\sqrt{(-13\sqrt{2} + 5)^2 + (12)^2 + (13)^2}} \begin{bmatrix} -13\sqrt{2} + 5 \\ 12 \\ 13 \end{bmatrix}$$

$$= \frac{1}{\sqrt{676 - 130\sqrt{2}}} \begin{bmatrix} -13\sqrt{2} + 5 \\ 12 \\ 13 \end{bmatrix}.$$

The reflector matrix is

$$\mathbf{M}_{ref} = \mathbf{I} - 2\mathbf{n}^{unit^T}\mathbf{n}^{unit}$$

$$= \begin{bmatrix} 1 & 0 & 0 \\ 0 & 1 & 0 \\ 0 & 0 & 1 \end{bmatrix} - 2 \cdot \frac{1}{\sqrt{676 - 130\sqrt{2}}} \cdot \frac{1}{\sqrt{676 - 130\sqrt{2}}}$$

$$\begin{bmatrix} -13\sqrt{2} + 5 \\ 12 \\ 13 \end{bmatrix} [-13\sqrt{2} + 5\ 12\ 13]$$

$$= \begin{bmatrix} 1 & 0 & 0 \\ 0 & 1 & 0 \\ 0 & 0 & 1 \end{bmatrix} - \frac{1}{338 - 65\sqrt{2}}$$

$$\begin{bmatrix} -130\sqrt{2} + 363 & -156\sqrt{2} + 60 & -169\sqrt{2} + 65 \\ -156\sqrt{2} + 60 & 144 & 156 \\ -169\sqrt{2} + 65 & 156 & 169 \end{bmatrix}$$

$$= \frac{1}{338 - 65\sqrt{2}} \begin{bmatrix} 25 + 65\sqrt{2} & -60 + 156\sqrt{2} & -65 + 169\sqrt{2} \\ -60 + 156\sqrt{2} & 194 - 65\sqrt{2} & -156 \\ -65 + 169\sqrt{2} & -156 & 169 - 65\sqrt{2} \end{bmatrix}.$$

11. We follow the strategy in Example 4.2, Section 4.2.

Since reflections preserve length (norm), the final vector must be (plus or minus) $\sqrt{1^2 + 1^2 + 1^2 + 1^2}[1\,0\,0\,0] = [\pm 2\,0\,0\,0]$. The normal to the mirror must then be parallel to the vector connecting the tips of (say) $[2\,0\,0\,0]$ and $[1\,1\,1\,1]$, that is,

$$\mathbf{n} = [1 - 1, -1 - 1].$$

Therefore

$$\mathbf{n}^{\text{unit}} = \frac{1}{\sqrt{1^2 + (-1)^2 + (-1)^2 + (-1)^2}}[1 - 1, -1 - 1]$$
$$= \frac{1}{2}[1 - 1, -1 - 1].$$

The reflector matrix is

$$\mathbf{M}_{\text{ref}} = \mathbf{I} - 2\mathbf{n}^{\text{unit}^T}\mathbf{n}^{\text{unit}}$$

$$= \begin{bmatrix} 1 & 0 & 0 & 0 \\ 0 & 1 & 0 & 0 \\ 0 & 0 & 1 & 0 \\ 0 & 0 & 0 & 1 \end{bmatrix} - 2 \cdot \frac{1}{2} \cdot \frac{1}{2} \begin{bmatrix} 1 \\ -1 \\ -1 \\ -1 \end{bmatrix} [1 - 1, -1 - 1]$$

$$= \begin{bmatrix} 1 & 0 & 0 & 0 \\ 0 & 1 & 0 & 0 \\ 0 & 0 & 1 & 0 \\ 0 & 0 & 0 & 1 \end{bmatrix} - \frac{1}{2} \begin{bmatrix} 1 & -1 & -1 & -1 \\ -1 & 1 & 1 & 1 \\ -1 & 1 & 1 & 1 \\ -1 & 1 & 1 & 1 \end{bmatrix}$$

$$= \begin{bmatrix} 1/2 & 1/2 & 1/2 & 1/2 \\ 1/2 & 1/2 & -1/2 & -1/2 \\ 1/2 & -1/2 & 1/2 & -1/2 \\ 1/2 & -1/2 & -1/2 & 1/2 \end{bmatrix}$$

13. We must find a reflector matrix $\mathbf{M}_{\text{ref}}$ which sends the second column vector to a vector along the x-axis. Since reflections preserve length (norm), the final vector must be (plus or minus) $\sqrt{2^2 + 1^2 + 3^2}\mathbf{i} = \sqrt{14}\mathbf{i}$. The normal to the mirror must then be parallel to the vector connecting the tips of (say) $\sqrt{14}\mathbf{i}$ and $[2\,1\,3]^T$, that is,

$$\mathbf{n} = \begin{bmatrix} 2 \\ 1 \\ 3 \end{bmatrix} - \begin{bmatrix} \sqrt{14} \\ 0 \\ 0 \end{bmatrix} = \begin{bmatrix} 2 - \sqrt{14} \\ 1 \\ 3 \end{bmatrix}.$$

Therefore

$$\mathbf{n}^{\text{unit}} = \frac{1}{\sqrt{(2 - \sqrt{14})^2 + (1)^2 + (3)^2}} \begin{bmatrix} 2 - \sqrt{14} \\ 1 \\ 3 \end{bmatrix}$$

$$= \frac{1}{\sqrt{28 - 4\sqrt{14}}} \begin{bmatrix} 2 - \sqrt{14} \\ 1 \\ 3 \end{bmatrix}.$$

The Householder reflector matrix is

$$\mathbf{M}_{\text{ref}} = \mathbf{I} - 2\mathbf{n}^{\text{unit}^T}\mathbf{n}^{\text{unit}}$$

$$= \begin{bmatrix} 1 & 0 & 0 \\ 0 & 1 & 0 \\ 0 & 0 & 1 \end{bmatrix} - 2 \cdot \frac{1}{\sqrt{28 - 4\sqrt{14}}} \cdot \frac{1}{\sqrt{28 - 4\sqrt{14}}}$$

$$\begin{bmatrix} 2 - \sqrt{14} \\ 1 \\ 3 \end{bmatrix} [2 - \sqrt{14}\ 1\ 3]$$

$$= \begin{bmatrix} 1 & 0 & 0 \\ 0 & 1 & 0 \\ 0 & 0 & 1 \end{bmatrix} - \frac{1}{14 - 2\sqrt{14}}$$

$$\begin{bmatrix} 18 - 4\sqrt{14} & 2 - \sqrt{14} & 6 - 3\sqrt{14} \\ 2 - \sqrt{14} & 1 & 3 \\ 6 - 3\sqrt{14} & 3 & 9 \end{bmatrix}$$

$$= \frac{1}{14 - 2\sqrt{14}} \begin{bmatrix} -4 + 2\sqrt{14} & -2 + \sqrt{14} & -6 + 3\sqrt{14} \\ -2 + \sqrt{14} & 13 - 2\sqrt{14} & -3 \\ -6 + 3\sqrt{14} & -3 & 5 - 2\sqrt{14} \end{bmatrix}$$

$$= \begin{bmatrix} \frac{\sqrt{14}}{7} & \frac{\sqrt{14}}{14} & \frac{3\sqrt{14}}{14} \\ \frac{\sqrt{14}}{14} & \frac{63 - \sqrt{14}}{70} & \frac{-21 - 3\sqrt{14}}{70} \\ \frac{3\sqrt{14}}{14} & \frac{-21 - 3\sqrt{14}}{70} & \frac{7 - 9\sqrt{14}}{70} \end{bmatrix}.$$

We show next that the zero pattern in the first column is not preserved when we multiply $\mathbf{A} = \begin{bmatrix} \text{col\#1} & \text{col\#2} & \text{col\#3} \\ \downarrow & \downarrow & \downarrow \end{bmatrix}$ to the left by $\mathbf{M}_{\text{ref}}$. It suffices to compute

$$\mathbf{M}_{\text{ref}} \begin{bmatrix} \text{col\#1} \\ \downarrow \end{bmatrix} = \begin{bmatrix} \frac{\sqrt{14}}{7} & \frac{\sqrt{14}}{14} & \frac{3\sqrt{14}}{14} \\ \frac{\sqrt{14}}{14} & \frac{63-\sqrt{14}}{70} & \frac{-21-3\sqrt{14}}{70} \\ \frac{3\sqrt{14}}{14} & \frac{-21-3\sqrt{14}}{70} & \frac{7-9\sqrt{14}}{70} \end{bmatrix} \begin{bmatrix} 4 \\ 0 \\ 0 \end{bmatrix} = \begin{bmatrix} \frac{4\sqrt{14}}{7} \\ \frac{2\sqrt{14}}{7} \\ \frac{6\sqrt{14}}{7} \end{bmatrix}.$$

15. The second column vector must be reflected to a vector of the form $[2\,a\,0]^T$, and since its length should be preserved, we get $a = \pm\sqrt{10}$. We choose $a = \sqrt{10}$. The normal to the mirror must then be

$$\mathbf{n} = \begin{bmatrix} 2 \\ 1 \\ 3 \end{bmatrix} - \begin{bmatrix} 2 \\ \sqrt{10} \\ 0 \end{bmatrix} = \begin{bmatrix} 0 \\ 1-\sqrt{10} \\ 3 \end{bmatrix}.$$

Therefore

$$\mathbf{n}^{\text{unit}} = \frac{1}{\sqrt{0^2 + (1-\sqrt{10})^2 + (3)^2}} \begin{bmatrix} 0 \\ 1-\sqrt{10} \\ 3 \end{bmatrix}$$

$$= \frac{1}{\sqrt{20 - 2\sqrt{10}}} \begin{bmatrix} 0 \\ 1-\sqrt{10} \\ 3 \end{bmatrix}.$$

The Householder reflector matrix is

$$\mathbf{M}_{\text{ref}} = \mathbf{I} - 2\mathbf{n}^{\text{unit}\,T}\mathbf{n}^{\text{unit}}$$

$$= \begin{bmatrix} 1 & 0 & 0 \\ 0 & 1 & 0 \\ 0 & 0 & 1 \end{bmatrix} - 2 \cdot \frac{1}{\sqrt{20 - 2\sqrt{10}}} \cdot \frac{1}{\sqrt{20 - 2\sqrt{10}}}$$

$$\begin{bmatrix} 0 \\ 1-\sqrt{10} \\ 3 \end{bmatrix} [0\ 1-\sqrt{10}\ 3]$$

$$= \begin{bmatrix} 1 & 0 & 0 \\ 0 & 1 & 0 \\ 0 & 0 & 1 \end{bmatrix} - \frac{1}{10-\sqrt{10}} \begin{bmatrix} 0 & 0 & 0 \\ 0 & 11-2\sqrt{10} & 3-3\sqrt{10} \\ 0 & 3-3\sqrt{10} & 9 \end{bmatrix}$$

$$= \frac{1}{10-\sqrt{10}} \begin{bmatrix} 1 & 0 & 0 \\ 0 & -1+\sqrt{10} & -3+3\sqrt{10} \\ 0 & -3+3\sqrt{10} & 1-\sqrt{10} \end{bmatrix}$$

$$= \begin{bmatrix} 1 & 0 & 0 \\ 0 & \sqrt{10}/10 & -3\sqrt{10}/10 \\ 0 & -3\sqrt{10}/10 & -\sqrt{10}/10 \end{bmatrix}.$$

We show next that the zero pattern in the first column is preserved when we multiply $\mathbf{A} = \begin{bmatrix} \text{col}\#1 & \text{col}\#2 & \text{col}\#3 \\ \downarrow & \downarrow & \downarrow \end{bmatrix}$ to the left by $\mathbf{M}_{\text{ref}}$. It suffices to compute

$$\mathbf{M}_{\text{ref}} \begin{bmatrix} \text{col}\#1 \\ \downarrow \end{bmatrix} = \begin{bmatrix} 1 & 0 & 0 \\ 0 & \sqrt{10}/10 & -3\sqrt{10}/10 \\ 0 & -3\sqrt{10}/10 & -\sqrt{10}/10 \end{bmatrix} \begin{bmatrix} 4 \\ 0 \\ 0 \end{bmatrix} = \begin{bmatrix} 4 \\ 0 \\ 0 \end{bmatrix}.$$

Hence, the unit normal vector to the mirror is

$$\mathbf{n} = \frac{1}{\sqrt{20 - 2\sqrt{10}}} \begin{bmatrix} 0 \\ 1 - \sqrt{10} \\ 3 \end{bmatrix}.$$

The Householder reflector is

$$\mathbf{M}_{\text{ref}} = \mathbf{I} - 2\mathbf{n}\mathbf{n}^T = \begin{bmatrix} 1 & 0 & 0 \\ 0 & \sqrt{10}/10 & -3\sqrt{10}/10 \\ 0 & -3\sqrt{10}/10 & -\sqrt{10}/10 \end{bmatrix}.$$

The zero pattern in the first column is preserved.

17. The reflection matrix multiplication is not commutative.

(a) Let $\mathbf{M_1} = \begin{bmatrix} 0 & 1 \\ 1 & 0 \end{bmatrix}$ be the matrix of the reflection about the line $y = x$, and let $\mathbf{M_2} = \begin{bmatrix} -1 & 0 \\ 0 & 1 \end{bmatrix}$ be the matrix of the reflection about the $y-$axis. We have

$$\mathbf{M_1 M_2} = \begin{bmatrix} 0 & 1 \\ -1 & 0 \end{bmatrix} \neq \begin{bmatrix} 0 & -1 \\ 1 & 0 \end{bmatrix} = \mathbf{M_2 M_1}$$

(b) Consider the point $[1\ 0]$. When reflecting it about the line $y = x$, it is sent to $[1\ 0]$, and the latter is sent to itself when reflecting it about the $y-$axis. However, when reflecting the point $[1\ 0]$ first about the $y-$axis, it is sent to $[-1\ 0]$, which is sent to $[0 - 1]$, when reflecting about the line $y = x$. Therefore, the composition of the two reflection is not commutative, and so the reflection matrix multiplication is not commutative in general.

19. (a) It suffice to show that the homogeneous linear system $(\mathbf{I}-\mathbf{A})\mathbf{v}=\mathbf{0}$ has only the trivial solution $\mathbf{v}=\mathbf{0}$. We compute

$$
\begin{aligned}
\|(\mathbf{I}-\mathbf{A})\mathbf{v}\|^2 &= ((\mathbf{I}-\mathbf{A})\mathbf{v})^T\,(\mathbf{I}-\mathbf{A})\mathbf{v} \\
&= \mathbf{v}^T(\mathbf{I}-\mathbf{A})^T(\mathbf{I}-\mathbf{A})\mathbf{v} \\
&= \mathbf{v}^T(\mathbf{I}-\mathbf{A}^T)(\mathbf{I}-\mathbf{A})\mathbf{v} \\
&= \mathbf{v}^T(\mathbf{I}-\mathbf{A}-\mathbf{A}^T-\mathbf{A}^T\mathbf{A})\mathbf{v} \\
&= \mathbf{v}^T(\mathbf{I}-\mathbf{A}^T\mathbf{A})\mathbf{v} \\
&= \mathbf{v}^T\mathbf{v}-\mathbf{v}^T\mathbf{A}^T\mathbf{A}\mathbf{v} \\
&= \|\mathbf{v}\|^2-(\mathbf{A}\mathbf{v})^T\,\mathbf{A}\mathbf{v} \\
&= \|\mathbf{v}\|^2+\|\mathbf{A}\mathbf{v}\|^2
\end{aligned}
$$

Now, if $\mathbf{v}$ is a solution of the homogeneous linear system $(\mathbf{I}-\mathbf{A})\mathbf{v}=\mathbf{0}$, then $\|(\mathbf{I}-\mathbf{A})\mathbf{v}\|^2=\mathbf{0}$, and so $\|\mathbf{v}\|^2+\|\mathbf{A}\mathbf{v}\|^2=0$. It follows that $\|\mathbf{v}\|=0$, and so $\mathbf{v}=\mathbf{0}$.

(b) Let $\mathbf{B}=(\mathbf{I}-\mathbf{A})^{-1}(\mathbf{I}+\mathbf{A})$. To show that $\mathbf{B}$ is orthogonal, it suffices to show that $\mathbf{B}\mathbf{B}^T=\mathbf{I}$. Since

$$
\begin{aligned}
\mathbf{B}^T &= \left((\mathbf{I}-\mathbf{A})^{-1}(\mathbf{I}+\mathbf{A})\right)^T = (\mathbf{I}+\mathbf{A})^T\left((\mathbf{I}-\mathbf{A})^{-1}\right)^T \\
&= (\mathbf{I}+\mathbf{A}^T)((\mathbf{I}-\mathbf{A})^T)^{-1},
\end{aligned}
$$

we can compute

$$
\begin{aligned}
\mathbf{B}\mathbf{B}^T &= (\mathbf{I}-\mathbf{A})^{-1}(\mathbf{I}+\mathbf{A})(\mathbf{I}+\mathbf{A}^T)((\mathbf{I}-\mathbf{A})^T)^{-1} \\
&= (\mathbf{I}-\mathbf{A})^{-1}(\mathbf{I}+\mathbf{A})(\mathbf{I}-\mathbf{A})((\mathbf{I}-\mathbf{A}^T))^{-1} \\
&= (\mathbf{I}-\mathbf{A})^{-1}(\mathbf{I}-\mathbf{A}^2)(\mathbf{I}+\mathbf{A})^{-1} \\
&= (\mathbf{I}-\mathbf{A})^{-1}(\mathbf{I}-\mathbf{A})(\mathbf{I}+\mathbf{A})(\mathbf{I}+\mathbf{A})^{-1} \\
&= \mathbf{I}.
\end{aligned}
$$

4.3   LEAST SQUARES

1. The linear system is $\mathbf{A}\mathbf{x} = \mathbf{b}$, where $\mathbf{A} = \begin{bmatrix} 1 & 3 \\ -1 & -1 \\ 1 & 3 \\ -1 & -1 \end{bmatrix}$, and
$\mathbf{b} = [1\,0\,0\,1]^T$. The column space of $\mathbf{A}$ is spanned by the vectors
$\mathbf{v_1} = [1\,-1\,1\,-1]^T$ and $\mathbf{v_2} = [3\,-1\,3\,-1]^T$. Using the solution
of Problem 7, Exercises 4.1, the Gram-Schmidt algorithm yields an
orthonormal basis for the column space of $\mathbf{A}$ consisting of the vectors
$\mathbf{w_1} = [1/2\,-1/2\,1/2\,-1/2]^T$ and $\mathbf{w_2} = [1/2\,1/2\,1/2\,1/2]^T$. The
projection of onto the column space of is given by formula (2) in
Section 4.3:

$$\mathbf{b}_{\text{col } \mathbf{A}} = (\mathbf{b} \cdot \mathbf{w_1})\mathbf{w_1} + (\mathbf{b} \cdot \mathbf{w_2})\mathbf{w_2}$$

$$= \left(\begin{bmatrix} 1 \\ 0 \\ 0 \\ 1 \end{bmatrix} \cdot \begin{bmatrix} 1/2 \\ -1/2 \\ 1/2 \\ -1/2 \end{bmatrix} \right) \begin{bmatrix} 1/2 \\ -1/2 \\ 1/2 \\ -1/2 \end{bmatrix} + \left(\begin{bmatrix} 1 \\ 0 \\ 0 \\ 1 \end{bmatrix} \cdot \begin{bmatrix} 1/2 \\ 1/2 \\ 1/2 \\ 1/2 \end{bmatrix} \right) \begin{bmatrix} 1/2 \\ 1/2 \\ 1/2 \\ 1/2 \end{bmatrix}$$

$$= (1/2 - 1/2) \begin{bmatrix} 1/2 \\ -1/2 \\ 1/2 \\ -1/2 \end{bmatrix} + (1/2 + 1/2) \begin{bmatrix} 1/2 \\ 1/2 \\ 1/2 \\ 1/2 \end{bmatrix}$$

$$= \begin{bmatrix} 1/2 \\ 1/2 \\ 1/2 \\ 1/2 \end{bmatrix}.$$

We solve next the equation $\mathbf{A}\mathbf{x} = \mathbf{b}_{\text{col } \mathbf{A}}$, which is equivalent to

$$\begin{bmatrix} 1 & 3 & \vdots & 1/2 \\ -1 & -1 & \vdots & 1/2 \\ 1 & 3 & \vdots & 1/2 \\ -1 & -1 & \vdots & 1/2 \end{bmatrix}.$$

The Gaussian elimination procedure reveals the solution

$$\mathbf{x} = \begin{bmatrix} -1 \\ 1/2 \end{bmatrix}.$$

The error is given by

$$\text{error} = \mathbf{b} - \mathbf{Ax} = \mathbf{b} - \mathbf{b}_{\text{col } A} = \begin{bmatrix} 1 \\ 0 \\ 0 \\ 1 \end{bmatrix} - \begin{bmatrix} 1/2 \\ 1/2 \\ 1/2 \\ 1/2 \end{bmatrix} = \begin{bmatrix} 1/2 \\ -1/2 \\ -1/2 \\ 1/2 \end{bmatrix}.$$

3. The system is of the form $\mathbf{Ax} = \mathbf{b}$, where $\mathbf{A} = \begin{bmatrix} 2 & 3 & 2 \\ 2 & 1 & 0 \\ -2 & -1 & -2 \\ -2 & -3 & -4 \end{bmatrix}$ and

$\mathbf{b} = [0\,0\,0\,1]^T$. The associated normal equations are $\mathbf{A}^T\mathbf{Ax} = \mathbf{A}^T\mathbf{b}$, which is equivalent to

$$\begin{bmatrix} 2 & 2 & -2 & -2 \\ 3 & 1 & -1 & -3 \\ 2 & 0 & -2 & -4 \end{bmatrix} \begin{bmatrix} 2 & 3 & 2 \\ 2 & 1 & 0 \\ -2 & -1 & -2 \\ -2 & -3 & -4 \end{bmatrix} \begin{bmatrix} x_1 \\ x_2 \\ x_3 \end{bmatrix}$$

$$= \begin{bmatrix} 2 & 2 & -2 & -2 \\ 3 & 1 & -1 & -3 \\ 2 & 0 & -2 & -4 \end{bmatrix} \begin{bmatrix} 0 \\ 0 \\ 0 \\ 1 \end{bmatrix},$$

or

$$\begin{bmatrix} 16 & 16 & 16 \\ 16 & 20 & 20 \\ 16 & 20 & 24 \end{bmatrix} \begin{bmatrix} x_1 \\ x_2 \\ x_3 \end{bmatrix} = \begin{bmatrix} -2 \\ -3 \\ -4 \end{bmatrix}.$$

The Gaussian elimination procedure exhibits the solution

$$\begin{bmatrix} x_1 \\ x_2 \\ x_3 \end{bmatrix} = \begin{bmatrix} 1/8 \\ 0 \\ -1/4 \end{bmatrix}.$$

5. The linear system is $\mathbf{Ax} = \mathbf{b}$, where

$$\mathbf{A} = \begin{bmatrix} -1 & 2 & 3 \\ 1 & 0 & 1 \\ -1 & 2 & 1 \\ 1 & 0 & -1 \end{bmatrix} \quad \text{and} \quad \mathbf{b} = [1\,0\,0\,1]^T.$$

The column space of $\mathbf{A}$ is spanned by the vectors $\mathbf{v_1} = [-1\,1\,-1\,1]^T$, $\mathbf{v_2} = [2\,0\,2\,0]^T$ and $\mathbf{v_3} = [3\,1\,1\,-1]^T$. Using the solution of Problem 10, Exercises 4.1, the Gram-Schmidt algorithm yields an orthonormal basis for the column space of $\mathbf{A}$ consisting of the vectors $\mathbf{w_1} = [-1/2\,1/2\,-1/2\,1/2]^T$, $\mathbf{w_2} = [1/2\,1/2\,1/2\,1/2]^T$, and $\mathbf{w_3} = [1/2\,1/2\,-1/2\,-1/2]^T$ The projection of onto the column space of is given by formula (2) in Section 4.3:

$$\mathbf{b}_{\text{col A}} = (\mathbf{b} \cdot \mathbf{w_1})\mathbf{w_1} + (\mathbf{b} \cdot \mathbf{w_2})\mathbf{w_2} + (\mathbf{b} \cdot \mathbf{w_3})\mathbf{w_3}$$

$$= \left(\begin{bmatrix} 1 \\ 0 \\ 0 \\ 1 \end{bmatrix} \cdot \begin{bmatrix} -1/2 \\ 1/2 \\ -1/2 \\ 1/2 \end{bmatrix} \right) \begin{bmatrix} -1/2 \\ 1/2 \\ -1/2 \\ 1/2 \end{bmatrix} + \left(\begin{bmatrix} 1 \\ 0 \\ 0 \\ 1 \end{bmatrix} \cdot \begin{bmatrix} 1/2 \\ 1/2 \\ 1/2 \\ 1/2 \end{bmatrix} \right) \begin{bmatrix} 1/2 \\ 1/2 \\ 1/2 \\ 1/2 \end{bmatrix}$$

$$+ \left(\begin{bmatrix} 1 \\ 0 \\ 0 \\ 1 \end{bmatrix} \cdot \begin{bmatrix} 1/2 \\ 1/2 \\ -1/2 \\ -1/2 \end{bmatrix} \right) \begin{bmatrix} 1/2 \\ 1/2 \\ -1/2 \\ -1/2 \end{bmatrix}$$

$$= (-1/2 + 1/2) \begin{bmatrix} -1/2 \\ 1/2 \\ -1/2 \\ 1/2 \end{bmatrix} + (1/2 + 1/2) \begin{bmatrix} 1/2 \\ 1/2 \\ 1/2 \\ 1/2 \end{bmatrix}$$

$$+ (1/2 - 1/2) \begin{bmatrix} 1/2 \\ 1/2 \\ -1/2 \\ -1/2 \end{bmatrix}$$

$$= \begin{bmatrix} 1/2 \\ 1/2 \\ 1/2 \\ 1/2 \end{bmatrix}.$$

We solve next the equation $\mathbf{Ax} = \mathbf{b}_{\text{col A}}$, which is equivalent to

$$
\begin{bmatrix}
-1 & 2 & 3 & \vdots & 1/2 \\
1 & 0 & 1 & \vdots & 1/2 \\
-1 & 2 & 1 & \vdots & 1/2 \\
1 & 0 & -1 & \vdots & 1/2
\end{bmatrix}.
$$

The Gaussian elimination procedure reveals the solution

$$
\mathbf{x} = \begin{bmatrix} 1/2 \\ 1/2 \\ 0 \end{bmatrix}.
$$

The error is given by

$$
\text{error} = \mathbf{b} - \mathbf{Ax} = \mathbf{b} - \mathbf{b}_{\text{col A}} = \begin{bmatrix} 1 \\ 0 \\ 0 \\ 1 \end{bmatrix} - \begin{bmatrix} 1/2 \\ 1/2 \\ 1/2 \\ 1/2 \end{bmatrix} = \begin{bmatrix} 1/2 \\ -1/2 \\ -1/2 \\ 1/2 \end{bmatrix}.
$$

7. Given a collection of pairs of numbers $\{(x_i, y_i)\}_{i=1}^{N}$, the equations for the slopes m of the straight line $y = mx$ through the origin that minimizes the sum of squares of errors

$$
\sum_{i=1}^{N} (y_i - mx_i)^2
$$

are the solutions of

$$
mx_1 = y_1,
$$
$$
mx_2 = y_2,
$$
$$
\vdots
$$
$$
mx_N = y_N
$$

In matrix form, this is

$$
\begin{bmatrix} x_1 \\ x_2 \\ \vdots \\ x_N \end{bmatrix} m = \begin{bmatrix} y_1 \\ y_2 \\ \vdots \\ y_N \end{bmatrix}.
$$

The normal equations for this system is

$$
\begin{bmatrix} x_1 & x_2 & \cdots & x_N \end{bmatrix} \begin{bmatrix} x_1 \\ x_2 \\ \vdots \\ x_N \end{bmatrix} m = \begin{bmatrix} x_1 & x_2 & \cdots & x_N \end{bmatrix} \begin{bmatrix} y_1 \\ y_2 \\ \vdots \\ y_N \end{bmatrix},
$$

or

$$
\left(\sum_{i=1}^{N} x_i^2 \right) m = \sum_{i=1}^{N} x_i y_i.
$$

Using the standard abbreviation $S_{xy} = \sum_{i=1}^{N} x_i y_i$, and $S_{x^2} = \sum_{i=1}^{N} x_i^2$, the solution is

$$
m = \frac{S_{xy}}{S_{x^2}}.
$$

9. Given a collection of pairs of numbers $\{(x_i, y_i)\}_{i=1}^{N}$, the equations for the slopes m of the straight line $y = mx + b_0$ with prescribed intercept b_0 that minimizes the sum of squares of errors

$$
\sum_{i=1}^{N} [y_i - (mx_i + b_0)]^2
$$

are the solutions of

$$
mx_1 + b_0 = y_1,
$$
$$
mx_2 + b_0 = y_2,
$$
$$
\vdots
$$
$$
mx_N + b_0 = y_N
$$

In matrix form, this is

$$
\begin{bmatrix} x_1 \\ x_2 \\ \vdots \\ x_N \end{bmatrix} m = \begin{bmatrix} y_1 - b_0 \\ y_2 - b_0 \\ \vdots \\ y_N - b_0 \end{bmatrix}.
$$

The normal equations for this system is

$$
\begin{bmatrix} x_1 & x_2 & \cdots & x_N \end{bmatrix} \begin{bmatrix} x_1 \\ x_2 \\ \vdots \\ x_N \end{bmatrix} m = \begin{bmatrix} x_1 & x_2 & \cdots & x_N \end{bmatrix} \begin{bmatrix} y_1 - b_0 \\ y_2 - b_0 \\ \vdots \\ y_N - b_0 \end{bmatrix},
$$

or

$$
\left(\sum_{i=1}^{N} x_i^2 \right) m = \sum_{i=1}^{N} x_i y_i - b_0 \sum_{i=1}^{N} x_i.
$$

Using the standard abbreviation $S_x = \sum_{i=1}^{N} x_i$, $S_{xy} = \sum_{i=1}^{N} x_i y_i$, and $S_{x^2} = \sum_{i=1}^{N} x_i^2$, the solution is

$$
\frac{S_{xy} - b_0 S_x}{S_{x^2}}.
$$

11. We compute

$$
\begin{aligned}
m\bar{x} + b &= \frac{S_x}{N} \frac{N S_{xy} - S_x S_y}{N S_{x^2} - S_x^2} + \frac{S_{x^2} S_y - S_x S_{xy}}{N S_{x^2} - S_x^2} \\
&= \frac{S_{xy} S_x - \frac{1}{N} S_x^2 S_y + S_{x^2} S_y - S_x S_{xy}}{N S_{x^2} - S_x^2} \\
&= \frac{S_y}{N} \frac{N S_{x^2} - S_x^2}{N S_{x^2} - S_x^2} \\
&= \frac{S_y}{N} \\
&= \bar{y}.
\end{aligned}
$$

13. (a) Consider the subspace $\text{Span}(\mathbf{v})$. An orthogonal basis of $\text{Span}(\mathbf{v})$ is given by the unit vector $\frac{1}{\|\mathbf{v}\|}\mathbf{v}$. Let now

$$\mathbf{b} = \mathbf{b}_\mathbf{v} + \mathbf{b}^\perp$$

be the orthogonal decomposition of the vector $\mathbf{b}$ with respect to this subspace. Since $\mathbf{b}_\mathbf{v}$ is in the one-dimensional subspace, we have $\mathbf{b}_\mathbf{v} = \frac{k}{\|\mathbf{v}\|}\mathbf{v}$, for some real number k. Taking the dot product in $\mathbf{v}$ with the unit vector $\frac{1}{\|\mathbf{v}\|}\mathbf{v}$, we compute

$$\frac{(\mathbf{b}\cdot\mathbf{v})}{\|\mathbf{v}\|} = k + 0 = k.$$

Hence

$$\mathbf{b}_\mathbf{v} = \frac{k}{\|\mathbf{v}\|}\mathbf{v} = \frac{(\mathbf{b}\cdot\mathbf{v})}{\|\mathbf{v}\|^2}\mathbf{v} = \frac{\mathbf{v}\mathbf{v}^T\mathbf{b}}{\mathbf{v}^T\mathbf{v}}.$$

(b) The normal equations for the least squares approximation problem $\mathbf{v}x = \mathbf{b}$ are $\mathbf{v}^T(\mathbf{b}-x\mathbf{v}) = 0$. This is the same as $\mathbf{v}^T\mathbf{v}x = \mathbf{v}^T\mathbf{b}$, which yields $x = \dfrac{\mathbf{v}^T\mathbf{b}}{\mathbf{v}^T\mathbf{v}}$.

15. (a) If $\mathbf{x}$ is in the null space of $\mathbf{A}$, then $\mathbf{A}\mathbf{x} = \mathbf{0}$, and so $\mathbf{A}^T\mathbf{A}\mathbf{x} = \mathbf{0}$, which means that $\mathbf{x}$ is in the null space of $\mathbf{A}^T\mathbf{A}$ as well.

(b) If $\mathbf{A}^T\mathbf{A}\mathbf{x} = \mathbf{0}$, then $\mathbf{x}^T\mathbf{A}^T\mathbf{A}\mathbf{x} = \mathbf{0}$. However

$$\mathbf{x}^T\mathbf{A}^T\mathbf{A}\mathbf{x} = (\mathbf{A}\mathbf{x})^T(\mathbf{A}\mathbf{x}) = \|\mathbf{A}\mathbf{x}\|,$$

and so $\|\mathbf{A}\mathbf{x}\| = 0$. But then $\mathbf{A}\mathbf{x} = \mathbf{0}$, which means that x is in the null space of $\mathbf{A}$.

(c) From (a) and (b) one can conclude that the null spaces of $\mathbf{A}$ and $\mathbf{A}^T\mathbf{A}$ coincide. In particular, the nullities of the two matrices coincide. From Theorem 6, Section 3.3, the result follows.

17. We must find the optimal exponential function $y = ce^{bx}$ that minimizes the sum of squares of errors

$$\sum_{i=1}^{N}(y_i - ce^{bx_i})^2.$$

Looking at the critical points, this amounts to solving the system of equation

$$ce^{bx_1} = y_1$$
$$ce^{bx_2} = y_2$$
$$\vdots$$
$$ce^{bx_N} = y_N$$

for the unknowns b and c. By taking logarithms and substituting C for $\log c$ and Y_i for $\log y_i$, we obtain the system

$$bx_1 + C = Y_1$$
$$bx_2 + C = Y_1$$
$$\vdots$$
$$bx_N + C = Y_N$$

for the unknowns b and C, which is the situation studied in Example 2, in Section 4.3. We obtain

$$b = \frac{1}{NS_{x^2} - S_x^2} \log \frac{y_1^{Nx_1} y_2^{Nx_2} \cdots y_N^{Nx_N}}{(y_1 y_2 \cdots y_N)^{S_x}}$$

$$\log c = \frac{1}{NS_{x^2} - S_x^2} \log \frac{(y_1 y_2 \cdots y_N)^{S_{x^2}}}{(y_1^{x_1} y_2^{x_2} \cdots y_N^{x_N})^{S_x}}.$$

19. Given a collection of pairs of numbers $\{(x_i, y_i)\}_{i=1}^{N}$, $y_i > 0$, $i = 1, 2, \ldots, N$ the equations for the least squares fit using a linear combinations of exponentials $y = c_0 + c_1 e^{-x} + c_2 e^{-2x} + \cdots + c_k e^{-kx}$ minimizes the sum of errors

$$\sum_{i=1}^{N} \left[y_i - (c_0 + c_1 e^{-x_i} + c_2 e^{-2x_i} + \cdots + c_k e^{-kx_i}) \right]^2.$$

Looking at the critical points, the unknowns $c_0, c_1, \ldots, c_k$ are solutions of

$$c_0 + c_1 e^{-x_1} + c_2 e^{-2x_1} + \cdots + c_k e^{-kx_1} = y_1,$$
$$c_0 + c_1 e^{-x_2} + c_2 e^{-2x_2} + \cdots + c_k e^{-kx_2} = y_2,$$
$$\vdots$$
$$c_0 + c_1 e^{-x_N} + c_2 e^{-2x_N} + \cdots + c_k e^{-kx_N} = y_N$$

In matrix form, this is

$$
\begin{bmatrix}
1 & e^{-x_1} & \cdots & e^{-kx_1} \\
1 & e^{-x_2} & \cdots & e^{-kx_2} \\
\vdots & \vdots & \ddots & \vdots \\
1 & e^{-x_N} & \cdots & e^{-kx_N}
\end{bmatrix}
\begin{bmatrix}
c_0 \\ c_1 \\ \vdots \\ c_k
\end{bmatrix}
=
\begin{bmatrix}
y_1 \\ y_2 \\ \vdots \\ y_N
\end{bmatrix}.
$$

The normal equations are

$$
\begin{bmatrix}
1 & 1 & \cdots & 1 \\
e^{-x_1} & e^{-x_2} & \cdots & e^{-x_N} \\
\vdots & \vdots & \ddots & \vdots \\
e^{-kx_1} & e^{-kx_2} & \cdots & e^{-kx_N}
\end{bmatrix}
\begin{bmatrix}
1 & e^{-x_1} & \cdots & e^{-kx_1} \\
1 & e^{-x_2} & \cdots & e^{-kx_2} \\
\vdots & \vdots & \ddots & \vdots \\
1 & e^{-x_N} & \cdots & e^{-kx_N}
\end{bmatrix}
\begin{bmatrix}
c_0 \\ c_1 \\ \vdots \\ c_N
\end{bmatrix}.
$$

$$
=
\begin{bmatrix}
1 & 1 & \cdots & 1 \\
e^{-x_1} & e^{-x_2} & \cdots & e^{-x_N} \\
\vdots & \vdots & \ddots & \vdots \\
e^{-kx_1} & e^{-x_N} & \cdots & e^{-kx_N}
\end{bmatrix}
\begin{bmatrix}
y_1 \\ y_2 \\ \vdots \\ y_N
\end{bmatrix}
$$

21. (a) Notice that SSE is the sum of squares of the entries of the column
vector $(\mathbf{b} - \mathbf{Ax})$. The jth entry of the column vector $(\mathbf{b} - \mathbf{Ax})$ is

$$
b_j - \sum_{k=1}^{n} A_{jk} x_k.
$$

Hence,

$$
\text{SSE} = \sum_{j=1}^{n} \left(b_j - \sum_{k=1}^{n} A_{jk} x_k \right)^2
$$

$$
= \sum_{j=1}^{n} b_j^2 - 2 \sum_{j=1}^{n} b_j \sum_{k=1}^{n} A_{jk} x_k + \sum_{j=1}^{n} \left(\sum_{k=1}^{n} A_{jk} x_k \right)^2
$$

$$
= \sum_{j=1}^{n} b_j^2 - 2 \sum_{j=1}^{n} \sum_{k=1}^{n} A_{jk} b_j x_k + \sum_{i=1}^{n} \sum_{j=1}^{n} \sum_{k=1}^{n} A_{ji} A_{jk} x_i x_k.
$$

(b) We compute partial derivatives using the formula obtained in
part (a), by analyzing the three terms separately. Notice that the

last one $\sum_{j=1}^{n} b_j^2$ is independent of x_p, so its derivative vanishes. For the second term, notice that the only terms containing the variable x_p is

$$-2 \left(\sum_{j=1}^{n} A_{jp} b_p \right) x_p,$$

and so its partial derivative with respect to x_p is $-2 \sum_{j=1}^{n} A_{jp} b_j$.

The first term is quadratic in x_p. Concerning the partial derivative with respect to x_p, we ignore the terms not containing x_p. The terms containing x_p are

$$\sum_{j=1}^{n} \sum_{k=1}^{n} A_{jp} A_{jk} x_k x_p + \sum_{i=1}^{n} \sum_{j=1}^{n} A_{ji} A_{jp} x_i x_p,$$

and the partial derivative is

$$\sum_{j=1}^{n} \sum_{k=1}^{n} A_{jp} A_{jk} x_k + \sum_{i=1}^{n} \sum_{j=1}^{n} A_{ji} A_{jp} x_i.$$

Collecting these observations, we get

$$\frac{\partial \text{SSE}}{\partial x_p} = \sum_{j=1}^{n} \sum_{k=1}^{n} A_{jp} A_{jk} x_k + \sum_{i=1}^{n} \sum_{j=1}^{n} A_{ji} A_{jp} x_i - 2 \sum_{j=1}^{n} A_{jp} b_j.$$

(c) We compute first the pth entry of $\mathbf{A}^T \mathbf{b}$, and we find it to be $\sum_{j=1}^{n} A_{jp} b_j$. On the other hand, we can also see that the pth entry of $2 \mathbf{A}^T \mathbf{A} \mathbf{x}$ is

$$2 \sum_{j=1}^{n} A_{jp} \left(\sum_{k=1}^{n} A_{jk} x_k \right) = \sum_{j=1}^{n} \sum_{k=1}^{n} A_{jp} A_{jk} x_k + \sum_{i=1}^{n} \sum_{j=1}^{n} A_{ji} A_{jp} x_i.$$

The conclusion follows immediately.

4.4 FUNCTION SPACES

1. (a) No. Notice that $(-1) \cdot 1 + 1 \cdot \sin^2 x + 1 \cdot \cos^2 x = 0$.

   (b) Yes. If we test for independence the relation $c_1 x + c_2 |x| = 0$ at the points $\{-1, 1\}$, we get $-c_1 + c + 2 = 0$, and $c_1 + c_2 = 0$. The only solution is $c_1 = c_2 = 0$.

   (c) No. On the interval $[-1, 0]$, we have $|x| = -x$, which can be interpreted as the linear dependence relation $1 \cdot x + 1 \cdot |x| = 0$.

   (d) Yes. If we test for independence at the points $\{-1, 0, 1\}$, the system (1) in Section 4.4 becomes

   $$\begin{bmatrix} -1 & 1 & -1 \\ 0 & 0 & 1 \\ 1 & 1 & -1 \end{bmatrix} \begin{bmatrix} c_1 \\ c_2 \\ c_3 \end{bmatrix} = \begin{bmatrix} 0 \\ 0 \\ 0 \end{bmatrix}.$$

   The determinant of the coefficient matrix is 2, so only the trivial solution is possible for $\{c_1, c_2, c_3\}$, implying the functions are independent.

   (e) Yes. If we test for independence at the points $\{-1, 0, 1\}$, the system (1) in Section 4.4 becomes

   $$\begin{bmatrix} -1 & e^{-1} & 1 \\ 0 & 0 & 1 \\ 1 & 1 & 1 \end{bmatrix} \begin{bmatrix} c_1 \\ c_2 \\ c_3 \end{bmatrix} = \begin{bmatrix} 0 \\ 0 \\ 0 \end{bmatrix}.$$

   The determinant of the coefficient matrix is $2 + e^{-1} \neq 0$, so only the trivial solution is possible for $\{c_1, c_2, c_3\}$, implying the functions are independent.

   (f) Yes. If we test for independence at the points $\{1, 4, 9\}$, the system (1) in Section 4.4 becomes

   $$\begin{bmatrix} 1 & 1 & 1 \\ 1/4 & 2 & 4 \\ 1/9 & 3 & 9 \end{bmatrix} \begin{bmatrix} c_1 \\ c_2 \\ c_3 \end{bmatrix} = \begin{bmatrix} 0 \\ 0 \\ 0 \end{bmatrix}.$$

   The determinant of the coefficient matrix is $85/18 \neq 0$, so only the trivial solution is possible for $\{c_1, c_2, c_3\}$, implying the functions are independent.

3. We proceed as in Example 4, Section 4.4. To respect the standard notations, let $v_1 = 1$, $v_2 = x$ and $v_3 = x^2$. We have

$$w_1 = v_1 = 1;$$
$$w_2 = v_2 - (v_2 \cdot w_1)w_1/\|w_1\|^2$$
$$= x - \left[\int_0^1 (x)(1)\, dx\right] 1/\left[\int_0^1 (1)^2\, dx\right] = x - \frac{1}{2};$$
$$w_3 = v_3 - (v_3 \cdot w_1)w_1/\|w_1\|^2 - (v_3 \cdot w_2)w_2/\|w_2\|^2$$
$$= x^2 - \left[\int_0^1 (x^2)(1)\, dx\right] 1/\left[\int_0^1 (1)^2\, dx\right]$$
$$- \left[\int_0^1 (x^2)(x - 1/2)\, dx\right] (x - 1/2)/\left[\int_0^1 (x - 1/2)^2\, dx\right]$$
$$= x^2 - x + \frac{1}{6}.$$

5. It suffices to show that the polynomials $\{1, (x - 1), (x - 1)^2, \ldots, (x - 1)^n\}$ are linearly independent. Suppose now that $c_0 + c_1(x - 1) + \cdots c_n(x - 1)^n = 0$ for every $x \in \mathbb{R}$. For $k = 0, 1, \ldots, n$, if we differentiate k times, we see that $k!c_k + \frac{(k+1)!}{1!}(x - 1) + \frac{(k+2)!}{2!}(x - 1)^2 + \cdots \frac{(n)!}{(n-k)!}(x - 1)^{n-k} = 0$. If we let $x = 1$, we see that $c_k = 0$, for every $k = 0, 1, \ldots, n$, and so the polynomials $\{1, (x - 1), (x - 1)^2, \ldots, (x - 1)^n\}$ are indeed linearly independent.

7. (a) The values of the three functions at $x = 1$ are $1, 1/2$ and $2/3$. Therefore, we need to rescale them by $1, 2$ and $3/2$, respectively. The new functions obtained this way are 1, $2(x - 1/2)$ and $3/2(x^2 - 1/3)$.

   (b) We consider the function $v_4 = x^3$, and we continue the Gram-Schmidt algorithm from Example 4, Section 4.4, to find the degree three orthogonal polynomial. We have

$$w_3 = v_4 - (v_4 \cdot w_1)w_1/\|w_1\|^2 - (v_4 \cdot w_2)w_2/\|w_2\|^2$$
$$- (v_4 \cdot w_2)w_3/\|w_2\|^2$$
$$= x^3 - \left[\int_{-1}^1 (x^3)(1)\, dx\right] 1/\left[\int_{-1}^1 (1)^2\, dx\right]$$

$$- \left[\int_{-1}^{1} (x^3)(x) \, dx \right] (x) \bigg/ \left[\int_{-1}^{1} (x)^2 \, dx \right]$$

$$- \left[\int_{-1}^{1} (x^3)(x^2 - 1/3) \, dx \right] (x^2 - 1/3) \bigg/$$

$$\left[\int_{-1}^{1} (x^2 - 1/3)^2 \, dx \right]$$

$$= x^3 - \frac{3}{5}x$$

Rescaling as in part (a) to take value 1 when $x = 1$, we find the degree three Legendre polynomial

$$\frac{1}{2}(5x^3 - 3x).$$

9. After differentiating $(n - 1)$ times we obtain the following linear system, with unknowns $\{c_1, c_2, \ldots, c_n\}$:

$$c_1 e^{\alpha_1 x} + c_2 e^{\alpha_2 x} + \cdots + c_n e^{\alpha_n x} = 0$$
$$c_1 \alpha_1 e^{\alpha_1 x} + c_2 \alpha_2 e^{\alpha_2 x} + \cdots + c_n \alpha_n e^{\alpha_n x} = 0$$
$$\vdots$$
$$c_1 \alpha_1^{n-1} e^{\alpha_1 x} + c_2 \alpha_2^{n-1} e^{\alpha_2 x} + \cdots + c_n \alpha_n^{n-1} e^{\alpha_n x} = 0.$$

In matrix for, this system is

$$\begin{bmatrix} e^{\alpha_1 x} & e^{\alpha_2 x} & \cdots & e^{\alpha_n x} \\ \alpha_1 e^{\alpha_1 x} & \alpha_2 e^{\alpha_2 x} & \cdots & \alpha_n e^{\alpha_n x} \\ \vdots & \vdots & \vdots & \vdots \\ \alpha_1^{n-1} e^{\alpha_1 x} & \alpha_2^{n-1} e^{\alpha_2 x} & \cdots & \alpha_n^{n-1} e^{\alpha_n x} \end{bmatrix} \begin{bmatrix} c_1 \\ c_2 \\ \vdots \\ c_n \end{bmatrix} = \begin{bmatrix} 0 \\ 0 \\ \vdots \\ 0 \end{bmatrix}.$$

We prove that the coefficients matrix factors is invertible, by noticing first that it factors as

$$\begin{bmatrix} 1 & 1 & \cdots & 1 \\ \alpha_1 & \alpha_2 & \cdots & \alpha_n \\ \vdots & \vdots & \ddots & \vdots \\ \alpha_1^{n-1} & \alpha_2^{n-1} & \cdots & \alpha_n^{n-1} \end{bmatrix} \begin{bmatrix} e^{\alpha_1 x} & 0 & \cdots & 0 \\ 0 & e^{\alpha_2 x} & \cdots & 0 \\ \vdots & \vdots & \ddots & \vdots \\ 0 & 0 & \cdots & e^{\alpha_n x} \end{bmatrix}.$$

It suffices show that the factors of this product matrix are invertible. This can be showed by proving that the determinants of the two matrices never vanish. The determinant of the first matrix is a Vandermonde determinant, and from Problem 23, Exercises 2.4 we know that

$$
\begin{vmatrix}
1 & 1 & \cdots & 1 \\
\alpha_1 & \alpha_2 & \cdots & \alpha_n \\
\vdots & \vdots & \ddots & \vdots \\
\alpha_1^{n-1} & \alpha_2^{n-1} & \cdots & \alpha_n^{n-1}
\end{vmatrix}
= \prod_{i<j}(\alpha_j - \alpha_i) \neq 0,
$$

as the constants α_i are distinct, while

$$
\begin{vmatrix}
e^{\alpha_1 x} & 0 & \cdots & 0 \\
0 & e^{\alpha_2 x} & \cdots & 0 \\
\vdots & \vdots & \ddots & \vdots \\
0 & 0 & \cdots & e^{\alpha_n x}
\end{vmatrix}
= e^{(\sum_i \alpha_i)x} \neq 0.
$$

11. (a) Suppose $c_1 + c_2 \ln x + c_3 e^x = 0$. Then

$$
c_3 = \lim_{x \to \infty} -\frac{c_1 + c_2 \ln x}{e^x} = 0,
$$

This implies $c_1 + c_2 \ln x = 0$, and by the same reasoning

$$
c_2 = \lim_{x \to \infty} -\frac{c_1}{\ln x} = 0.
$$

It follows immediately that $c_1 = 0$, as well.

(b) Suppose $c_1 e^x + c_2 x e^x + c_3 x^2 e^x = 0$. Then $c_1 + c_2 x + c_3 x^2 = 0$, and so

$$
c_3 = \lim_{x \to \infty} -\frac{c_1 + c_2 x}{x^2} = 0,
$$

This implies $c_1 + c_2 x = 0$, and by the same reasoning

$$
c_2 = \lim_{x \to \infty} -\frac{c_1}{x} = 0.
$$

It follows immediately that $c_1 = 0$, as well.

13. The normal equations are

$$c_1 \langle \sin x, \sin x \rangle + c_2 \langle \sin x, \cos x \rangle = \langle \sin x, \cos^3 x \rangle$$
$$c_1 \langle \sin x, \cos x \rangle + c_2 \langle \cos x, \cos x \rangle = \langle \cos x, \cos^3 x \rangle.$$

We have $c_1 = c_2 = 3/4$, and so the least squares approximation of $\cos^3 x$ by a linear combination of $\sin x$ and $\cos x$, over the interval $(0, 2\pi)$ is $(3 \cos x)/4$.

15. (a) Since the functions $\{1, x\}$ form a basis for the vector space of linear functions, the normal equations found in Problem 12c, Exercises 4.4 become

$$m\langle 1, x \rangle + b\langle 1, 1 \rangle = \langle 1, f(x) \rangle$$
$$m\langle x, x \rangle + b\langle x, 1 \rangle = \langle x, f(x) \rangle,$$

where

$$\langle g(x), h(x) \rangle = \int_{x_1}^{x_2} g(x)h(x)\, dx.$$

(b) We find that the straight line that provides the least squares approximation of x^2, over the interval $(0, 1)$ is the function

$$y = x - 1/6.$$

(c) Following the hint provided, we notice that the linear Chebyshev approximation of x^2 over the interval $(0, 1)$ is the straight line

$$y = x - 1/8.$$

17. We have $\langle 1, \cos kx \rangle = \displaystyle\int_0^{2\pi} \cos kx\, dx = -\left.\frac{\sin kx}{k}\right|_0^{2\pi} = 0$. A similar computation shows $\langle 1, \sin kx \rangle = \displaystyle\int_0^{2\pi} \sin kx\, dx = 0$. For every $i \neq j$, compute now

$$\langle \sin ix, \sin jx \rangle = \int_0^{2\pi} \sin(ix)\sin(jx)dx$$

$$= \frac{\cos(i-j)x - \cos(i+j)x}{2}\Big|_0^{2\pi} = 0$$

$$\langle \sin ix, \cos jx \rangle = \int_0^{2\pi} \sin(ix)\cos(jx)dx$$

$$= \frac{\sin(i+j)x - \sin(i+j)x}{2}\Big|_0^{2\pi} = 0$$

$$\langle \cos ix, \cos jx \rangle = \int_0^{2\pi} \cos(ix)\cos(jx)dx$$

$$= \frac{\cos(i-j)x + \cos(i+j)x}{2}\Big|_0^{2\pi} = 0.$$

19. (a) It suffices to find n linearly independent polynomials of degree n. Consider the following set of degree n polynomials

$$\{1 + x + \cdots + x^n, \ x + x^2 + \cdots + x^n, \ldots, x^{n-1} + x^n, x^n\}.$$

To show that these polynomials are linearly independent, assume

$$c_1(1 + x + \cdots + x^n) + c_2(x + x^2 + \cdots + x^n) + \cdots c_n x^n = 0.$$

This is equivalent to

$$c_1 + (c_1 + c_2)x + \cdots (c_1 + x_2 + \cdots c_n)x^n = 0.$$

Since the set of polynomials $\{1, x, \ldots, x^n\}$ form a basis of $\mathbb{P}_n$, we obtain the following system of linear equations:

$$c_1 + c_2 + \cdots + c_n = 0$$
$$c_2 + \cdots + c_n = 0$$
$$\vdots$$
$$c_n = 0$$

A simple inspection shows that the linear system has only the trivial solution $c_1 = c_2 = \cdots = c_n = 0$.

(b) The set of polynomials $\{x, x^2, \ldots, x^n\}$ consists of linearly independent polynomials taking the value zero at $x = 0$. To show it forms a basis, it suffices to show it spans the subspace of $\mathbb{P}_n$ consisting of polynomials taking the value zero at $x = 0$. However, for a polynomial

$$f(x) = c_n x^n + c_{n-1} x^{n-1} + \cdots + c_1 x + c_0$$

to live in this space, it means c_0, so it can be written as a linear combination of $\{x, x^2, \ldots, x^n\}$.

(c) The set of polynomials $\{(x-1), (x^2-1), \ldots, (x^n-1)\}$ consists of linearly independent polynomials taking the value zero at $x = 1$. To show it forms a basis, it suffices to show it spans the subspace of $\mathbb{P}_n$ consisting of polynomials taking the value zero at $x = 0$. However, for a polynomial

$$f(x) = c_n x^n + c_{n-1} x^{n-1} + \cdots + c_1 x + c_0$$

to live in this space, it means $c_n + c_{n-1} + \cdots + c_1 + c_0 = 0$. Therefore,

$$\begin{aligned}
f(x) &= f(x) - (c_n + c_{n-1} + \cdots + c_1 + c_0) \\
&= c_n x^n + c_{n-1} x^{n-1} + \cdots + c_1 x + c_0 - (c_n + c_{n-1} + \cdots + c_1 + c_0) \\
&= c_n (x^n - 1) + c_{n-1}(x^{n-1} - 1) + \cdots + c_1(x-1).
\end{aligned}$$

Hence, every polynomial in the given subspace can be written as a linear combination of polynomials in the set $\{(x-1), (x^2-1), \ldots, (x^n-1)\}$.

21. Since $\dim \mathbb{P}_n = n + 1$, it suffices to show that the polynomials p_k, $k = 0, 1, \ldots, n$ are linearly independent. Suppose $c_0 p_0 + c_1 p_1 + \cdots + c_n p_n = 0$. One can use the growth rate method to show that $c_0 = c_1 = \cdots = c_n = 0$. We have

$$c_n = \lim_{x \to \infty} -\frac{c_0 p_0 + c_1 p_1 + \cdots + c_{n-1} p_{n-1}}{p_n} = 0,$$

as the degree of the denominator is larger that the degree of the numerator. Hence we have $c_0 p_0 + c_1 p_1 + \cdots + c_{n-1} p_{n-1} = 0$. Iterating the above procedure, we get $c_{n-1} = \cdots = c_1 = 0$, which forces $c_0 = 0$, as well.

PART II

REVIEW PROBLEMS FOR PART II

1. (a) We will solve the problem for $\mathbf{A}$ in $\mathbf{R}^{5,5}$. The general case (including a rectangular $\mathbf{A}$) can be obtained by a similar argument.

First observe that multiplying $\mathbf{A}$ by $\begin{bmatrix} 1 & 1 & 1 & 1 & 1 \end{bmatrix}^T$ on the right gives the vector $\mathbf{v}$ of its row sums (of length 5). It therefore remains to verify all of its terms are equal, in other words, that $v_1 - v_j = 0$, $2 \leq j \leq 5$ with v_j being the j-th element of $\mathbf{v}$. This system of 4 equations can be written as

$$
\begin{bmatrix}
-1 & 1 & 0 & 0 & 0 \\
-1 & 0 & 1 & 0 & 0 \\
-1 & 0 & 0 & 1 & 0 \\
-1 & 0 & 0 & 0 & 1
\end{bmatrix}
\mathbf{v} =
\begin{bmatrix}
0 \\
0 \\
0 \\
0
\end{bmatrix}.
$$

It remains to recall how $\mathbf{v}$ was obtained in order to write down the condition of equality of row sums:

$$
\begin{bmatrix}
-1 & 1 & 0 & 0 & 0 \\
-1 & 0 & 1 & 0 & 0 \\
-1 & 0 & 0 & 1 & 0 \\
-1 & 0 & 0 & 0 & 1
\end{bmatrix}
\mathbf{A}
\begin{bmatrix}
1 \\
1 \\
1 \\
1 \\
1
\end{bmatrix}
=
\begin{bmatrix}
0 \\
0 \\
0 \\
0
\end{bmatrix}.
$$

Solutions Manual to Accompany Fundamentals of Matrix Analysis with Applications,
First Edition. Edward Barry Saff and Arthur David Snider.
© 2016 John Wiley & Sons, Inc. Published 2016 by John Wiley & Sons, Inc.

(b) Since equality of column sums of matrix $\mathbf{A}$ is equivalent to equality of row sums of matrix $\mathbf{A}^T$, we can write the required condition as the above display with $\mathbf{A}^T$ in place of $\mathbf{A}$. Then transposing both sides gives

$$
\begin{bmatrix} 1 & 1 & 1 & 1 & 1 \end{bmatrix} \mathbf{A}
\begin{bmatrix}
-1 & -1 & -1 & -1 \\
1 & 0 & 0 & 0 \\
0 & 1 & 0 & 0 \\
0 & 0 & 1 & 0 \\
0 & 0 & 0 & 1
\end{bmatrix} = \begin{bmatrix} 0 & 0 & 0 & 0 \end{bmatrix}.
$$

(c) In the pictured "magic square" the row and column sums as well as the sums on the main diagonal and the reverse diagonal are all equal.

3. The argument follows strategy of Exercise 9, see below. We will first solve the homogeneous system with the given vectors as columns to obtain a non-trivial linear combination equal to zero. The system

$$
\begin{bmatrix}
1 & 1 & 2 & 2 & \vdots & 0 \\
1 & 0 & 13 & 1 & \vdots & 0 \\
2 & 1 & 23 & 33 & \vdots & 0
\end{bmatrix}
\quad \text{is reduced to} \quad
\begin{bmatrix}
1 & 0 & 5 & -29 & \vdots & 0 \\
0 & 1 & 5 & 61 & \vdots & 0 \\
0 & 0 & 8 & 30 & \vdots & 0
\end{bmatrix}
$$

and has, for example, solution $[191 \quad -169 \quad -15 \quad 4]^T$. Rearranging terms in the resulting linear combination gives the vector

$$
\begin{bmatrix} 22 \\ 191 \\ 213 \end{bmatrix} = 191 \begin{bmatrix} 1 \\ 1 \\ 2 \end{bmatrix} - 169 \begin{bmatrix} 1 \\ 0 \\ 1 \end{bmatrix} = 15 \begin{bmatrix} 2 \\ 13 \\ 23 \end{bmatrix} - 4 \begin{bmatrix} 2 \\ 1 \\ 33 \end{bmatrix},
$$

common to both spans.

5. (a) Not a subspace: the zero matrix is clearly not orthogonal since its rows are not unit vectors.

(b) Is a subspace: for any pair of diagonal matrices $\mathbf{A}$, $\mathbf{B}$ and any scalars α, β, the linear combination $\alpha\mathbf{A} + \beta\mathbf{B}$ is again diagonal.

(c) Not a subspace: the zero matrix is not a reflection matrix. Indeed, according to Example 3 of Section 2.2, a reflection matrix has the form $\mathbf{M}_{\text{ref}} = \mathbf{E} - 2\mathbf{v}\mathbf{v}^T$, where $\mathbf{E}$ is the identity matrix, and

$\mathbf{v}$ is the unit normal column-vector. In particular, the zero matrix from $\mathbf{R}^{4,4}$ cannot be represented in this form since rank$(\mathbf{E})$=4 but rank$(2\mathbf{v}\mathbf{v}^T) = 1$.

(d) Not a subspace: does not contain the zero matrix. As the Gauss elimination operators are invertible, this set is a subset of (e) below.

(e) Not a subspace: does not contain the zero matrix.

7. One could take vectors [1 0 0 0] and [0 1 0 0] since the matrix

$$\begin{bmatrix} 1 & 0 & 0 & 0 \\ 0 & 1 & 0 & 0 \\ 1 & 2 & 3 & 4 \\ 4 & 3 & 2 & 1 \end{bmatrix} \quad \text{has the row echelon form} \quad \begin{bmatrix} 1 & 0 & 0 & 0 \\ 0 & 1 & 0 & 0 \\ 0 & 0 & 1 & 3 \\ 0 & 0 & 0 & 5 \end{bmatrix},$$

and so its rows form a basis of $\mathbf{R}^4_{\text{row}}$.

9. Let vectors $\mathbf{v}_1$, $\mathbf{v}_2$, $\mathbf{v}_3$ and $\mathbf{w}_1$, $\mathbf{w}_2$, $\mathbf{w}_3$ be bases for the two subspaces. Any six vectors in a five-dimensional space are linearly dependent, so there exist scalars c_j, $1 \le j \le 6$, at least one of which is distinct from zero, such that

$$c_1\mathbf{v}_1 + c_2\mathbf{v}_2 + c_3\mathbf{v}_3 + c_4\mathbf{w}_1 + c_5\mathbf{w}_2 + c_6\mathbf{w}_3 = \mathbf{0}.$$

Let us write the above display as

$$c_1\mathbf{v}_1 + c_2\mathbf{v}_2 + c_3\mathbf{v}_3 = -c_4\mathbf{w}_1 - c_5\mathbf{w}_2 - c_6\mathbf{w}_3.$$

Observe that it is an equality of two non-zero vectors (otherwise all c_j are zeros). Furthermore, the vector on the left is an element of the one three-dimensional subspace, and the vector on the right is an element of another. We have therefore constructed a common nonzero vector of the two three-dimensional subspaces.

11. In view of the identity $\cos^2 t = \cos 2t + \sin^2 t$, the function $\cos^2 t$ may be discarded. We will show that $\sin 2t$, $\cos 2t$, $\sin^2 t$ are linearly independent. Indeed, as in Section 4.4, it is sufficient to pick 3 points and show that the matrix of values of the above functions in these points has full rank. We will use $\{0, \pi/4, \pi/2\}$. The resulting matrix is

$$\begin{array}{c} \\ 0 \\ \pi/4 \\ \pi/2 \end{array} \begin{array}{ccc} \sin 2t & \cos 2t & \sin^2 t \\ \begin{bmatrix} 0 & 1 & 0 \\ 1 & 0 & 1/2 \\ 0 & -1 & 1 \end{bmatrix}, \end{array}$$

which is easily seen to have full rank. Thus $\sin 2t$, $\cos 2t$, $\sin^2 t$ is a basis of the given space.

13. Recall that by Problem 37 to Section 3.3, it holds $\text{rank}(\mathbf{AB}) \leq \min\{\text{rank}(\mathbf{A}), \text{rank}(\mathbf{B})\}$, so $\text{rank}(\mathbf{AB}) \leq \text{rank}(\mathbf{A}) = 3$. Since $\mathbf{AB}$ is 5-by-5, it is singular.

15. Yes. Writing the given elements of $\mathbf{R}^{2,2}$ as vectors in $\mathbf{R}^4_{\text{row}}$, we have

$$\begin{bmatrix} 1 & 1 & 1 & 0 \\ 2 & 3 & 0 & 1 \\ 0 & 2 & 0 & -1 \\ -2 & 3 & 0 & 0 \end{bmatrix}, \text{ in the row echelon form } \begin{bmatrix} 1 & 1 & 1 & 0 \\ 0 & 1 & -2 & 1 \\ 0 & 0 & 4 & -3 \\ 0 & 0 & 0 & 4 \end{bmatrix},$$

so the matrix has rank 4. Because the dimension of space $\mathbf{R}^{2,2}$ is 4, it must be that the four original matrices constitute a basis.

17. We will argue for an $\mathbf{A}$ in $\mathbf{R}^{k,k}$ and assume that i, j vary in the range $1, 2, \ldots, k$. Observe that for the j-th canonical base vector e_j, $\mathbf{A}e_j$ is the j-th column of $\mathbf{A}$. This implies further that $A_{jj} = e_j^T \mathbf{A} e_j = e_j^T e_j = 1$. Let now $u = e_j + e_i$, $i \neq j$, a vector that has "1" in exactly two coordinates and "0" in the rest of them. We have

$$2 = u^T u = (e_j^T + e_i^T)\mathbf{A}(e_j + e_i) = e_j^T \mathbf{A} e_j + e_j^T \mathbf{A} e_i + e_i^T \mathbf{A} e_j + e_i^T \mathbf{A} e_i.$$

Using the above expression for A_{jj}, this can be reduced to

$$0 = e_j^T \mathbf{A} e_i + e_i^T \mathbf{A} e_j.$$

As we noted earlier, $\mathbf{A}e_i$ is the i-th column of $\mathbf{A}$. It is therefore not hard to see that $e_j^T \mathbf{A} e_i = A_{ji}$. Similarly, $e_i^T \mathbf{A} e_j = A_{ij}$. Recall that matrix $\mathbf{A}$ is symmetric, so $A_{ji} = A_{ij}$. Hence by the previous display, $A_{ij} = 0$, $i \neq j$. Together with $A_{jj} = 1$ this means $\mathbf{A}$ is the identity matrix.

19. Recall that by definition $|v|^2 = v \cdot v$. Similarly, $4 = |u + v|^2 = (u + v) \cdot (u + v) = u \cdot u + 2u \cdot v + v \cdot v = 8 + 2u \cdot v$. Hence $u \cdot v = -2$.

5

EIGENVECTORS AND EIGENVALUES

5.1 EIGENVECTOR BASICS

1. The Example 2, Section 2.2 is a particular case of the more general discussion about the orthogonal subspace decomposition in Section 4.3.

   Recall that if V is a vector space (say $V = \mathbb{R}^n$) and W a subspace, the V admits an orthogonal decomposition $V = W \oplus W^\perp$. With respect to such a decomposition, every vector $\mathbf{v}$ in V can be written (uniquely!) as $\mathbf{v} = \mathbf{w} + \mathbf{w}^\perp$, where $\mathbf{w}$ is in W, and $\mathbf{w}^\perp$ is in $W^\perp$ (in particular, $\mathbf{w}$ and $\mathbf{w}^\perp$ are orthogonal). With respect to such an orthogonal decomposition, the orthogonal projector matrix acts as

   $$\mathbf{M}_{\text{proj}}\mathbf{v} = \mathbf{M}_{\text{proj}}(\mathbf{w} + \mathbf{w}^\perp) = \mathbf{w}$$

   Using now the uniqueness of the orthogonal subspace decomposition, we can see now that $W^\perp$ and W are eigenspaces of $\mathbf{M}_{\text{proj}}$ corresponding to eigenvalues 0 and 1, respectively.

   Moreover, there are no other eigenspaces. Any other eigenspace would contain vectors $\mathbf{u}$ of the form $\mathbf{w} + \mathbf{w}^\perp$ with $\mathbf{w}, \mathbf{w}^\perp \neq \mathbf{0}$. If $\mathbf{u}$ corresponds to an eigenvalue r, then $\mathbf{M}_{\text{proj}}\mathbf{u} = r\mathbf{u} = r(\mathbf{w} + \mathbf{w}^\perp)$, and since $\mathbf{M}_{\text{proj}}\mathbf{u} = \mathbf{w}$, we get $r(\mathbf{w} + \mathbf{w}^\perp) = \mathbf{w}$, which is equivalent to

Solutions Manual to Accompany Fundamentals of Matrix Analysis with Applications, First Edition. Edward Barry Saff and Arthur David Snider.
© 2016 John Wiley & Sons, Inc. Published 2016 by John Wiley & Sons, Inc.

$r\mathbf{w}^{\perp} = (1 - r)\mathbf{w}$. By taking the dot product on both sides by $\mathbf{w} \neq 0$, we see that $r = 1$. But taking the dot product by $\mathbf{w}^{\perp}$ yields $r = 0$, contradiction!

3. We notice that if in the vector $\mathbf{v}$ we keep the first entry equal to one, but change the position on which -1, is we do not change the value of $\mathbf{F}\mathbf{v}^{T} = -\mathbf{v}^{T}$. Therefore, we obtain the following 6 linearly independent eigenvectors, corresponding to the eigenvalue -1.

$$\left\{ \begin{bmatrix} 1 \\ -1 \\ 0 \\ 0 \\ 0 \\ 0 \\ 0 \end{bmatrix}, \begin{bmatrix} 1 \\ 0 \\ -1 \\ 0 \\ 0 \\ 0 \\ 0 \end{bmatrix}, \begin{bmatrix} 1 \\ 0 \\ 0 \\ -1 \\ 0 \\ 0 \\ 0 \end{bmatrix}, \begin{bmatrix} 1 \\ 0 \\ 0 \\ 0 \\ -1 \\ 0 \\ 0 \end{bmatrix}, \begin{bmatrix} 1 \\ 0 \\ 0 \\ 0 \\ 0 \\ -1 \\ 0 \end{bmatrix}, \begin{bmatrix} 1 \\ 0 \\ 0 \\ 0 \\ 0 \\ 0 \\ -1 \end{bmatrix} \right\}.$$

The second eigenvalue is 6 and an eigenvector is $[1\,1\,1\ 1\,1\,1\,1]^{T}$, as it can be easily checked by a simple inspection.

5. The Gauss-elimination matrix that multiplies the ith row of an $m \times n$ matrix by a constant c is the $m \times m$ matrix

$$\begin{bmatrix} 1 & & & & & & \\ & \ddots & & & & & \\ & & 1 & & & & \\ & & & c & & & \\ & & & & 1 & & \\ & & & & & \ddots & \\ & & & & & & 1 \end{bmatrix},$$

where the entries off the main diagonal are zero. One eigenvalues is c with the eigenvector $[0 \ldots 0\,1\,0 \ldots 0]^{T}$, where the only non-zero entry is in the ith row. Another eigenvalue is 1. A set of linearly independent eigenvectors consist of the vectors of the form $[0 \ldots 0\,1\,0 \ldots 0]^{T}$, where the non-zero entry can be placed on any row, except the ith. Counting the number of linearly independent vectors of the matrix, we conclude that 1 and c are the only eigenvalues.

7. Looking for an eigenvector of the form $\mathbf{x} = [x\,1\,0\,0]^T$ corresponding to the eigenvalue 2 amounts to solving the equation $\mathbf{Ux} = 2\mathbf{x}$, which is equivalent to the equation $x + 1 = 2x$. Therefore, an eigenvector corresponding to the eigenvalue 2 is $[1\,1\,0\,0]^T$. Looking now for eigenvectors of the form $[y\,z\,1\,0]^T$ corresponding to the eigenvalue 2 amounts to solving the system of equations

$$y + z + 1 = 2y$$
$$2z + \alpha = 2z.$$

The second equation is consistent if and only if $\alpha = 0$. In this case, we obtain the eigenvector $[1\,0\,1\,0]^T$, which is linearly independent from $[1\,1\,0\,0]^T$.

9. The eigenvector interpretation is that $\mathbf{u}_4$ is an eigenvector of $\mathbf{B}$ corresponding to the eigenvalue $2i$. Indeed, we can see by a direct computation that

$$
\begin{bmatrix}
0 & 1 & 0 & -1 \\
-1 & 0 & 1 & 0 \\
0 & -1 & 0 & 1 \\
1 & 0 & -1 & 0
\end{bmatrix}
\begin{bmatrix}
i \\ -1 \\ -i \\ 1
\end{bmatrix}
= -
\begin{bmatrix}
i \\ -1 \\ -i \\ 1
\end{bmatrix}.
$$

11. (a) Let $\mathbf{u}_1 = \begin{bmatrix} -1 & 1 & -1 & 1 \end{bmatrix}^T$. We compute

$$
\mathbf{Au}_1 =
\begin{bmatrix}
0 & 1 & 0 & 0 \\
0 & 0 & 1 & 0 \\
0 & 0 & 0 & 1 \\
1 & 0 & 0 & 0
\end{bmatrix}
\begin{bmatrix}
-1 \\ 1 \\ -1 \\ 1
\end{bmatrix}
$$

$$
=
\begin{bmatrix}
1 \\ -1 \\ 1 \\ -1
\end{bmatrix}
= (-1)
\begin{bmatrix}
-1 \\ 1 \\ -1 \\ 1
\end{bmatrix},
$$

$$\mathbf{Bu_1} = \begin{bmatrix} 0 & 1 & 0 & -1 \\ -1 & 0 & 1 & 0 \\ 0 & -1 & 0 & 1 \\ 1 & 0 & -1 & 0 \end{bmatrix} \begin{bmatrix} -1 \\ 1 \\ -1 \\ 1 \end{bmatrix}$$

$$= \begin{bmatrix} 0 \\ 0 \\ 0 \\ 0 \end{bmatrix} = 0 \begin{bmatrix} -1 \\ 1 \\ -1 \\ 1 \end{bmatrix},$$

$$\mathbf{Cu_1} = \begin{bmatrix} 0 & 1 & 0 & 1 \\ 1 & 0 & 1 & 0 \\ 0 & 1 & 0 & 1 \\ 1 & 0 & 1 & 0 \end{bmatrix} \begin{bmatrix} -1 \\ 1 \\ -1 \\ 1 \end{bmatrix}$$

$$= \begin{bmatrix} 2 \\ -2 \\ 2 \\ -2 \end{bmatrix} = (-2) \begin{bmatrix} -1 \\ 1 \\ -1 \\ 1 \end{bmatrix}.$$

Hence, the corresponding eigenvalues of $\mathbf{u_2}$ are $-1, 0$, and -2, respectively.

(b) Let $\mathbf{u_2} = [-i \ -1 \ i \ 1]^T$. We compute:

$$\mathbf{Au_1} = \begin{bmatrix} 0 & 1 & 0 & 0 \\ 0 & 0 & 1 & 0 \\ 0 & 0 & 0 & 1 \\ 1 & 0 & 0 & 0 \end{bmatrix} \begin{bmatrix} -i \\ -1 \\ i \\ 1 \end{bmatrix}$$

$$= \begin{bmatrix} -1 \\ i \\ 1 \\ -i \end{bmatrix} = (-i) \begin{bmatrix} -i \\ -1 \\ i \\ 1 \end{bmatrix},$$

$$\mathbf{Bu_1} = \begin{bmatrix} 0 & 1 & 0 & -1 \\ -1 & 0 & 1 & 0 \\ 0 & -1 & 0 & 1 \\ 1 & 0 & -1 & 0 \end{bmatrix} \begin{bmatrix} -i \\ -1 \\ i \\ 1 \end{bmatrix}$$

$$= \begin{bmatrix} -2 \\ 2i \\ 2 \\ -2i \end{bmatrix} = (-2i) \begin{bmatrix} -i \\ -1 \\ i \\ 1 \end{bmatrix},$$

$$Cu_1 = \begin{bmatrix} 0 & 1 & 0 & 1 \\ 1 & 0 & 1 & 0 \\ 0 & 1 & 0 & 1 \\ 1 & 0 & 1 & 0 \end{bmatrix} \begin{bmatrix} -i \\ -1 \\ i \\ 1 \end{bmatrix}$$

$$= \begin{bmatrix} 0 \\ 0 \\ 0 \\ 0 \end{bmatrix} = 0 \begin{bmatrix} -i \\ -1 \\ i \\ 1 \end{bmatrix}.$$

Hence u_2 is an eigenvector for all of the matrices **A, B** and **C** of eigenvalues $-i, -2i$ and 0, respectively.

Let $u_3 = [1\ 1\ 1\ 1]^T$. We compute:

$$Au_1 = \begin{bmatrix} 0 & 1 & 0 & 0 \\ 0 & 0 & 1 & 0 \\ 0 & 0 & 0 & 1 \\ 1 & 0 & 0 & 0 \end{bmatrix} \begin{bmatrix} 1 \\ 1 \\ 1 \\ 1 \end{bmatrix} \qquad = \begin{bmatrix} 1 \\ 1 \\ 1 \\ 1 \end{bmatrix} = 1 \begin{bmatrix} 1 \\ 1 \\ 1 \\ 1 \end{bmatrix},$$

$$Bu_1 = \begin{bmatrix} 0 & 1 & 0 & -1 \\ -1 & 0 & 1 & 0 \\ 0 & -1 & 0 & 1 \\ 1 & 0 & -1 & 0 \end{bmatrix} \begin{bmatrix} 1 \\ 1 \\ 1 \\ 1 \end{bmatrix} \qquad = \begin{bmatrix} 0 \\ 0 \\ 0 \\ 0 \end{bmatrix} = 0 \begin{bmatrix} 1 \\ 1 \\ 1 \\ 1 \end{bmatrix},$$

$$Cu_1 = \begin{bmatrix} 0 & 1 & 0 & 1 \\ 1 & 0 & 1 & 0 \\ 0 & 1 & 0 & 1 \\ 1 & 0 & 1 & 0 \end{bmatrix} \begin{bmatrix} 1 \\ 1 \\ 1 \\ 1 \end{bmatrix} \qquad = \begin{bmatrix} 2 \\ 2 \\ 2 \\ 2 \end{bmatrix} = 2 \begin{bmatrix} 1 \\ 1 \\ 1 \\ 1 \end{bmatrix}.$$

Hence u_3 is an eigenvector for all of the matrices **A, B** and **C** of eigenvalues $1, 0$ and 2, respectively.

13. The matrices are

$$A = \begin{bmatrix} 0 & 1 & 0 & 0 & 0 & 0 \\ 0 & 0 & 1 & 0 & 0 & 0 \\ 0 & 0 & 0 & 1 & 0 & 0 \\ 0 & 0 & 0 & 0 & 1 & 0 \\ 0 & 0 & 0 & 0 & 0 & 1 \\ 1 & 0 & 0 & 0 & 0 & 0 \end{bmatrix},$$

$$
\mathbf{B} = \begin{bmatrix}
0 & 1 & 0 & -1 & 0 & 1 \\
1 & 0 & 1 & 0 & -1 & 0 \\
0 & 1 & 0 & 1 & 0 & -1 \\
-1 & 0 & 1 & 0 & 1 & 0 \\
0 & -1 & 0 & 1 & 0 & 1 \\
1 & 0 & -1 & 0 & 1 & 0
\end{bmatrix}
$$

$$
\mathbf{C} = \begin{bmatrix}
0 & 1 & 0 & 1 & 0 & 1 \\
1 & 0 & 1 & 0 & 1 & 0 \\
0 & 1 & 0 & 1 & 0 & 1 \\
1 & 0 & 1 & 0 & 1 & 0 \\
0 & 1 & 0 & 1 & 0 & 1 \\
1 & 0 & 1 & 0 & 1 & 0
\end{bmatrix},
$$

$$
\mathbf{D} = \begin{bmatrix}
d_1 & d_2 & d_3 & d_4 & d_5 & d_6 \\
d_6 & d_1 & d_2 & d_3 & d_4 & d_5 \\
d_5 & d_6 & d_1 & d_2 & d_3 & d_4 \\
d_4 & d_5 & d_6 & d_1 & d_2 & d_3 \\
d_3 & d_4 & d_5 & d_6 & d_1 & d_2 \\
d_2 & d_3 & d_4 & d_5 & d_6 & d_1
\end{bmatrix}
$$

We discuss and perform the computations for the matrix $\mathbf{D}$ only, as $\mathbf{A}$, $\mathbf{B}$ and $\mathbf{C}$ are special cases. For convenience, let $f(x) = d_1 + d_2 x + d_3 x^2 + d_4 x^3 + d_5 x^4 + d_6 x^5$.

Let $\mathbf{u_6} = \begin{bmatrix} \omega & \omega^2 & \omega^3 & \omega^4 & \omega^5 & \omega^6 \end{bmatrix}^T$. Then

$$
\mathbf{Du_6} = \begin{bmatrix}
d_1 & d_2 & d_3 & d_4 & d_5 & d_6 \\
d_6 & d_1 & d_2 & d_3 & d_4 & d_5 \\
d_5 & d_6 & d_1 & d_2 & d_3 & d_4 \\
d_4 & d_5 & d_6 & d_1 & d_2 & d_3 \\
d_3 & d_4 & d_5 & d_6 & d_1 & d_2 \\
d_2 & d_3 & d_4 & d_5 & d_6 & d_1
\end{bmatrix}
\begin{bmatrix}
\omega \\ \omega^2 \\ \omega^3 \\ \omega^4 \\ \omega^5 \\ \omega^6
\end{bmatrix}
$$

$$
= \begin{bmatrix}
d_1\omega + d_2\omega^2 + d_3\omega^3 + d_4\omega^4 + d_5\omega^5 + d_6\omega^6 \\
d_1\omega^2 + d_2\omega^3 + d_3\omega^4 + d_4\omega^5 + d_5\omega^6 + d_6\omega \\
d_1\omega^3 + d_2\omega^4 + d_3\omega^5 + d_4\omega^6 + d_5\omega + d_6\omega^2 \\
d_1\omega^4 + d_2\omega^5 + d_3\omega^6 + d_4\omega + d_5\omega^2 + d_6\omega^3 \\
d_1\omega^5 + d_2\omega^6 + d_3\omega + d_4\omega^2 + d_5\omega^3 + d_6\omega^4 \\
d_1\omega^6 + d_2\omega + d_3\omega^2 + d_4\omega^3 + d_5\omega^4 + d_6\omega^5
\end{bmatrix}
$$

$$
= \begin{bmatrix} f(\omega)\omega \\ f(\omega)\omega^2 \\ f(\omega)\omega^3 \\ f(\omega)\omega^4 \\ f(\omega)\omega^5 \\ f(\omega)\omega^6 \end{bmatrix} = f(\omega) \begin{bmatrix} \omega \\ \omega^2 \\ \omega^3 \\ \omega^4 \\ \omega^5 \\ \omega^6 \end{bmatrix}.
$$

Hence, the vector $\mathbf{u_6}$ is an eigenvector of $\mathbf{D}$ corresponding to the eigenvalue $f(\omega)$.

Specializing to the matrices matrices $\mathbf{A}, \mathbf{B}$ and $\mathbf{C}$, we conclude that $\mathbf{u_6}$ is a common eigenvector corresponding to the eigenvalues $\omega, \omega - \omega^3 + \omega^5 - 2$, and $\omega + \omega^3 + \omega^5 - 0$, respectively.

Let $\mathbf{u_1} = \begin{bmatrix} \omega^2\,\omega^4\,\omega^6\,\omega^8\,\omega^{10}\,\omega^{12} \end{bmatrix}^T$. We get

$$
\mathbf{Du_1} = \begin{bmatrix} d_1 & d_2 & d_3 & d_4 & d_5 & d_6 \\ d_6 & d_1 & d_2 & d_3 & d_4 & d_5 \\ d_5 & d_6 & d_1 & d_2 & d_3 & d_4 \\ d_4 & d_5 & d_6 & d_1 & d_2 & d_3 \\ d_3 & d_4 & d_5 & d_6 & d_1 & d_2 \\ d_2 & d_3 & d_4 & d_5 & d_6 & d_1 \end{bmatrix} \begin{bmatrix} \omega^2 \\ \omega^4 \\ \omega^6 \\ \omega^8 \\ \omega^{10} \\ \omega^{12} \end{bmatrix}.
$$

$$
= \begin{bmatrix} d_1\omega^2 + d_2\omega^4 + d_3\omega^6 + d_4\omega^8 + d_5\omega^{10} + d_6\omega^{12} \\ d_1\omega^4 + d_2\omega^6 + d_3\omega^8 + d_4\omega^{10} + d_5\omega^{12} + d_6\omega^2 \\ d_1\omega^6 + d_2\omega^8 + d_3\omega^{10} + d_4\omega^{12} + d_5\omega^2 + d_6\omega^4 \\ d_1\omega^8 + d_2\omega^{10} + d_3\omega^{12} + d_4\omega^2 + d_5\omega^4 + d_6\omega^6 \\ d_1\omega^{10} + d_2\omega^{12} + d_3\omega^2 + d_4\omega^4 + d_5\omega^6 + d_6\omega^8 \\ d_1\omega^{12} + d_2\omega^2 + d_3\omega^4 + d_4\omega^6 + d_5\omega^8 + d_6\omega^{10} \end{bmatrix}
$$

$$
= \begin{bmatrix} f(\omega^2)\omega^2 \\ f(\omega^2)\omega^4 \\ f(\omega^2)\omega^4 \\ f(\omega^2)\omega^6 \\ f(\omega^2)\omega^5 \\ f(\omega^2)\omega^6 \end{bmatrix} = f(\omega^2) \begin{bmatrix} \omega^2 \\ \omega^4 \\ \omega^6 \\ \omega^8 \\ \omega^{10} \\ \omega^{12} \end{bmatrix}.
$$

Hence, the vector

$$
\mathbf{u_1} = \begin{bmatrix} \omega^2\,\omega^4\,\omega^6\,\omega^8\,\omega^{10}\,\omega^{12} \end{bmatrix}^T = \begin{bmatrix} \omega^2\,\omega^4\,1\,\omega^2\,\omega^4\,1 \end{bmatrix}^T
$$

is an eigenvector of $\mathbf{D}$ corresponding to the eigenvalue $f(\omega^2)$.

Specializing to the matrices matrices **A, B** and **C**, we conclude that $\mathbf{u_1}$ is a common eigenvector corresponding to the eigenvalues ω^2, $\omega^2 - \omega^4 + \omega^6 = -2\omega^4 = 2\omega$, and $\omega^2 + \omega^4 + \omega^6 = 0$, respectively.

Let $\mathbf{u_2} = \begin{bmatrix} \omega^3 & \omega^6 & \omega^9 & \omega^{12} & \omega^{15} & \omega^{18} \end{bmatrix}^T$. We get

$$
\mathbf{Du_2} = \begin{bmatrix}
d_1 & d_2 & d_3 & d_4 & d_5 & d_6 \\
d_6 & d_1 & d_2 & d_3 & d_4 & d_5 \\
d_5 & d_6 & d_1 & d_2 & d_3 & d_4 \\
d_4 & d_5 & d_6 & d_1 & d_2 & d_3 \\
d_3 & d_4 & d_5 & d_6 & d_1 & d_2 \\
d_2 & d_3 & d_4 & d_5 & d_6 & d_1
\end{bmatrix}
\begin{bmatrix}
\omega^3 \\ \omega^6 \\ \omega^9 \\ \omega^{12} \\ \omega^{15} \\ \omega^{18}
\end{bmatrix}
$$

$$
= \begin{bmatrix}
d_1\omega^3 + d_2\omega^6 + d_3\omega^9 + d_4\omega^{12} + d_5\omega^{15} + d_6\omega^{18} \\
d_1\omega^6 + d_2\omega^9 + d_3\omega^{12} + d_4\omega^{15} + d_5\omega^{18} + d_6\omega^3 \\
d_1\omega^9 + d_2\omega^{12} + d_3\omega^{15} + d_4\omega^{18} + d_5\omega^3 + d_6\omega^6 \\
d_1\omega^{12} + d_2\omega^{15} + d_3\omega^{18} + d_4\omega^3 + d_5\omega^6 + d_6\omega^9 \\
d_1\omega^{15} + d_2\omega^{18} + d_3\omega^3 + d_4\omega^6 + d_5\omega^9 + d_6\omega^{12} \\
d_1\omega^{18} + d_2\omega^3 + d_3\omega^6 + d_4\omega^9 + d_5\omega^{12} + d_6\omega^{15}
\end{bmatrix}
$$

$$
= \begin{bmatrix}
f(\omega^3)\omega^3 \\
f(\omega^3)\omega^6 \\
f(\omega^3)\omega^9 \\
f(\omega^3)\omega^{12} \\
f(\omega^3)\omega^{15} \\
f(\omega^3)\omega^{18}
\end{bmatrix}
= f(\omega^3)
\begin{bmatrix}
\omega^3 \\ \omega^6 \\ \omega^9 \\ \omega^{12} \\ \omega^{15} \\ \omega^{18}
\end{bmatrix}.
$$

Hence, the vector

$$
\mathbf{u_2} = \begin{bmatrix} \omega^3 & \omega^6 & \omega^9 & \omega^{12} & \omega^{15} & \omega^{18} \end{bmatrix}^T = \begin{bmatrix} -1 & 1 & -1 & 1 & -1 & 1 \end{bmatrix}^T
$$

is an eigenvector of **D** corresponding to the eigenvalue $f(\omega^3)$.

Specializing to the matrices matrices **A, B** and **C**, we conclude that $\mathbf{u_2}$ is a common eigenvector corresponding to the eigenvalues $\omega^3 = -1, \omega^3 - \omega^6 + \omega^3 = -3$, and $\omega^3 + \omega^6 + \omega^9 = -1$, respectively.

Let $\mathbf{u_3} = \begin{bmatrix} \omega^4 & \omega^8 & \omega^{12} & \omega^{16} & \omega^{20} & \omega^{24} \end{bmatrix}^T$. We get

$$
\mathbf{Du_3} = \begin{bmatrix}
d_1 & d_2 & d_3 & d_4 & d_5 & d_6 \\
d_6 & d_1 & d_2 & d_3 & d_4 & d_5 \\
d_5 & d_6 & d_1 & d_2 & d_3 & d_4 \\
d_4 & d_5 & d_6 & d_1 & d_2 & d_3 \\
d_3 & d_4 & d_5 & d_6 & d_1 & d_2 \\
d_2 & d_3 & d_4 & d_5 & d_6 & d_1
\end{bmatrix}
\begin{bmatrix}
\omega^4 \\ \omega^8 \\ \omega^{12} \\ \omega^{16} \\ \omega^{20} \\ \omega^{24}
\end{bmatrix}
$$

$$
= \begin{bmatrix}
d_1\omega^4 + d_2\omega^8 + d_3\omega^{12} + d_4\omega^{16} + d_5\omega^{20} + d_6\omega^{24} \\
d_1\omega^8 + d_2\omega^{12} + d_3\omega^{16} + d_4\omega^{20} + d_5\omega^{24} + d_6\omega^4 \\
d_1\omega^{12} + d_2\omega^{16} + d_3\omega^{20} + d_4\omega^{24} + d_5\omega^4 + d_6\omega^8 \\
d_1\omega^{16} + d_2\omega^{20} + d_3\omega^{24} + d_4\omega^4 + d_5\omega^8 + d_6\omega^{12} \\
d_1\omega^{20} + d_2\omega^{24} + d_3\omega^4 + d_4\omega^8 + d_5\omega^{12} + d_6\omega^{16} \\
d_1\omega^{24} + d_2\omega^4 + d_3\omega^8 + d_4\omega^{12} + d_5\omega^{16} + d_6\omega^{20}
\end{bmatrix}
$$

$$
= \begin{bmatrix}
f(\omega^4)\omega^4 \\
f(\omega^4)\omega^8 \\
f(\omega^4)\omega^{12} \\
f(\omega^4)\omega^{16} \\
f(\omega^4)\omega^{20} \\
f(\omega^4)\omega^{24}
\end{bmatrix}
= f(\omega^4)
\begin{bmatrix}
\omega^4 \\ \omega^8 \\ \omega^{12} \\ \omega^{16} \\ \omega^{20} \\ \omega^{24}
\end{bmatrix}.
$$

Hence, the vector $\mathbf{u_3} = \begin{bmatrix} \omega^4 & \omega^8 & \omega^{12} & \omega^{16} & \omega^{20} & \omega^{24} \end{bmatrix}^T = \begin{bmatrix} \omega^4 & \omega^2 & 1 & \omega^4 & \omega^2 & 1 \end{bmatrix}^T$ is an eigenvector of $\mathbf{D}$ corresponding to the eigenvalue $f(\omega^4)$.

Specializing to the matrices matrices $\mathbf{A}$, $\mathbf{B}$ and $\mathbf{C}$, we conclude that $\mathbf{u_3}$ is a common eigenvector corresponding to the eigenvalues $\omega^4 = -\omega$, $\omega^4 - \omega^8 + \omega^{12} = -2\omega^2$, and $\omega^4 + \omega^8 + \omega^{12} = 0$, respectively.

Let $\mathbf{u_4} = \begin{bmatrix} \omega^5 & \omega^{10} & \omega^{15} & \omega^{20} & \omega^{25} & \omega^{30} \end{bmatrix}^T$. We get

$$
\mathbf{Du_4} = \begin{bmatrix}
d_1 & d_2 & d_3 & d_4 & d_5 & d_6 \\
d_6 & d_1 & d_2 & d_3 & d_4 & d_5 \\
d_5 & d_6 & d_1 & d_2 & d_3 & d_4 \\
d_4 & d_5 & d_6 & d_1 & d_2 & d_3 \\
d_3 & d_4 & d_5 & d_6 & d_1 & d_2 \\
d_2 & d_3 & d_4 & d_5 & d_6 & d_1
\end{bmatrix}
\begin{bmatrix}
\omega^5 \\ \omega^{10} \\ \omega^{15} \\ \omega^{20} \\ \omega^{25} \\ \omega^{30}
\end{bmatrix}
$$

$$
= \begin{bmatrix} d_1\omega^5 + d_2\omega^{10} + d_3\omega^{15} + d_4\omega^{20} + d_5\omega^{25} + d_6\omega^{30} \\ d_1\omega^{10} + d_2\omega^{15} + d_3\omega^{20} + d_4\omega^{25} + d_5\omega^{30} + d_6\omega^{5} \\ d_1\omega^{15} + d_2\omega^{20} + d_3\omega^{25} + d_4\omega^{30} + d_5\omega^{5} + d_6\omega^{10} \\ d_1\omega^{20} + d_2\omega^{25} + d_3\omega^{30} + d_4\omega^{5} + d_5\omega^{10} + d_6\omega^{15} \\ d_1\omega^{25} + d_2\omega^{30} + d_3\omega^{5} + d_4\omega^{10} + d_5\omega^{15} + d_6\omega^{20} \\ d_1\omega^{30} + d_2\omega^{5} + d_3\omega^{10} + d_4\omega^{15} + d_5\omega^{20} + d_6\omega^{25} \end{bmatrix}
$$

$$
= \begin{bmatrix} f(\omega^5)\omega^5 \\ f(\omega^5)\omega^{10} \\ f(\omega^5)\omega^{15} \\ f(\omega^5)\omega^{20} \\ f(\omega^5)\omega^{25} \\ f(\omega^5)\omega^{30} \end{bmatrix} = f(\omega^5) \begin{bmatrix} \omega^5 \\ \omega^{10} \\ \omega^{15} \\ \omega^{20} \\ \omega^{25} \\ \omega^{30} \end{bmatrix}.
$$

Hence, the vector

$$
\mathbf{u_4} = \begin{bmatrix} \omega^5 \ \omega^{10} \ \omega^{15} \ \omega^{20} \ \omega^{25} \ \omega^{30} \end{bmatrix}^T = \begin{bmatrix} -\omega^2 \ -\omega \ -1 \ \omega^2 \ \omega \ 1 \end{bmatrix}^T
$$

is an eigenvector of $\mathbf{D}$ corresponding to the eigenvalue $f(\omega^5)$.

Specializing to the matrices matrices $\mathbf{A}$, $\mathbf{B}$ and $\mathbf{C}$, we conclude that $\mathbf{u_4}$ is a common eigenvector corresponding to the eigenvalues $\omega^5 = -\omega^2$, $\omega^5 - \omega^{10} + \omega^{15} = 0$, and $\omega^5 + \omega^{10} + \omega^{15} = -2\omega$, respectively.

Let $\mathbf{u_5} = \begin{bmatrix} \omega^6 \ \omega^{12} \ \omega^{18} \ \omega^{24} \ \omega^{30} \ \omega^{36} \end{bmatrix}^T$. We get

$$
\mathbf{Du_5} = \begin{bmatrix} d_1 & d_2 & d_3 & d_4 & d_5 & d_6 \\ d_6 & d_1 & d_2 & d_3 & d_4 & d_5 \\ d_5 & d_6 & d_1 & d_2 & d_3 & d_4 \\ d_4 & d_5 & d_6 & d_1 & d_2 & d_3 \\ d_3 & d_4 & d_5 & d_6 & d_1 & d_2 \\ d_2 & d_3 & d_4 & d_5 & d_6 & d_1 \end{bmatrix} \begin{bmatrix} \omega^6 \\ \omega^{12} \\ \omega^{18} \\ \omega^{24} \\ \omega^{30} \\ \omega^{36} \end{bmatrix}
$$

$$
= \begin{bmatrix} d_1\omega^6 + d_2\omega^{12} + d_3\omega^{18} + d_4\omega^{24} + d_5\omega^{30} + d_6\omega^{36} \\ d_1\omega^{12} + d_2\omega^{18} + d_3\omega^{24} + d_4\omega^{30} + d_5\omega^{36} + d_6\omega^{6} \\ d_1\omega^{18} + d_2\omega^{24} + d_3\omega^{30} + d_4\omega^{36} + d_5\omega^{6} + d_6\omega^{12} \\ d_1\omega^{24} + d_2\omega^{30} + d_3\omega^{36} + d_4\omega^{6} + d_5\omega^{12} + d_6\omega^{18} \\ d_1\omega^{30} + d_2\omega^{36} + d_3\omega^{6} + d_4\omega^{12} + d_5\omega^{18} + d_6\omega^{24} \\ d_1\omega^{36} + d_2\omega^{6} + d_3\omega^{12} + d_4\omega^{18} + d_5\omega^{24} + d_6\omega^{30} \end{bmatrix}
$$

$$= \begin{bmatrix} f(\omega^6)\omega^6 \\ f(\omega^6)\omega^{12} \\ f(\omega^6)\omega^{18} \\ f(\omega^6)\omega^{24} \\ f(\omega^6)\omega^{30} \\ f(\omega^6)\omega^{36} \end{bmatrix} = f(\omega^6) \begin{bmatrix} \omega^6 \\ \omega^{12} \\ \omega^{18} \\ \omega^{24} \\ \omega^{30} \\ \omega^{36} \end{bmatrix}.$$

Hence, the vector

$$\mathbf{u_5} = \begin{bmatrix} \omega^6 & \omega^{12} & \omega^{18} & \omega^{24} & \omega^{30} & \omega^{36} \end{bmatrix}^T = \begin{bmatrix} 1 & 1 & 1 & 1 & 1 & 1 \end{bmatrix}^T$$

is an eigenvector of $\mathbf{D}$ corresponding to the eigenvalue $f(\omega^6)$.

Specializing to the matrices matrices $\mathbf{A}, \mathbf{B}$ and $\mathbf{C}$, we conclude that $\mathbf{u_5}$ is a common eigenvector corresponding to the eigenvalues $\omega^6 = 1, \omega^{12} - \omega^{24} + \omega^{36} = 1$ and $\omega^{12} + \omega^{24} + \omega^{36} = 3$, respectively.

15. If $\mathbf{u}$ is an eigenvector of $(\mathbf{AB})$ corresponding to the eigenvalue r_0, then

$$(\mathbf{BA})\mathbf{Bu} = (\mathbf{BAB})\mathbf{u} = \mathbf{B}(\mathbf{AB})\mathbf{u} = r_0\mathbf{Bu}.$$

This shows not only that $\mathbf{Bu}$ is an eigenvector of $(\mathbf{BA})$, but it also shows that if r_0 is an eigenvalue of $(\mathbf{AB})$, it is an eigenvalue of $(\mathbf{BA})$ as well. Reverting the roles of $\mathbf{A}$ and $\mathbf{B}$, we see that $(\mathbf{AB})$ and $(\mathbf{BA})$ have the same eigenvalues.

17. If 1 is not an eigenvalue of $\mathbf{A}$, then equation $\mathbf{Ax} = \mathbf{x}$ has only the trivial solution, which is the same thing as saying that the equation $(\mathbf{I} - \mathbf{A})\mathbf{x} = \mathbf{0}$ has only the trivial solution. This is equivalent to the invertibility of the matrix $\mathbf{I} - \mathbf{A}$.

19. (a) We compute

$$(\mathbf{A} + \mathbf{u}_1\mathbf{b}^T)\mathbf{u}_1 = \mathbf{A}\mathbf{u}_1 + \mathbf{u}_1\mathbf{b}^T\mathbf{u}_1 = r_1\mathbf{u}_1 + \mathbf{u}_1\mathbf{b}^T\mathbf{u}_1$$
$$= (r_1 + \mathbf{b}^T\mathbf{u}_1)\mathbf{u}_1.$$

(b) For every $i = 2, 3, \ldots, m$, we compute

$$(\mathbf{A} + \mathbf{u}_1\mathbf{b}^T)\left(\mathbf{u}_i + \frac{\mathbf{b}^T\mathbf{u}_i}{r_i - r_1'}\mathbf{u}_1\right) = (\mathbf{A} + \mathbf{u}_1\mathbf{b}^T)\mathbf{u}_i$$

$$+ (\mathbf{A} + \mathbf{u}_1\mathbf{b}^T)\left(\frac{\mathbf{b}^T\mathbf{u}_i}{r_i - r_1'}\mathbf{u}_1\right)$$

$$= \mathbf{A}\mathbf{u}_i + \mathbf{u}_1\mathbf{b}^T\mathbf{u}_i + \frac{\mathbf{b}^T\mathbf{u}_i}{r_i - r_1'}r_1'\mathbf{u}_1$$

$$= r_i\mathbf{u}_i + \left(\mathbf{b}^T\mathbf{u}_i + \frac{r_1'\mathbf{b}^T\mathbf{u}_i}{r_i - r_1'}\right)\mathbf{u}_1$$

$$= r_i\left(\mathbf{u}_i + \frac{\mathbf{b}^T\mathbf{u}_i}{r_i - r_1'}\mathbf{u}_1\right).$$

5.2 CALCULATING EIGENVALUES AND EIGENVECTORS

1. The characteristic equation for the matrix $\mathbf{A} = \begin{bmatrix} 9 & -3 \\ -2 & 4 \end{bmatrix}$ is

$$|\mathbf{A} - r\mathbf{I}| = \begin{vmatrix} 9 - r & -3 \\ -2 & 4 - r \end{vmatrix}$$
$$= (9 - r)(4 - r) - 6$$
$$= r^2 - 13r + 30$$
$$= (r - 3)(r - 10) = 0.$$

Hence the eigenvalues of $\mathbf{A}$ are $r_1 = 3$ and $r_2 = 10$. To find the eigenvectors corresponding to $r_1 = 3$, we must solve $(\mathbf{A} - r_1\mathbf{I})\mathbf{u} = \mathbf{0}$. Substituting for $\mathbf{A}$ and r_1 gives

$$\begin{bmatrix} 6 & -3 \\ -2 & 1 \end{bmatrix}\begin{bmatrix} u_1 \\ u_2 \end{bmatrix} = \begin{bmatrix} 0 \\ 0 \end{bmatrix}.$$

Notice that this matrix equation is equivalent to the single scalar equation $-2u_1 + u_2 = 0$, whose solutions are obtained by assigning an arbitrary value for u_1 (say $u_1 = s$) and setting $u_2 = 2s$. Consequently, the eigenvectors associated to $r_1 = 3$ and be expressed as

$$\mathbf{u_1} = s \begin{bmatrix} 1 \\ 2 \end{bmatrix}.$$

For $r_2 = 10$, the equation $(\mathbf{A} - r_2\mathbf{I})\mathbf{u} = \mathbf{0}$ becomes

$$\begin{bmatrix} -1 & -3 \\ -2 & -6 \end{bmatrix} \begin{bmatrix} u_1 \\ u_2 \end{bmatrix} = \begin{bmatrix} 0 \\ 0 \end{bmatrix}.$$

Solving, we obtain $u_1 = 3t$ and $u_2 = t$. Therefore, the eigenvectors associated with the eigenvalue $r_2 = 10$ are

$$\mathbf{u_2} = t \begin{bmatrix} 3 \\ -1 \end{bmatrix}.$$

3. The characteristic equation for the matrix $\mathbf{A} = \begin{bmatrix} 1 & -2 \\ -2 & 1 \end{bmatrix}$ is

$$|\mathbf{A} - r\mathbf{I}| = \begin{vmatrix} 1-r & -2 \\ -2 & 1-r \end{vmatrix} = (1-r)^2 - 4$$
$$= (-1-r)(3-r) = 0.$$

Hence the eigenvalues of $\mathbf{A}$ are $r_1 = -1$ and $r_2 = 3$. To find the eigenvectors corresponding to $r_1 = -1$, we must solve $(\mathbf{A} - r_1\mathbf{I})\mathbf{u} = \mathbf{0}$. Substituting for $\mathbf{A}$ and r_1 gives

$$\begin{bmatrix} 2 & -2 \\ -2 & 2 \end{bmatrix} \begin{bmatrix} u_1 \\ u_2 \end{bmatrix} = \begin{bmatrix} 0 \\ 0 \end{bmatrix}.$$

Notice that this matrix equation is equivalent to the single scalar equation $-2u_1 + 2u_2 = 0$, whose solutions are obtained by assigning an arbitrary value for u_1 (say $u_1 = s$) and setting $u_2 = s$. Consequently, the eigenvectors associated to $r_1 = -1$ and be expressed as

$$\mathbf{u_1} = s \begin{bmatrix} 1 \\ 1 \end{bmatrix}.$$

For $r_2 = 3$, the equation $(\mathbf{A} - r_2\mathbf{I})\mathbf{u} = \mathbf{0}$ becomes

$$\begin{bmatrix} -2 & -2 \\ -2 & -2 \end{bmatrix} \begin{bmatrix} u_1 \\ u_2 \end{bmatrix} = \begin{bmatrix} 0 \\ 0 \end{bmatrix}.$$

Solving, we obtain $u_1 = t$ and $u_2 = -t$. Therefore, the eigenvectors associated with the eigenvalue $r_2 = 3$ are

$$\mathbf{u_2} = t \begin{bmatrix} 1 \\ -1 \end{bmatrix}.$$

5. The characteristic equation for the matrix $\mathbf{A} = \begin{bmatrix} 1 & 1 \\ 1 & 1 \end{bmatrix}$ is

$$|\mathbf{A} - r\mathbf{I}| = \begin{vmatrix} 1 - r & 1 \\ 1 & 1 - r \end{vmatrix} = (1 - r)^2 - 1 = -r(2 - r) = 0.$$

Hence the eigenvalues of $\mathbf{A}$ are $r_1 = 0$ and $r_2 = 2$. To find the eigenvectors corresponding to $r_1 = 0$, we must solve $(\mathbf{A} - r_1\mathbf{I})\mathbf{u} = \mathbf{0}$. Substituting for $\mathbf{A}$ and r_1 gives

$$\begin{bmatrix} 1 & 1 \\ 1 & 1 \end{bmatrix} \begin{bmatrix} u_1 \\ u_2 \end{bmatrix} = \begin{bmatrix} 0 \\ 0 \end{bmatrix}.$$

This matrix equation is equivalent to the scalar equation $u_1 + u_2 = 0$, whose solutions are obtained by assigning an arbitrary value for u_1 (say $u_1 = s$) and setting $u_2 = -s$. Consequently, the eigenvectors associated to $r_1 = 0$ and be expressed as

$$\mathbf{u_1} = s \begin{bmatrix} 1 \\ -1 \end{bmatrix}.$$

For $r_2 = 2$, the equation $(\mathbf{A} - r_2\mathbf{I})\mathbf{u} = \mathbf{0}$ becomes

$$\begin{bmatrix} -1 & 1 \\ 1 & -1 \end{bmatrix} \begin{bmatrix} u_1 \\ u_2 \end{bmatrix} = \begin{bmatrix} 0 \\ 0 \end{bmatrix}.$$

Solving, we obtain $u_1 = t$ and $u_2 = t$. Therefore, the eigenvectors associated with the eigenvalue $r_2 = 2$ are

$$\mathbf{u_2} = t \begin{bmatrix} 1 \\ 1 \end{bmatrix}.$$

7. The characteristic equation for the matrix $\mathbf{A} = \begin{bmatrix} 4 & -1 & -2 \\ 2 & 1 & -2 \\ 1 & -1 & 1 \end{bmatrix}$ is

$$|\mathbf{A} - r\mathbf{I}| = \begin{vmatrix} 4-r & -1 & -2 \\ 2 & 1-r & -2 \\ 1 & -1 & 1-r \end{vmatrix} = 0,$$

which simplifies to $(r - 1)(r - 2)(r - 3) = 0$. Hence, the eigenvalues of $\mathbf{A}$ are $r_1 = 1, r_2 = 2$ and $r_2 = 3$. To find the eigenvectors corresponding to $r_1 = 1$, we set $r = 1$ in $(\mathbf{A} - r\mathbf{I})\mathbf{u} = \mathbf{0}$. This gives

$$\begin{bmatrix} 3 & -1 & -2 \\ 2 & 0 & -2 \\ 1 & -1 & 0 \end{bmatrix} \begin{bmatrix} u_1 \\ u_2 \\ u_3 \end{bmatrix} = \begin{bmatrix} 0 \\ 0 \\ 0 \end{bmatrix}.$$

Using elementary row operations (Gaussian elimination), this is equivalent to solving

$$
\begin{aligned}
2u_1 \quad - \quad\quad 2u_3 &= 0, \\
u_1 \quad - \quad u_2 \quad\quad &= 0.
\end{aligned}
$$

Setting $u_1 = s$, we get $u_2 = u_3 = s$. Consequently, the eigenvectors associated to $r_1 = 1$ and be expressed as

$$\mathbf{u_1} = s \begin{bmatrix} 1 \\ 1 \\ 1 \end{bmatrix}.$$

For $r_2 = 2$, we solve

$$\begin{bmatrix} 2 & -1 & -2 \\ 2 & -1 & -2 \\ 1 & -1 & -1 \end{bmatrix} \begin{bmatrix} u_1 \\ u_2 \\ u_3 \end{bmatrix} = \begin{bmatrix} 0 \\ 0 \\ 0 \end{bmatrix}.$$

Solving the linear system in a similar fashion, we obtain the eigenvectors

$$\mathbf{u}_2 = s \begin{bmatrix} 1 \\ 0 \\ 1 \end{bmatrix}.$$

Finally, for $r_3 = 3$, we solve

$$\begin{bmatrix} 1 & -1 & -2 \\ 2 & -2 & -2 \\ 1 & -1 & -2 \end{bmatrix} \begin{bmatrix} u_1 \\ u_2 \\ u_3 \end{bmatrix} = \begin{bmatrix} 0 \\ 0 \\ 0 \end{bmatrix}.$$

and get the eigenvectors

$$\mathbf{u}_3 = s \begin{bmatrix} 1 \\ 1 \\ 0 \end{bmatrix}.$$

9. The characteristic equation for the matrix $\mathbf{A} = \begin{bmatrix} 0 & -1 \\ 1 & 0 \end{bmatrix}$ is

$$|\mathbf{A} - r\mathbf{I}| = \begin{vmatrix} -r & -1 \\ 1 & -r \end{vmatrix} = r^2 + 1 = 0.$$

Hence the eigenvalues of $\mathbf{A}$ are $r_1 = i$ and $r_2 = -i$. To find the eigenvectors corresponding to $r_1 = i$, we must solve $(\mathbf{A} - r_1\mathbf{I})\mathbf{u} = \mathbf{0}$. Substituting for $\mathbf{A}$ and r_1 gives

$$\begin{bmatrix} -i & -1 \\ 1 & -i \end{bmatrix} \begin{bmatrix} u_1 \\ u_2 \end{bmatrix} = \begin{bmatrix} 0 \\ 0 \end{bmatrix}.$$

This matrix equation is equivalent to the scalar equation $-iu_1 - u_2 = 0$, whose solutions are obtained by assigning an arbitrary value for $u_1 = s$

(say $u_1 = s$) and setting $u_2 = -is$. Consequently, the eigenvectors associated to $r_1 = i$ are expressed as

$$\mathbf{u_1} = s \begin{bmatrix} 1 \\ -i \end{bmatrix}.$$

For $r_2 = -i$, the equation $(\mathbf{A} - r_2\mathbf{I})\mathbf{u} = \mathbf{0}$ becomes

$$\begin{bmatrix} i & -1 \\ 1 & i \end{bmatrix} \begin{bmatrix} u_1 \\ u_2 \end{bmatrix} = \begin{bmatrix} 0 \\ 0 \end{bmatrix},$$

which is equivalent to the scalar equation $iu_1 - u_2 - 0$. As above, we see that the eigenvectors associated with the eigenvalue $r_2 = -i$ are

$$\mathbf{u_2} = t \begin{bmatrix} 1 \\ i \end{bmatrix}.$$

11. The characteristic equation for the matrix $\mathbf{A} = \begin{bmatrix} 1 & 2i \\ -1 & 3+i \end{bmatrix}$ is

$$|\mathbf{A} - r\mathbf{I}| = \begin{vmatrix} 1 - r & 2i \\ -1 & (3+i) - r \end{vmatrix} = 0.$$

This simplifies to $r^2 - (4 + i)r + 3(1 + i) = 0$, which is equivalent to $(r - 3)(r - 1 - i) = 0$. Hence the eigenvalues of $\mathbf{A}$ are $r_1 = 3$ and $r_2 = 1 + i$. To find the eigenvectors corresponding to $r_1 = i$, we must solve $(\mathbf{A} - r_1\mathbf{I})\mathbf{u} = \mathbf{0}$. Substituting for $\mathbf{A}$ and r_1 gives

$$\begin{bmatrix} -2 & 2i \\ -1 & i \end{bmatrix} \begin{bmatrix} u_1 \\ u_2 \end{bmatrix} = \begin{bmatrix} 0 \\ 0 \end{bmatrix}.$$

This matrix equation is equivalent to the scalar equation $-u_1 + iu_2 = 0$, whose solutions are obtained by assigning $u_2 = s$ and setting $u_1 = is$. Consequently, the eigenvectors associated to $r_1 = 3$ and be expressed as

$$\mathbf{u_1} = s \begin{bmatrix} i \\ 1 \end{bmatrix}.$$

For $r_2 = i + 1$, the equation $(\mathbf{A} - r_2\mathbf{I})\mathbf{u} = \mathbf{0}$ becomes

$$\begin{bmatrix} -i & 2i \\ -1 & 2 \end{bmatrix} \begin{bmatrix} u_1 \\ u_2 \end{bmatrix} = \begin{bmatrix} 0 \\ 0 \end{bmatrix},$$

which is equivalent to the scalar equation $-u_1 + 2u_2 = 0$. It follows that the eigenvectors associated with the eigenvalue $r_2 = i + 1$ are

$$\mathbf{u}_2 = t \begin{bmatrix} 2 \\ 1 \end{bmatrix}.$$

13. The characteristic equation for the matrix $\mathbf{A} = \begin{bmatrix} 10 & -4 \\ 4 & 2 \end{bmatrix}$ is

$$|\mathbf{A} - r\mathbf{I}| = \begin{vmatrix} 10 - r & -4 \\ 4 & 2 - r \end{vmatrix} = 0$$

which is equivalent to $r^2 - 12r + 36 = 0$, or $(r - 6)^2 = 0$. Hence the matrix $\mathbf{A}$ has only one eigenvalue $r = 6$. To find the eigenvectors corresponding to $r = 6$, we must solve $(\mathbf{A} - 6\mathbf{I})\mathbf{u} = \mathbf{0}$. This gives

$$\begin{bmatrix} 4 & -4 \\ 4 & -4 \end{bmatrix} \begin{bmatrix} u_1 \\ u_2 \end{bmatrix} = \begin{bmatrix} 0 \\ 0 \end{bmatrix}.$$

This matrix equation is equivalent to the scalar equation $4u_1 - 4u_2 = 0$. Consequently, the eigenvectors associated to $r = 6$ are the vectors of the form

$$\mathbf{u} = s \begin{bmatrix} 1 \\ 1 \end{bmatrix}.$$

15. The characteristic equation for the matrix $\mathbf{A} = \begin{bmatrix} 3 & 2 \\ -2 & 7 \end{bmatrix}$ is

$$|\mathbf{A} - r\mathbf{I}| = \begin{vmatrix} 3 - r & 2 \\ -2 & 7 - r \end{vmatrix} = 0$$

which is equivalent to $r^2 - 10r + 25 = 0$, or $(r - 5)^2 = 0$. Hence the matrix $\mathbf{A}$ has only one eigenvalue $r = 5$, with algebraic multiplicity 2.

To find the eigenvectors corresponding to $r = 5$, we must solve $(\mathbf{A} - 5\mathbf{I})\mathbf{u} = \mathbf{0}$. This gives

$$\begin{bmatrix} -2 & 2 \\ -2 & 2 \end{bmatrix} \begin{bmatrix} u_1 \\ u_2 \end{bmatrix} = \begin{bmatrix} 0 \\ 0 \end{bmatrix}.$$

This matrix equation is equivalent to the scalar equation $-2u_1 + 2u_2 = 0$. Consequently, the eigenvectors associated to $r = 5$ are the vectors of the form

$$\mathbf{u} = s \begin{bmatrix} 1 \\ 1 \end{bmatrix}.$$

17. The characteristic equation for the matrix $\mathbf{A} = \begin{bmatrix} 0 & 1 & 1 \\ 1 & 0 & 1 \\ 1 & 1 & 0 \end{bmatrix}$ is

$$|\mathbf{A} - r\mathbf{I}| = \begin{vmatrix} -r & 1 & 1 \\ 1 & -r & 1 \\ 1 & 1 & -r \end{vmatrix} = 0,$$

which is equivalent to $-r^3 + 3r + 2 = 0$, or $-(r + 1)^2(r + 2) = 0$. Hence, the eigenvalues of $\mathbf{A}$ are $r_1 = -1$ with algebraic multiplicity 2, and $r_2 = 2$, with algebraic multiplicity 1. To find the eigenvectors corresponding to $r_1 = -1$, amounts to finding the solutions of the equation $(\mathbf{A} + \mathbf{I})\mathbf{u} = \mathbf{0}$. This is equivalent to the equation

$$\begin{bmatrix} 1 & 1 & 1 \\ 1 & 1 & 1 \\ 1 & 1 & 1 \end{bmatrix} \begin{bmatrix} u_1 \\ u_2 \\ u_3 \end{bmatrix} = \begin{bmatrix} 0 \\ 0 \\ 0 \end{bmatrix}.$$

Using elementary row operations (Gaussian elimination), this is equivalent to solving

$$u_1 + u_2 + u_3 = 0.$$

Setting $u_2 = -s$ and $u_3 = -t$, we get $u_1 = s + t$. Consequently, the eigenspace associated to $r_1 = -1$ is two dimensional

$$\mathbf{u} = s \begin{bmatrix} 1 \\ -1 \\ 0 \end{bmatrix} + t \begin{bmatrix} 1 \\ 0 \\ -1 \end{bmatrix}.$$

For $r_2 = 2$, we solve the equation $(\mathbf{A} - 2\mathbf{I})\mathbf{u} = \mathbf{0}$. This is equivalent to the equation

$$\begin{bmatrix} -2 & 1 & 1 \\ 1 & -2 & 1 \\ 1 & 1 & -2 \end{bmatrix} \begin{bmatrix} u_1 \\ u_2 \\ u_3 \end{bmatrix} = \begin{bmatrix} 0 \\ 0 \\ 0 \end{bmatrix}.$$

Using elementary row operations (Gaussian elimination), this is equivalent to solving

$$u_1 - 2u_2 + u_3 = 0,$$
$$3u_2 - 3u_3 = 0.$$

We assign $u_2 = s$, and obtain $u_3 = t$ and $u_1 = 1$. Hence, the eigenvectors corresponding to the eigenvalue $r_2 = 2$ are the vectors of the form

$$\mathbf{v} = s \begin{bmatrix} 1 \\ 1 \\ 1 \end{bmatrix}.$$

19. The characteristic equation for the matrix $\mathbf{A} = \begin{bmatrix} 1 & 0 & 0 \\ 0 & 2 & 3 \\ 1 & 0 & 2 \end{bmatrix}$ is

$$|\mathbf{A} - r\mathbf{I}| = \begin{vmatrix} 1-r & 0 & 0 \\ 0 & 2-r & 3 \\ 1 & 0 & 2-r \end{vmatrix} = 0,$$

which is equivalent to $(1 - r)(2 - r)^2 = 0$. Hence, the eigenvalues of $\mathbf{A}$ are $r_1 = 1$ with algebraic multiplicity 1, and $r_2 = 2$, with algebraic multiplicity 2. To find the eigenvectors corresponding to $r_1 = 1$, amounts to finding the solutions of the equation $(\mathbf{A} - \mathbf{I})\mathbf{u} = \mathbf{0}$. This is equivalent to the equation

$$\begin{bmatrix} 0 & 0 & 0 \\ 0 & 1 & 3 \\ 1 & 0 & 1 \end{bmatrix} \begin{bmatrix} u_1 \\ u_2 \\ u_3 \end{bmatrix} = \begin{bmatrix} 0 \\ 0 \\ 0 \end{bmatrix}.$$

This is equivalent to solving

$$\begin{aligned} u_1 + \quad & u_3 = 0, \\ u_2 + \quad & 3u_3 = 0. \end{aligned}$$

Assigning $u_1 = s$, we get $u_3 = -s$, and $u_2 = 3s$. Consequently, the eigenvectors associated to $r_1 = 1$ are of the form

$$\mathbf{u} = s \begin{bmatrix} 1 \\ 3 \\ -1 \end{bmatrix}.$$

For $r_2 = 2$, we solve the equation $(\mathbf{A} - 2\mathbf{I})\mathbf{u} = \mathbf{0}$. This is equivalent to the equation

$$\begin{bmatrix} -1 & 0 & 0 \\ 0 & 0 & 3 \\ 1 & 0 & 0 \end{bmatrix} \begin{bmatrix} u_1 \\ u_2 \\ u_3 \end{bmatrix} = \begin{bmatrix} 0 \\ 0 \\ 0 \end{bmatrix}.$$

The solution of the corresponding linear system is $u_1 = u_3 = 0$, while $u_2 = t$. Hence, the eigenvectors corresponding to the eigenvalue $r_2 = 2$ are the vectors of the form

$$\mathbf{v} = t \begin{bmatrix} 0 \\ 1 \\ 0 \end{bmatrix}.$$

21. (a) Recall that the highest eigenvalue of the matrix $\mathbf{A} = \begin{bmatrix} 9 & -3 \\ -2 & 4 \end{bmatrix}$ is 10, with $\mathbf{u_1} = [3 \; -1]^T$ a corresponding eigenvector. According

to Problem 19 of Exercises 5.1, we choose next a vector $\mathbf{b} = [b_1\ b_2]^T$ such that

$$0 = 10 + \mathbf{b}^T \mathbf{u}_1.$$

An immediate choice is $\mathbf{b} = [-3\ 1]^T$. We compute now the deflated matrix

$$
\mathbf{A} + \mathbf{u}_1 \mathbf{b}^T = \begin{bmatrix} 9 & -3 \\ -2 & 4 \end{bmatrix} + \begin{bmatrix} 3 \\ -1 \end{bmatrix} \begin{bmatrix} -3 & 1 \end{bmatrix}
$$

$$
= \begin{bmatrix} 9 & -3 \\ -2 & 4 \end{bmatrix} + \begin{bmatrix} -9 & 3 \\ 3 & -1 \end{bmatrix}
$$

$$
= \begin{bmatrix} 0 & 0 \\ 1 & 3 \end{bmatrix}.
$$

It is a triangular matrix, and its eigenvalues are the its diagonal entries, 0 and 3.

(b) The characteristic equation for the matrix $\mathbf{A} = \begin{bmatrix} 1 & -2 & 1 \\ -1 & 2 & -1 \\ 0 & 1 & 1 \end{bmatrix}$ is

$$
|\mathbf{A} - r\mathbf{I}| = \begin{vmatrix} 1-r & -2 & 1 \\ -1 & 2-r & -1 \\ 0 & 1 & 1-r \end{vmatrix} = 0,
$$

which is equivalent to $-r^3 + 4r^2 - 4r = 0$, or $-r(r-2)^2 = 0$ Hence, the eigenvalues of A are $r_1 = 0$ with algebraic multiplicity 1, and $r_2 = 2$, with algebraic multiplicity 2. The computation of the eigenvectors corresponding to the highest eigenvalue, $r_2 = 2$, amounts to finding the solutions of the equation $(\mathbf{A} - 2\mathbf{I})\mathbf{u} = \mathbf{0}$. This is equivalent to the equation

$$
\begin{bmatrix} -1 & -2 & 1 \\ -1 & 0 & -1 \\ 0 & 1 & -1 \end{bmatrix} \begin{bmatrix} u_1 \\ u_2 \\ u_3 \end{bmatrix} = \begin{bmatrix} 0 \\ 0 \\ 0 \end{bmatrix}.
$$

This is equivalent to solving

$$
\begin{aligned}
-u_1 \quad - \quad u_3 &= 0, \\
u_2- \quad u_3 &= 0.
\end{aligned}
$$

Assigning $u_1 = s$, we get $u_3 = -s$, and $u_2 = -s$. Consequently, an eigenvector associated to $r_2 = 2$ is

$$
\mathbf{u} = \begin{bmatrix} 1 \\ -1 \\ -1 \end{bmatrix}.
$$

obtained for $s = 1$. As before, we look next for a vector $\mathbf{b} = [b_1 \, b_2 \, b_3]^T$ such that

$$
0 = 2 + \mathbf{b}^T\mathbf{u},
$$

which is equivalent to the scalar equation $b_1 - b_2 - b_3 = -2$. An immediate choice is $\mathbf{b} = [0 \, 1 \, 1]^T$.

We compute now the deflated matrix

$$
\begin{aligned}
\mathbf{A} + \mathbf{u}\mathbf{b}^T &= \begin{bmatrix} 1 & -2 & 1 \\ -1 & 2 & -1 \\ 0 & 1 & 1 \end{bmatrix} + \begin{bmatrix} 1 \\ -1 \\ -1 \end{bmatrix} [0 \, 1 \, 1] \\
&= \begin{bmatrix} 1 & -2 & 1 \\ -1 & 2 & -1 \\ 0 & 1 & 1 \end{bmatrix} + \begin{bmatrix} 0 & 1 & 1 \\ 0 & -1 & -1 \\ 0 & -1 & -1 \end{bmatrix} \\
&= \begin{bmatrix} 1 & -1 & 2 \\ -1 & 1 & -2 \\ 0 & 0 & 0 \end{bmatrix}
\end{aligned}
$$

We compute now the eigenvalues of the deflated matrix $\begin{bmatrix} 1 & -1 & 2 \\ -1 & 1 & -2 \\ 0 & 0 & 0 \end{bmatrix}$. The characteristic polynomial is

$$
\begin{vmatrix} 1-r & -1 & 2 \\ -1 & 1-r & -2 \\ 0 & 0 & -r \end{vmatrix} = 0,
$$

which is equivalent to $r^2(2 - r) = 0$. Notice that in this case, the higher eigenvalue $r = 2$ is now an eigenvalue of algebraic multiplicity one, and the eigenvalue $r = 0$, has now multiplicity 2.

23. (a) Let $p^T(r) = \det(\mathbf{A}^T - r\mathbf{I})$ be the characteristic polynomial of $\mathbf{A}^T$. We have

$$p^T(r) = \det(\mathbf{A}^T - r\mathbf{I}) = \det\left((\mathbf{A} - r\mathbf{I})^T\right) = \det(\mathbf{A} - r\mathbf{I}).$$

However, $p(r) = \det(\mathbf{A} - r\mathbf{I})$ is the characteristic polynomial of $\mathbf{A}$. Therefore, the matrices $\mathbf{A}$ and $\mathbf{A}^T$ have the same characteristic polynomial. In particular, they have the same eigenvalues.

(b) Let r_1 be an eigenvalue of $\mathbf{A}^T$ with the correspond eigenvector $\mathbf{v}_1$, and r_2 be an eigenvalue of $\mathbf{A}$ with the corresponding eigenvector $\mathbf{v}_2$, where $r_1 \neq r_2$. We have $\mathbf{v}_1^T(\mathbf{A}\mathbf{v}_2) = \mathbf{v}_1^T(r_2\mathbf{v}_2) = r_2\mathbf{v}_1^T\mathbf{v}_2$. Similarly, have $(\mathbf{A}^T\mathbf{v}_1)\mathbf{v}_2 = r_1\mathbf{v}_1^T\mathbf{v}_2$. Since $\mathbf{v}_1^T(\mathbf{A}\mathbf{v}_2) = (\mathbf{A}^T\mathbf{v}_1)\mathbf{v}_2$, it follows that $r_2\mathbf{v}_1^T\mathbf{v}_2 = r_1\mathbf{v}_1^T\mathbf{v}_2$, which yields $\mathbf{v}_1^T\mathbf{v}_2 = 0$.

(c) (i) The eigenvalues and the corresponding eigenvectors of $\mathbf{A} = \begin{bmatrix} 3 & 2 \\ -1 & 0 \end{bmatrix}$ can be easily computed as above. The eigenvalues are $r_1 = 1$, with the corresponding eigenvector $\mathbf{v}_1 = [1 - 1]^T$ and $r_2 = 2$, with the corresponding eigenvector $\mathbf{v}_2 = [2 - 1]^T$. The eigenvalues of $\mathbf{A}^T$ are $r_1 = 1$, with the corresponding eigenvector $\mathbf{w}_1 = [1\,2]^T$ and $r_2 = 2$, with the corresponding eigenvector $\mathbf{w}_2 = [1\,1]^T$. Notice that the vectors $\mathbf{v}_1^T\mathbf{w}_2 = \mathbf{v}_2^T\mathbf{w}_1 = 0$.

(ii) Similarly, the eigenvalues of $\mathbf{A} = \begin{bmatrix} -3 & -2 & 2 \\ 2 & 5 & -4 \\ 1 & 5 & -4 \end{bmatrix}$ are $r_1 = -1$, with the corresponding eigenvector $\mathbf{v}_1 = [1\,1\,2]^T$, $r_2 = -2$, with the corresponding eigenvector $\mathbf{v}_2 = [2\,0\,1]^T$, and $r_3 = 1$, with the corresponding eigenvector $\mathbf{v}_3 = [0\,1\,1]^T$. The eigenvalues of $\mathbf{A}^T$ are $r_1 = -1$, with the corresponding eigenvector $\mathbf{w}_1 = [1\,2 - 2]^T$, $r_2 = -2$, with the corresponding eigenvector $\mathbf{w}_2 = [1\,1 - 1]^T$, and $r_3 = 1$, with the corresponding eigenvector $\mathbf{w}_3 = [1\,3 - 2]^T$. A direct computation shows that

$$\mathbf{v}_1^T\mathbf{w}_2 = \mathbf{v}_1^T\mathbf{w}_3 = \mathbf{v}_2^T\mathbf{w}_1 = \mathbf{v}_2^T\mathbf{w}_3 = \mathbf{v}_3^T\mathbf{w}_1 = \mathbf{v}_3^T\mathbf{w}_2 = 0.$$

25. Notice that if $f(x) = a_n x^n + a_{n-1} x^{n-1} + \cdots + a_1 x + a_0$ is an arbitrary polynomial of degree n, then $a_0 = f(0)$. In our case,

$$c_0 = p(0) = \det(\mathbf{A} - 0\mathbf{I}) = \det(\mathbf{A}).$$

27. The characteristic equation of a 2×2 matrix $\mathbf{A} = \begin{bmatrix} a & b \\ c & d \end{bmatrix}$ is

$$|\mathbf{A} - r\mathbf{I}| = \begin{vmatrix} a - r & b \\ c & d - r \end{vmatrix} = 0.$$

This simplifies as $r^2 - (a + d)r + (ad - bc) = 0$, or $r^2 - \text{Trace}(\mathbf{A}) + \det(\mathbf{A}) = 0$. In our case, $\text{Trace}(\mathbf{A}) = 2$, and $\det(\mathbf{A}) = -8$, and so the characteristic equation is $r^2 - 2r - 8 = 0$, which is equivalent to $(r + 2)(r - 4) = 0$. Therefore, the eigenvalues are -2 and 4.

5.3   SYMMETRIC AND HERMITIAN MATRICES

1. The characteristic equation for the matrix $\mathbf{A} = \begin{bmatrix} -7 & 6 \\ 6 & 2 \end{bmatrix}$ is

$$|\mathbf{A} - r\mathbf{I}| = \begin{vmatrix} -7 - r & 6 \\ 6 & 2 - r \end{vmatrix} = 0$$

which is equivalent to $r^2 + 5r - 50 = (r + 10)(r - 5) = 0$. Hence the eigenvalues of $\mathbf{A}$ are $r_1 = -10$ and $r_2 = 5$. To find the eigenvectors corresponding to $r_1 = -10$, we must solve the equation $(\mathbf{A} + 10\mathbf{I})\mathbf{u} = \mathbf{0}$, which means

$$\begin{bmatrix} 3 & 6 \\ 6 & 12 \end{bmatrix} \begin{bmatrix} u_1 \\ u_2 \end{bmatrix} = \begin{bmatrix} 0 \\ 0 \end{bmatrix}.$$

Notice that this matrix equation is equivalent to the single scalar equation $6u_1 + 12u_2 = 0$, whose solutions are obtained by assigning an arbitrary value for u_2 (say $u_2 = s$) and setting $u_1 = -2s$. A particular such eigenvector is obtained for $s = -1$

$$\mathbf{u_1} = \begin{bmatrix} 2 \\ -1 \end{bmatrix}.$$

For $r_2 = 5$, we must solve the equation $(\mathbf{A} - 5\mathbf{I})\mathbf{u} = \mathbf{0}$ becomes

$$\begin{bmatrix} -12 & 6 \\ 6 & -3 \end{bmatrix} \begin{bmatrix} u_1 \\ u_2 \end{bmatrix} = \begin{bmatrix} 0 \\ 0 \end{bmatrix}.$$

Solving, we obtain $u_1 = t$ and $u_2 = 2t$. A particular eigenvector associated with the eigenvalue $r_2 = 5$ is obtained by setting $t = 1$:

$$\mathbf{u_2} = \begin{bmatrix} 1 \\ 2 \end{bmatrix}.$$

We can directly check now that the dot product

$$\mathbf{u_1} \cdot \mathbf{u_2} = 2 \cdot 1 + (-1) \cdot 2 = 0,$$

which means the vectors $\mathbf{u_1}$ and $\mathbf{u_2}$ are orthogonal.

3. The characteristic equation for the matrix $\mathbf{A} = \begin{bmatrix} 1 & 4 & 3 \\ 4 & 1 & 0 \\ 3 & 0 & 1 \end{bmatrix}$ is

$$|\mathbf{A} - r\mathbf{I}| = \begin{vmatrix} 1-r & 4 & 3 \\ 4 & 1-r & 0 \\ 3 & 0 & 1-r \end{vmatrix} = 0,$$

which simplifies to $(1 - r)(r + 4)(r - 6) = 0$. Hence, the eigenvalues of $\mathbf{A}$ are $r_1 = 1, r_2 = -4$ and $r_2 = 6$. To find the eigenvectors corresponding to $r_1 = 1$, we set $r = 1$ in $(\mathbf{A} - r\mathbf{I})\mathbf{u} = \mathbf{0}$. This gives

$$\begin{bmatrix} 0 & 4 & 3 \\ 4 & 0 & 0 \\ 3 & 0 & 0 \end{bmatrix} \begin{bmatrix} u_1 \\ u_2 \\ u_3 \end{bmatrix} = \begin{bmatrix} 0 \\ 0 \\ 0 \end{bmatrix}.$$

We see immediately that $u_1 = 0$, and $4u_2 + 3u_3 = 0$. A particular solution is $u_2 = 3$ and $u_3 = -4$, and the eigenvectors associated to $r_1 = 1$ are of the form

$$\mathbf{u_1} = s \begin{bmatrix} 0 \\ 3 \\ -4 \end{bmatrix}.$$

For $r_2 = -4$, we solve

$$\begin{bmatrix} 5 & 4 & 3 \\ 4 & 5 & 0 \\ 3 & 0 & 5 \end{bmatrix} \begin{bmatrix} u_1 \\ u_2 \\ u_3 \end{bmatrix} = \begin{bmatrix} 0 \\ 0 \\ 0 \end{bmatrix}.$$

Solving the linear system, we obtain the eigenvectors

$$\mathbf{u_2} = s \begin{bmatrix} 5 \\ -4 \\ -3 \end{bmatrix}.$$

Finally, for $r_3 = 6$, we solve

$$\begin{bmatrix} -5 & 4 & 3 \\ 4 & -5 & 0 \\ 3 & 0 & -5 \end{bmatrix} \begin{bmatrix} u_1 \\ u_2 \\ u_3 \end{bmatrix} = \begin{bmatrix} 0 \\ 0 \\ 0 \end{bmatrix}.$$

and get the eigenvectors

$$\mathbf{u_3} = s \begin{bmatrix} 5 \\ 4 \\ 3 \end{bmatrix}.$$

As the matrix $\mathbf{A}$ is symmetric, the vectors $\mathbf{u_1}$, $\mathbf{u_2}$ and $\mathbf{u_3}$ corresponding to distinct eigenvalues are orthogonal. We can check their orthogonality directly, by computing their dot products for $s = 1$:

$$\mathbf{u_1} \cdot \mathbf{u_2} = [0\,3 - 4]^T \cdot [5 - 4 - 3]^T$$
$$= 0 \cdot 5 + 3 \cdot (-4) + (-4) \cdot (-3) = 0,$$
$$\mathbf{u_1} \cdot \mathbf{u_3} = [0\,3 - 4]^T \cdot [5\,4\,3]^T = 0 \cdot 5 + 3 \cdot 4 + (-4) \cdot 3 = 0,$$
$$\mathbf{u_2} \cdot \mathbf{u_3} = [5 - 4 - 3]^T \cdot [5\,4\,3]^T = 5 \cdot 5 + (-4) \cdot 4 + (-3) \cdot 3 = 0.$$

5. The characteristic equation for the matrix $\mathbf{A} = \begin{bmatrix} 2 & 1 & 1 \\ 1 & 2 & -1 \\ 1 & -1 & 2 \end{bmatrix}$ is

$$|\mathbf{A} - r\mathbf{I}| = \begin{vmatrix} 2-r & 1 & 1 \\ 1 & 2-r & -1 \\ 1 & -1 & 2-r \end{vmatrix} = 0,$$

which simplifies to $r(r - 3)^2 = 0$. Hence, the eigenvalues of $\mathbf{A}$ are $r_1 = 0$, with algebraic multiplicity 1, and $r_2 = 3$, and with algebraic multiplicity 2. To find the eigenvectors corresponding to $r_1 = 0$, we set $r = 0$ in $(\mathbf{A} - r\mathbf{I})\mathbf{u} = \mathbf{0}$. This gives

$$\begin{bmatrix} 2 & 1 & 1 \\ 1 & 2 & -1 \\ 1 & -1 & 2 \end{bmatrix} \begin{bmatrix} u_1 \\ u_2 \\ u_3 \end{bmatrix} = \begin{bmatrix} 0 \\ 0 \\ 0 \end{bmatrix}.$$

After the Gaussian elimination, this is equivalent to

$$u_1 + u_2 \qquad = 0,$$
$$u_1 + \qquad u_3 = 0.$$

A particular solution is $u_1 = 1$, $u_2 = -1$ and $u_3 = -1$, and the eigenvectors associated to $r_1 = 0$ are of the form

$$\mathbf{u_1} = s \begin{bmatrix} 1 \\ -1 \\ -1 \end{bmatrix}.$$

For $r_2 = 3$, we solve

$$\begin{bmatrix} -1 & 1 & 1 \\ 1 & -1 & -1 \\ 1 & -1 & -1 \end{bmatrix} \begin{bmatrix} u_1 \\ u_2 \\ u_3 \end{bmatrix} = \begin{bmatrix} 0 \\ 0 \\ 0 \end{bmatrix},$$

which is equivalent to $u_1 - u_2 - u_3 = 0$. The eigenspace corresponding to the eigenvalue $r_2 = 3$ is therefore two-dimensional, and any corresponding eigenvector is of the form $\mathbf{u} = s\mathbf{v}_1 + t\mathbf{v}_2$, where $\mathbf{v}_1 = [1\ 1\ 0]^T$, and $\mathbf{v}_1 = [1\ 0\ 1]^T$.

Since the matrix $\mathbf{A}$ is symmetric, the vector $\mathbf{u}_1$ is orthogonal to both $\mathbf{v}_1$ and $\mathbf{v}_2$ above, as they correspond to distinct eigenvalues. However $\mathbf{v}_1$ and $\mathbf{v}_2$ are not orthogonal (their dot product equals to 1). To find three orthogonal eigenvectors, let $\mathbf{u}_1 = [1\ -1\ -1]^T$ as above, $\mathbf{u}_2 = \mathbf{v}_1 = [1\ 1\ 0]^T$, and we look for a third eigenvector $\mathbf{u}_3$ of the form $s\mathbf{v}_1 + t\mathbf{v}_2$ orthogonal to to $\mathbf{u}_2$ (it is automatically orthogonal to $\mathbf{u}_1$). solving the equation $\mathbf{u} \cdot \mathbf{u}_2 = 0$, we notice that a possible solution is obtain for $s = -1$ and $t = 2$. Therefore, a set of orthogonal eigenvectors for the given matrix consists of the vectors

$$\left\{ [1\ -1\ -1]^T, [1\ 1\ 0]^T, [1\ -1\ 2]^T \right\}.$$

7. The characteristic equation for the matrix $\mathbf{A} = \begin{bmatrix} 0 & i \\ -i & 0 \end{bmatrix}$ is

$$|\mathbf{A} - r\mathbf{I}| = \begin{vmatrix} -r & i \\ -i & -r \end{vmatrix} = 0$$

which is equivalent to $r^2 - 1 = (r + 1)(r - 1) = 0$. Hence the eigenvalues of $\mathbf{A}$ are $r_1 = -1$ and $r_2 = 1$. To find the eigenvectors corresponding to $r_1 = -1$, we must solve the equation $(\mathbf{A} + \mathbf{I})\mathbf{u} = \mathbf{0}$, which means

$$\begin{bmatrix} 1 & i \\ -i & 1 \end{bmatrix} \begin{bmatrix} u_1 \\ u_2 \end{bmatrix} = \begin{bmatrix} 0 \\ 0 \end{bmatrix}.$$

Notice that this matrix equation is equivalent to the single scalar equation $u_1 + iu_2 = 0$. A particular solution is obtained by assigning an arbitrary value for u_2 (say $u_2 = 1$) and setting $u_1 = i$. A particular such eigenvector is

$$\mathbf{u_1} = \begin{bmatrix} i \\ -1 \end{bmatrix}.$$

For $r_2 = -1$, we must solve the equation $(\mathbf{A} - \mathbf{I})\mathbf{u} = \mathbf{0}$ becomes

$$\begin{bmatrix} -1 & i \\ -i & -1 \end{bmatrix} \begin{bmatrix} u_1 \\ u_2 \end{bmatrix} = \begin{bmatrix} 0 \\ 0 \end{bmatrix}.$$

As before, a particular solution is the eigenvector

$$\mathbf{u_2} = \begin{bmatrix} i \\ 1 \end{bmatrix}.$$

We can directly check now that the complex inner product

$$\mathbf{u_1} \cdot \mathbf{u_2} = [-i - 1] \cdot \begin{bmatrix} i \\ 1 \end{bmatrix} = (-i) \cdot i + (-1) \cdot 1 = 1 - 1 = 0,$$

which means the vectors $\mathbf{u_1}$ and $\mathbf{u_2}$ are orthogonal.

9. The characteristic equation for the matrix $\mathbf{A} = \begin{bmatrix} -1 & 0 & -1+i \\ 0 & -1 & 0 \\ -1-i & 0 & 0 \end{bmatrix}$ is

$$|\mathbf{A} - r\mathbf{I}| = \begin{vmatrix} -1-r & 0 & -1+i \\ 0 & -1-r & 0 \\ -1-i & 0 & -r \end{vmatrix} = 0,$$

which simplifies to $(r^2-1)(r+2) = 0$. Hence, the eigenvalues of $\mathbf{A}$ are $r_1 = 1$, $r_2 = -1$ and $r_2 = -2$. To find the eigenvectors corresponding to $r_1 = 1$, we set $r = 1$ in $(\mathbf{A} - r\mathbf{I})\mathbf{u} = \mathbf{0}$. This gives

$$\begin{bmatrix} -2 & 0 & -1+i \\ 0 & -1 & 0 \\ -1-i & 0 & -1 \end{bmatrix} \begin{bmatrix} u_1 \\ u_2 \\ u_3 \end{bmatrix} = \begin{bmatrix} 0 \\ 0 \\ 0 \end{bmatrix}.$$

This is equivalent to $u_2 = 0$, and $(1 + i)u_1 + u_3 = 0$. A particular eigenvectors associated to $r_1 = 1$ is

$$\mathbf{u_1} = \begin{bmatrix} 1 \\ 0 \\ -(i+1) \end{bmatrix}.$$

For $r_2 = -1$, we solve

$$\begin{bmatrix} 0 & 0 & -1+i \\ 0 & 0 & 0 \\ -1-i & 0 & 1 \end{bmatrix} \begin{bmatrix} u_1 \\ u_2 \\ u_3 \end{bmatrix} = \begin{bmatrix} 0 \\ 0 \\ 0 \end{bmatrix},$$

which yields $u_1 = u_3 = 0$. A particular eigenvectors associated to $r_2 = -1$ is

$$\mathbf{u_2} = \begin{bmatrix} 0 \\ 1 \\ 0 \end{bmatrix}.$$

Finally, for $r_3 = -2$, we solve

$$\begin{bmatrix} 1 & 0 & -1+i \\ 0 & 1 & 0 \\ -1-i & 0 & 2 \end{bmatrix} \begin{bmatrix} u_1 \\ u_2 \\ u_3 \end{bmatrix} = \begin{bmatrix} 0 \\ 0 \\ 0 \end{bmatrix},$$

This is equivalent to $u_2 = 0$, and $u_1 + (-1 + i)u_3 = 0$. A particular eigenvectors associated to $r_3 = -2$ is

$$\mathbf{u_3} = \begin{bmatrix} (i-1) \\ 0 \\ -1 \end{bmatrix}.$$

Since the matrix $\mathbf{A}$ is Hermitian, the vectors

$$\left\{ [1\ 0 - (i+1)]^T, [0\ 1\ 0]^T, [(i-1)\ 0 - 1]^T \right\}$$

are mutually orthogonal, as they correspond to distinct eigenvalues.

11. The characteristic equation for the matrix $\mathbf{A} = \begin{bmatrix} 0 & 0 & -i \\ 0 & 0 & i \\ i & -i & 0 \end{bmatrix}$ is

$$|\mathbf{A} - r\mathbf{I}| = \begin{bmatrix} -r & 0 & -i \\ 0 & -r & i \\ i & -i & -r \end{bmatrix},$$

which simplifies to $r(r^2 - 2) = 0$. Hence, the eigenvalues of $\mathbf{A}$ are $r_1 = 0$, $r_2 = -\sqrt{2}$ and $r_2 = \sqrt{2}$. To find the eigenvectors corresponding to $r_1 = 0$, we set $r = 0$ in $(\mathbf{A} - r\mathbf{I})\mathbf{u} = \mathbf{0}$. This gives

$$\begin{bmatrix} 0 & 0 & -i \\ 0 & 0 & i \\ i & -i & 0 \end{bmatrix} \begin{bmatrix} u_1 \\ u_2 \\ u_3 \end{bmatrix} = \begin{bmatrix} 0 \\ 0 \\ 0 \end{bmatrix}.$$

This is equivalent to $u_3 = 0$, and $u_1 - u_2 = 0$. Hence, particular solution is $u_1 = 1$, $u_2 = 1$ and $u_3 = 1$, and so an eigenvector associated to $r_1 = 0$ is

$$\mathbf{u}_1 = \begin{bmatrix} 1 \\ 1 \\ 0 \end{bmatrix}.$$

For $r_2 = -\sqrt{2}$, we solve

$$\begin{bmatrix} \sqrt{2} & 0 & -i \\ 0 & \sqrt{2} & i \\ i & -i & \sqrt{2} \end{bmatrix} \begin{bmatrix} u_1 \\ u_2 \\ u_3 \end{bmatrix} = \begin{bmatrix} 0 \\ 0 \\ 0 \end{bmatrix}.$$

After the Gaussian elimination procedure, this is equivalent to

$$\sqrt{2}u_1 - \qquad iu_3 = 0$$
$$\sqrt{2}u_2 + iu_3 = 0.$$

An eigenvector associated to $r_2 = -\sqrt{2}$ is

$$\mathbf{u}_2 = \begin{bmatrix} i \\ -i \\ \sqrt{2} \end{bmatrix}.$$

Similarly, we find that an eigenvector associated to $r_3 = \sqrt{2}$ is

$$\mathbf{u}_3 = \begin{bmatrix} i \\ -i \\ -\sqrt{2} \end{bmatrix}.$$

Since the matrix $\mathbf{A}$ is Hermitian, the vectors

$$\left\{ [1\ 1\ 0]^T, \left[i - i\sqrt{2} \right]^T, \left[i - i - \sqrt{2} \right]^T \right\},$$

are mutually orthogonal, as they correspond to distinct eigenvalues.

13. For $\mathbf{A} = \begin{bmatrix} 2 & 1 \\ 1+i & 1+2i \end{bmatrix}$, the characteristic polynomial is

$$|\mathbf{A} - r\mathbf{I}| = \begin{vmatrix} 2-r & 1 \\ 1+i & 1+2i-r \end{vmatrix} = 0$$

which is equivalent to $(r - i - 1)(r - i - 2) = 0$. Hence the eigenvalues are $r_1 = i + 1$, and $r_2 = i + 2$. An eigenvector corresponding to the eigenvalue $r_1 = 1 + i$ is $\mathbf{v}_1 = [1\ i - 1]^T$, while $\mathbf{v}_2 = [1\ i]^T$ is an eigenvector corresponding to the eigenvalue $r_2 = 2 + i$. The matrix

$$\mathbf{A}^H = \begin{bmatrix} 2 & 1-i \\ 1 & 1-2i \end{bmatrix}$$

has the eigenvalues $r_1 = 1 - i$ with the corresponding eigenvector $\mathbf{w}_1 = [1 - i]^T$, and $r_2 = 2 - i$ with the corresponding eigenvector $\mathbf{w}_2 = [2\ 1 - i]^T$. Notice that $\mathbf{A}$ and $\mathbf{A}^H$ have no common eigenvalues, and the eigenvectors of $\mathbf{A}$ are not orthogonal to those of $\mathbf{A}^H$.

For matrices $\mathbf{A}$ with only real eigenvalues, the matrices $\mathbf{A}$ and $\mathbf{A}^H$ have the same eigenvalues, and the eigenvectors corresponding to distinct eigenvalues are orthogonal.

15. Using Problem 14, Exercises 5.3 we know that changing the dot products to complex inner products in the Gram-Schmidt algorithm results in a set of vectors that are orthogonal in the sense of Definition 2.

In our case, let $\mathbf{v_1} = [i\,0\,0]$, $\mathbf{v_2} = [i\,1\,1]$ and $\mathbf{v_3} = [i\,i\,1]$. Notice that

$$\mathbf{v_1}\mathbf{v_1^H} = [i\,0\,0]\begin{bmatrix} -i \\ 0 \\ 0 \end{bmatrix} = i \cdot (-i) = 1,$$

so $\mathbf{v_1}$ is a unit vector. Let $\mathbf{w_1} = \mathbf{v_1}$ and consider the vector

$$\mathbf{w_2} = \mathbf{v_2} - (\mathbf{v_2}\mathbf{w_1^H})\mathbf{w_1} = [i\,1\,1] - [i\,0\,0] = [0\,1\,1],$$

as

$$\mathbf{v_2}\mathbf{w_1^H} = [i\,1\,1]\begin{bmatrix} -i \\ 0 \\ 0 \end{bmatrix} = 1.$$

Then $\mathbf{w_2}^{\text{unit}} = \frac{1}{\sqrt{2}}[0\,1\,1]$. Let

$$\mathbf{w_2} = \mathbf{v_3} - (\mathbf{v_3}\mathbf{w_1^H})\mathbf{w_1} - (\mathbf{v_3}\mathbf{w_2}^{\text{unit H}})\mathbf{w_2}^{\text{unit}}.$$

We compute

$$\mathbf{v_3}\mathbf{w_1^H} = [i\,i\,1]\begin{bmatrix} -i \\ 0 \\ 0 \end{bmatrix} = 1$$

and

$$\mathbf{v_3}\mathbf{w_2}^{\text{unit H}} = \frac{1}{\sqrt{2}}[i\,i\,1]\begin{bmatrix} 0 \\ 1 \\ 1 \end{bmatrix} = \frac{i+1}{\sqrt{2}}.$$

Hence

$$\mathbf{w}_2 = \mathbf{v}_3 - \mathbf{w}_1 - \frac{i+1}{\sqrt{2}}\mathbf{w}_2^{\text{unit}} = [i\,i\,1] - [i\,0\,0] - \frac{i+2}{2}[0\,1\,1]$$

$$= \left[0\,\frac{i+1}{2}\,\frac{1-i}{2}\right].$$

Notice now that $\mathbf{w}_2$ is a unit vector:

$$\mathbf{w}_2\mathbf{w}_2^{\text{H}} = \left[0\,\frac{i+1}{2}\,\frac{1-i}{2}\right]\begin{bmatrix} 0 \\ \dfrac{-i+1}{2} \\ \dfrac{1+i}{2} \end{bmatrix} = 0 + 1/2 + 1/2 = 1.$$

Therefore, an orthonormal set spanning $\mathbb{C}^3_{\text{row}}$ is

$$\left\{[i\,0\,0], \frac{1}{\sqrt{2}}[0\,1\,1], \frac{1}{\sqrt{3}}\left[0\,\frac{(i+1)}{2}\,\frac{(1-i)}{2}\right]\right\}.$$

17. The characteristic equation for the matrix $\mathbf{A} = \begin{bmatrix} 6i & -1-5i \\ -1-5i & 4i \end{bmatrix}$ is

$$|\mathbf{A} - r\mathbf{I}| = \begin{vmatrix} 6i - r & -1-5i \\ -1-5i & 4i - r \end{vmatrix} = 0,$$

which is equivalent to $r^2 - (10i)r - 10i = 0$ (see Problem 27, Exercises 5.2). This shows that

(a) The eigenvalues are not real. If the eigenvalues were two real numbers a and b, then the characteristic polynomial would be $(r - a)(r - b) = r^2 - (a + b)r + ab$, a polynomial with real coefficients, unlike $r^2 - (10i)r - 10i$.

(b) The eigenvalues are not complex conjugate. If the eigenvalues a pair of complex conjugate numbers $a \pm bi$, with a, b real numbers, then the characteristic polynomial would be $(r - a - bi)$ $(r - a + bi) = r^2 - 2ar + a^2 + b^2$, a polynomial with real coefficients, unlike $r^2 - (10i)r - 10i$.

(c) Let the eigenvalues of A be $r_{1,2} = a_{1,2} + b_{1,2}i$. Since the characteristic equation is $r^2 - (10i)r - 10i = 0$, we have $(a_1 + b_1 i) + (a_2 + b_2 i) = 10i$ and $(a_1 + b_1 i)(a_2 + b_2 i) = 10i$, which is equivalent to

$$a_1 + a_2 = 0, \ b_1 + b_2 = 10, \ a_1 a_2 - b_1 b_2 = 0 \text{ and}$$
$$a_1 b_2 + a_2 b_1 = -10. \tag{5.1}$$

To find the eigenvector corresponding to $r_1 = a_1 + b_1 i$, we must solve the equation $(A + (a_1 + b_1 i)I)u = 0$, which means

$$\begin{bmatrix} -a_1 + (6 - b_1)i & -1 - 5i \\ -1 - 5i & -a_1 + (4 - b_1)i \end{bmatrix} \begin{bmatrix} u_1 \\ u_2 \end{bmatrix} = \begin{bmatrix} 0 \\ 0 \end{bmatrix}.$$

As usual, this matrix equation is equivalent to the single scalar equation $(-a_1 + (6 - b_1)i)u_1 - (1 + 5i)u_2 = 0$. A particular solution is

$$v_1 = \begin{bmatrix} 1 + 5i \\ -a_1 + (6 - b_1)i \end{bmatrix}.$$

Similarly, an eigenvector corresponding to the eigenvalue $r_2 = a_2 + b_2 i$ is

$$v_2 = \begin{bmatrix} 1 + 5i \\ -a_1 + (6 - b_1)i \end{bmatrix}.$$

We compute now

$$v_1^H v_2 = \begin{bmatrix} 1 - 5i & -a_1 - (6 - b_1)i \end{bmatrix} \begin{bmatrix} 1 + 5i \\ -a_2 + (6 - b_2)i \end{bmatrix}$$
$$= (1 - 5i) \cdot (1 + 5i)$$
$$\quad + (-a_1 - (6 - b_1)i) \cdot (-a_2 + (6 - b_2)i)$$
$$= 26 + a_1 a_2 + (6 - b_1)(6 - b_2)$$
$$\quad + (-a_1(6 - b_2) + a_2(6 - b_1))\, i$$
$$= 2 + a_1 a_2 + b_1 b_2 + (-6(a_1 - a_2) + (a_1 b_2 - b_1 a_2))\, i.$$

If $v_1^H v_2 = 0$, its real and imaginary parts computed above must vanish. Using (5.1), the vanishing of the imaginary part is equivalent to $a_1 = 0$. But then r_1 is purely imaginary, and since $r_1 r_2 = -10i$ it follows that r_2 is real, which is impossible.

(d) Since the matrix $\mathbf{A}$ is symmetric, the proof of formula (2) in Section 5.3 remains valid, regardless the matrix is complex or not. As a consequence, the relation $\mathbf{v_1}^T \mathbf{v_2}$ holds true.

6

SIMILARITY

6.1 SIMILARITY TRANSFORMATIONS AND DIAGONALIZABILITY

1. If $\mathbf{A} = \mathbf{P}^{-1}\mathbf{B}\mathbf{P}$, then

$$
\begin{aligned}
\mathbf{A}^n &= (\mathbf{P}^{-1}\mathbf{B}\mathbf{P})(\mathbf{P}^{-1}\mathbf{B}\mathbf{P})\cdots(\mathbf{P}^{-1}\mathbf{B}\mathbf{P}) \\
&= \mathbf{P}^{-1}\mathbf{B}(\mathbf{P}\mathbf{P}^{-1})\mathbf{B}(\mathbf{P}\mathbf{P}^{-1})\cdots(\mathbf{P}\mathbf{P}^{-1})\mathbf{B}\mathbf{P} \\
&= \mathbf{P}^{-1}\mathbf{B}\mathbf{I}\mathbf{B}\mathbf{I}\cdots\mathbf{I}\mathbf{B}\mathbf{P} \\
&= \mathbf{P}^{-1}\mathbf{B}^n\mathbf{P}.
\end{aligned}
$$

3. Let $\mathbf{A} = \mathbf{P}^{-1}\mathbf{D}\mathbf{P}$, where $\mathbf{D}$

$$
\mathbf{D} = \begin{bmatrix} a_1 & 0 & \cdots & 0 \\ 0 & a_2 & \ddots & \vdots \\ \vdots & \ddots & \ddots & 0 \\ 0 & \cdots & 0 & a_n \end{bmatrix}
$$

is a diagonal matrix. Let a be the only eigenvalue of $\mathbf{A}$. Since $\mathbf{A}$ and $\mathbf{D}$ are similar, they have the same eigenvalues. Notice now that the

Solutions Manual to Accompany Fundamentals of Matrix Analysis with Applications,
First Edition. Edward Barry Saff and Arthur David Snider.
© 2016 John Wiley & Sons, Inc. Published 2016 by John Wiley & Sons, Inc.

eigenvalues of $\mathbf{D}$ are $a_1, a_2, \ldots, a_n$, and so $a_1 = a_2 = \cdots = a_n = a$. This means $\mathbf{D} = a\mathbf{I}$. In this case, we can see that $\mathbf{A} = \mathbf{P}^{-1}\mathbf{D}\mathbf{P} = \mathbf{P}^{-1}(a\mathbf{I})\mathbf{P} = a\mathbf{P}^{-1}\mathbf{I}\mathbf{P} = a\mathbf{I}$.

5. We notice first that any two diagonal matrices with the same diagonal entries are similar. To prove this claim, it suffices to observe that multiplication to the left and to the right by the elementary matrix obtained from $\mathbf{I}$ by swapping two rows, rearranges the diagonal entries. Now, if $\mathbf{A}$ and $\mathbf{B}$ are two diagonalizable matrices with the same eigenvalues, their associated diagonal matrices $\mathbf{D}_1$ and $\mathbf{D}_2$ have the same eigenvalues. In particular, $\mathbf{D}_1$ and $\mathbf{D}_2$ have the same diagonal entries, and so they are similar. By transitivity, $\mathbf{A}$ and $\mathbf{B}$ are similar.

7. If $\mathbf{A}$ is invertible, then

$$\mathbf{BA} = (\mathbf{A}^{-1}\mathbf{A})\mathbf{BA} = \mathbf{A}^{-1}(\mathbf{AB})\mathbf{A},$$

which shows that $\mathbf{AB}$ and $\mathbf{BA}$ are similar. If $\mathbf{B}$ is invertible, then

$$\mathbf{AB} = (\mathbf{B}^{-1}\mathbf{B})\mathbf{AB} = \mathbf{B}^{-1}(\mathbf{BA})\mathbf{B},$$

which shows that $\mathbf{AB}$ and $\mathbf{BA}$ are similar.

If neither $\mathbf{A}$ or $\mathbf{B}$ is invertible, pick

$$\mathbf{A} = \begin{bmatrix} 0 & 1 \\ 0 & 0 \end{bmatrix} \quad \text{and} \quad \mathbf{B} = \begin{bmatrix} 1 & 0 \\ 0 & 0 \end{bmatrix}.$$

Then $\mathbf{AB} = \begin{bmatrix} 0 & 1 \\ 0 & 0 \end{bmatrix}\begin{bmatrix} 1 & 0 \\ 0 & 0 \end{bmatrix} = \begin{bmatrix} 0 & 0 \\ 0 & 0 \end{bmatrix}$, while $\mathbf{BA} = \begin{bmatrix} 1 & 0 \\ 0 & 0 \end{bmatrix}\begin{bmatrix} 0 & 1 \\ 0 & 0 \end{bmatrix} = \begin{bmatrix} 0 & 1 \\ 0 & 0 \end{bmatrix}$. We notice immediately that the matrices $\begin{bmatrix} 0 & 0 \\ 0 & 0 \end{bmatrix}$ and $\begin{bmatrix} 0 & 1 \\ 0 & 0 \end{bmatrix}$ are not similar.

9. The eigenvalues $r_1 = 3$ and $r_2 = 10$ were computed in Problem 1, Exercises 5.2, together with their corresponding eigenvectors. By setting $s = t = 1$ we obtain the following eigenvectors eigenvectors:

$$\mathbf{u}_1 = \begin{bmatrix} 1 \\ 2 \end{bmatrix} \quad \text{and} \quad \mathbf{u}_2 = \begin{bmatrix} 3 \\ -1 \end{bmatrix}$$

These vectors are the columns of the similarity transformation matrix

$$\mathbf{P} = \begin{bmatrix} 1 & 3 \\ 2 & -1 \end{bmatrix}.$$

The diagonal matrix D has $r_1 = 2$ and $r_2 = 10$ as the entries on the main diagonal

$$\mathbf{D} = \begin{bmatrix} 3 & 0 \\ 0 & 10 \end{bmatrix}.$$

We have now:

$$\begin{bmatrix} 9 & -3 \\ -2 & 4 \end{bmatrix}^{-1} = \begin{bmatrix} 1 & 3 \\ 2 & -1 \end{bmatrix} \begin{bmatrix} 1/3 & 0 \\ 0 & 1/10 \end{bmatrix} \begin{bmatrix} 1 & 3 \\ 2 & -1 \end{bmatrix}^{-1}$$

$$e^{\begin{bmatrix} 9 & -3 \\ -2 & 4 \end{bmatrix}} = \begin{bmatrix} 1 & 3 \\ 2 & -1 \end{bmatrix} \begin{bmatrix} e^3 & 0 \\ 0 & e^{10} \end{bmatrix} \begin{bmatrix} 1 & 3 \\ 2 & -1 \end{bmatrix}^{-1}.$$

11. The eigenvalues $r_1 = -1$ and $r_2 = 3$ were computed in Problem 3, Exercises 5.2, together with their corresponding eigenvectors. By setting $s = t = 1$ we obtain the following eigenvectors eigenvectors:

$$\mathbf{u_1} = \begin{bmatrix} 1 \\ 1 \end{bmatrix} \quad \text{and} \quad \mathbf{u_2} = \begin{bmatrix} 1 \\ -1 \end{bmatrix}$$

These vectors are the columns of the similarity transformation matrix

$$\mathbf{P} = \begin{bmatrix} 1 & 1 \\ 1 & -1 \end{bmatrix}.$$

The diagonal matrix D has $r_1 = -1$ and $r_2 = 3$ as the entries on the main diagonal

$$\mathbf{D} = \begin{bmatrix} -1 & 0 \\ 0 & 3 \end{bmatrix}.$$

We have now:

$$\begin{bmatrix} 1 & -2 \\ -2 & 1 \end{bmatrix}^{-1} = \begin{bmatrix} 1 & 1 \\ 1 & -1 \end{bmatrix} \begin{bmatrix} -1 & 0 \\ 0 & 1/3 \end{bmatrix} \begin{bmatrix} 1 & 1 \\ 1 & -1 \end{bmatrix}^{-1}$$

$$e^{\begin{bmatrix} 1 & -2 \\ -2 & 1 \end{bmatrix}} = \begin{bmatrix} 1 & 1 \\ 1 & -1 \end{bmatrix} \begin{bmatrix} e^{-1} & 0 \\ 0 & e^3 \end{bmatrix} \begin{bmatrix} 1 & 1 \\ 1 & -1 \end{bmatrix}^{-1}.$$

13. The eigenvalues $r_1 = 0$ and $r_2 = 2$ were computed in Problem 5, Exercises 5.2, together with their corresponding eigenvectors. By setting $s = t = 1$ we obtain the following eigenvectors eigenvectors:

$$\mathbf{u_1} = \begin{bmatrix} 1 \\ -1 \end{bmatrix} \quad \text{and} \quad \mathbf{u_2} = \begin{bmatrix} 1 \\ 1 \end{bmatrix}$$

These vectors are the columns of the similarity transformation matrix
$\mathbf{P} = \begin{bmatrix} 1 & 1 \\ -1 & 1 \end{bmatrix}$.

The diagonal matrix D has $r_1 = 0$ and $r_2 = 2$ as the entries on the main diagonal

$$\mathbf{D} = \begin{bmatrix} 0 & 0 \\ 0 & 2 \end{bmatrix}.$$

Notice that the given matrix is singular (0 is one of its eigenvalues!). We have now:

$$e^{\begin{bmatrix} 1 & 1 \\ 1 & 1 \end{bmatrix}} = \begin{bmatrix} 1 & 1 \\ -1 & 1 \end{bmatrix} \begin{bmatrix} 1 & 0 \\ 0 & e^2 \end{bmatrix} \begin{bmatrix} 1 & 1 \\ -1 & 1 \end{bmatrix}^{-1}.$$

15. The eigenvalues $r_1 = 1$, $r_2 = 2$ and $r_3 = 3$ were computed in Problem 7, Exercises 5.2, together with their corresponding eigenvectors. By setting $s = 1$ we obtain the following eigenvectors eigenvectors:

$$\mathbf{u_1} = \begin{bmatrix} 1 \\ 1 \\ 1 \end{bmatrix}, \quad \mathbf{u_2} = \begin{bmatrix} 1 \\ 0 \\ 1 \end{bmatrix} \quad \text{and} \quad \mathbf{u_3} = \begin{bmatrix} 1 \\ 1 \\ 0 \end{bmatrix}.$$

These vectors are the columns of the similarity transformation matrix

$$\mathbf{P} = \begin{bmatrix} 1 & 1 & 1 \\ 1 & 0 & 1 \\ 1 & 1 & 0 \end{bmatrix}.$$

The diagonal matrix D has $r_1 = 1$, $r_2 = 2$ and $r_3 = 3$ as the entries on the main diagonal

$$D = \begin{bmatrix} 1 & 0 & 0 \\ 0 & 2 & 0 \\ 0 & 0 & 3 \end{bmatrix}.$$

We have now:

$$\begin{bmatrix} 4 & -1 & -2 \\ 2 & 1 & -2 \\ 1 & -1 & 1 \end{bmatrix}^{-1}$$

$$= \begin{bmatrix} 1 & 1 & 1 \\ 1 & 0 & 1 \\ 1 & 1 & 0 \end{bmatrix} \begin{bmatrix} 1 & 0 & 0 \\ 0 & 1/2 & 0 \\ 0 & 0 & 1/3 \end{bmatrix} \begin{bmatrix} 1 & 1 & 1 \\ 1 & 0 & 1 \\ 1 & 1 & 0 \end{bmatrix}^{-1}.$$

$$e^{\begin{bmatrix} 4 & -1 & -2 \\ 2 & 1 & -2 \\ 1 & -1 & 1 \end{bmatrix}^{-1}}$$

$$= \begin{bmatrix} 1 & 1 & 1 \\ 1 & 0 & 1 \\ 1 & 1 & 0 \end{bmatrix} \begin{bmatrix} e & 0 & 0 \\ 0 & e^2 & 0 \\ 0 & 0 & e^3 \end{bmatrix} \begin{bmatrix} 1 & 1 & 1 \\ 1 & 0 & 1 \\ 1 & 1 & 0 \end{bmatrix}^{-1}.$$

17. The eigenvalues $r_1 = i$ and $r_2 = -i$ were computed in Problem 9, Exercises 5.2, together with their corresponding eigenvectors. By setting $s = t = 1$ we obtain the following eigenvectors eigenvectors:

$$\mathbf{u_1} = \begin{bmatrix} 1 \\ -i \end{bmatrix} \quad \text{and} \quad \mathbf{u_2} = \begin{bmatrix} 1 \\ i \end{bmatrix}$$

These vectors are the columns of the similarity transformation matrix
$$\mathbf{P} = \begin{bmatrix} 1 & 1 \\ -i & i \end{bmatrix}.$$
The diagonal matrix D has $r_1 = i$ and $r_2 = -i$ as the entries on the main diagonal

$$\begin{bmatrix} i & 0 \\ 0 & -i \end{bmatrix}.$$

We have now:

$$\begin{bmatrix} 0 & -1 \\ 1 & 0 \end{bmatrix}^{-1} = \begin{bmatrix} 1 & 1 \\ -i & i \end{bmatrix} \begin{bmatrix} -i & 0 \\ 0 & i \end{bmatrix} \begin{bmatrix} 1 & 1 \\ -i & i \end{bmatrix}^{-1}$$

$$e^{\begin{bmatrix} 0 & -1 \\ 1 & 0 \end{bmatrix}} = \begin{bmatrix} 1 & 1 \\ -i & i \end{bmatrix} \begin{bmatrix} e^i & 0 \\ 0 & e^{-i} \end{bmatrix} \begin{bmatrix} 1 & 1 \\ -i & i \end{bmatrix}^{-1}.$$

19. The eigenvalues $r_1 = 3$ and $r_2 = i + 1$ were computed in Problem 9, Exercises 5.2, together with their corresponding eigenvectors. By setting $s = t = 1$ we obtain the following eigenvectors eigenvectors:

$$\mathbf{u_1} = \begin{bmatrix} i \\ 1 \end{bmatrix} \quad \text{and} \quad \mathbf{u_2} = \begin{bmatrix} 2 \\ 1 \end{bmatrix}$$

These vectors are the columns of the similarity transformation matrix $\mathbf{P} = \begin{bmatrix} i & 2 \\ 1 & 1 \end{bmatrix}$.

The diagonal matrix D has $r_1 = 3$ and $r_2 = i + 1$ as the entries on the main diagonal

$$\mathbf{D} = \begin{bmatrix} 3 & 0 \\ 0 & i+1 \end{bmatrix}.$$

We have now:

$$\begin{bmatrix} 1 & 2i \\ -1 & 3+i \end{bmatrix}^{-1} = \begin{bmatrix} i & 2 \\ 1 & 1 \end{bmatrix} \begin{bmatrix} 1/3 & 0 \\ 0 & 1/(i+1) \end{bmatrix} \begin{bmatrix} i & 2 \\ 1 & 1 \end{bmatrix}^{-1}$$

$$e^{\begin{bmatrix} 1 & 2i \\ -1 & 3+i \end{bmatrix}} = \begin{bmatrix} i & 2 \\ 1 & 1 \end{bmatrix} \begin{bmatrix} e^3 & 0 \\ 0 & e^{i+1} \end{bmatrix} \begin{bmatrix} i & 2 \\ 1 & 1 \end{bmatrix}^{-1}.$$

21. Let $\mathbf{A} = \begin{bmatrix} 0 & a \\ 0 & 0 \end{bmatrix}$. We have

$$\begin{bmatrix} 0 & a \\ 0 & 0 \end{bmatrix} \begin{bmatrix} 0 & a \\ 0 & 0 \end{bmatrix} = \begin{bmatrix} 0 & 0 \\ 0 & 0 \end{bmatrix}$$

This shows that $\mathbf{A}$ is a square root of the zero matrix for every value of a.

Suppose now the matrix $\mathbf{A}$ has a square root $\mathbf{B}$, and $a \neq 0$. Let $\mathbf{B} = \begin{bmatrix} x & y \\ z & t \end{bmatrix}$. The equation $\mathbf{B}^2 = \mathbf{A}$ yields

$$\begin{cases} x^2 + yz = 0 \\ t^2 + yz = 0 \\ z(x+t) = 0 \\ y(x+t) = 0. \end{cases}$$

From the first two equations, we see that $x^2 = t^2$, that means $x = \pm t$. If $x = -t$, from the last equation, we get $0 = a$, which is impossible. If $x = t$, the third equation implies $zx = 0$. If $x = 0$, then $t = 0$, and the last equation yields again, $0 = a$, which is impossible. Finally, if $z = 0$, the first equation implies $x = 0$, and we reach a contradiction, as above.

23. We compute

$$\mathbf{A} = \mathbf{Q}\mathbf{D}\mathbf{Q}^{\mathrm{T}} = \mathbf{Q} \begin{bmatrix} r_1 & 0 & \cdots & 0 \\ 0 & r_2 & \cdots & 0 \\ \vdots & \vdots & \ddots & \vdots \\ 0 & 0 & \cdots & r_n \end{bmatrix} \mathbf{Q}^{\mathrm{T}}$$

$$= \begin{bmatrix} \mathbf{Q} \begin{bmatrix} r_1 \\ 0 \\ \vdots \\ 0 \end{bmatrix} & \mathbf{Q} \begin{bmatrix} 0 \\ r_2 \\ \vdots \\ 0 \end{bmatrix} \cdots \mathbf{Q} \begin{bmatrix} 0 \\ 0 \\ \vdots \\ r_n \end{bmatrix} \end{bmatrix} \mathbf{Q}^{\mathrm{T}}$$

$$= [r_1\mathbf{u_1} \ r_2\mathbf{u_2} \ \cdots \ r_n\mathbf{u_1}] \, \mathbf{Q}^{\mathrm{T}}$$

$$= r_1\mathbf{u_1}\mathbf{u_1^{\mathrm{T}}} + r_2\mathbf{u_2}\mathbf{u_2^{\mathrm{T}}} + \cdots r_n\mathbf{u_n}\mathbf{u_n^{\mathrm{T}}}.$$

For Hermitian matrices, the same proof yields

$$\mathbf{A} = r_1\mathbf{u_1}\bar{\mathbf{u}}_1^{\mathrm{T}} + r_1\mathbf{u_1}\bar{\mathbf{u}}_1^{\mathrm{T}} + \cdots + r_1\mathbf{u_1}\bar{\mathbf{u}}_1^{\mathrm{T}}.$$

The result is not valid for non-hermitian matrices. One can easily see that the matrix from exercise 9, Section 6.1 does not have such property.

25. If $\mathbf{v_i}$ is the ith column of the Vandermonde matrix $\mathbf{V}_q$, we compute

$$
\mathbf{C}_q\mathbf{v_i} = \begin{bmatrix} 0 & 1 & 0 & \cdots & 0 \\ 0 & 0 & 1 & \cdots & 0 \\ & & \vdots & & \\ 0 & 0 & 0 & \cdots & 1 \\ -a_0 & -a_1 & -a_2 & \cdots & -a_{n-1} \end{bmatrix} \begin{bmatrix} 1 \\ r_i \\ r_i^2 \\ \vdots \\ r_i^{n-1} \end{bmatrix}
$$

$$
= \begin{bmatrix} r_i \\ r_i^2 \\ r_i^3 \\ \vdots \\ -\sum_{i=0}^{n-1} a_i r_i^{n-1} \end{bmatrix}
$$

$$
= \begin{bmatrix} r_i \\ r_i^2 \\ r_i^3 \\ \vdots \\ r_i^n \end{bmatrix} = r_i\mathbf{v_i}.
$$

This means not only that the vectors $\mathbf{v_i}$ are eigenvectors of $\mathbf{C}_q$ corresponding to the eigenvalues r_i for every $i = 1, \ldots, n$, but we also have (see Lemma 1 in Section 6.1) $\mathbf{C}_q\mathbf{V}_q = \mathbf{V}_q\mathbf{D}$, where

$$
\mathbf{D} = \begin{bmatrix} r_1 & 0 & \cdots & 0 \\ 0 & r_2 & & 0 \\ \vdots & & \ddots & \vdots \\ 0 & 0 & \cdots & r_n \end{bmatrix},
$$

which yields $\mathbf{V}_q^{-1}\mathbf{C}_q\mathbf{V}_q = \mathbf{D}$.

27. We apply the result from Exercise 26, Section 6.1. We have

$$
\|\mathbf{Q_1AQ_2}\|_{\text{Frob}} = \|\mathbf{Q_1(AQ_2)}\|_{\text{Frob}} = \|\mathbf{AQ_2}\|_{\text{Frob}}
$$
$$
= \|\mathbf{A(Q_2^T)^T}\|_{\text{Frob}} = \|\mathbf{A}\|_{\text{Frob}}.
$$

The last equality follows from the fact that $\mathbf{Q_2^T}$ is an orthogonal matrix, as well.

The complex version of this statement is that the Hermitian Frobenius norm is preserved under left and right multiplication by a unitary matrix.

6.2 PRINCIPAL AXES AND NORMAL MODES

1. (a)

$$[x \quad y] \begin{bmatrix} a & c/2 \\ c/2 & b \end{bmatrix} \begin{bmatrix} x \\ y \end{bmatrix} = [ax + (c/2)y \quad (c/2)x + by] \begin{bmatrix} x \\ y \end{bmatrix}$$

$$= ax^2 + (c/2)xy + (c/2)xy + by^2$$

$$= ax^2 + by^2 + cxy.$$

(b)

$$\mathbf{y}^T \begin{bmatrix} 1 & 2 \\ 2 & 3 \end{bmatrix} \mathbf{y} - 3$$

$$= (\mathbf{x}^T + [0 \quad 1]) \begin{bmatrix} 1 & 2 \\ 2 & 3 \end{bmatrix} \left(\mathbf{x} + \begin{bmatrix} 0 \\ 1 \end{bmatrix} \right) - 3$$

$$= \mathbf{x}^T \begin{bmatrix} 1 & 2 \\ 2 & 3 \end{bmatrix} \mathbf{x} + \mathbf{x}^T \begin{bmatrix} 1 & 2 \\ 2 & 3 \end{bmatrix} \begin{bmatrix} 0 \\ 1 \end{bmatrix}$$

$$+ [0 \quad 1] \begin{bmatrix} 1 & 2 \\ 2 & 3 \end{bmatrix} \mathbf{x} + [0 \quad 1] \begin{bmatrix} 1 & 2 \\ 2 & 3 \end{bmatrix} \begin{bmatrix} 0 \\ 1 \end{bmatrix} - 3$$

$$= \mathbf{x}^T \begin{bmatrix} 1 & 2 \\ 2 & 3 \end{bmatrix} \mathbf{x} + \mathbf{x}^T \begin{bmatrix} 2 \\ 3 \end{bmatrix} + [2 \quad 3] \mathbf{x} + \begin{bmatrix} 2 \\ 3 \end{bmatrix} \begin{bmatrix} 0 \\ 1 \end{bmatrix} - 3$$

$$= \mathbf{x}^T \begin{bmatrix} 1 & 2 \\ 2 & 3 \end{bmatrix} \mathbf{x} + (2x_1 + 3x_2) + (2x_1 + 3x_2) + 3 - 3$$

$$= [x_1 \quad 2] \begin{bmatrix} 1 & 2 \\ 2 & 3 \end{bmatrix} \begin{bmatrix} x_1 \\ x_2 \end{bmatrix} + [4 \quad 6] \begin{bmatrix} x_1 \\ x_2 \end{bmatrix}$$

3. The symmetric matrix corresponding to the quadratic form is

$$\mathbf{A} = \begin{bmatrix} -7 & 6 \\ 6 & 2 \end{bmatrix}.$$

In Problem 1, Exercises 5.3, we found that $\mathbf{A}$ has eigenvalues -10 and 5, with the corresponding vectors $\mathbf{u}_1 = [2 \; -1]^T$, and $\mathbf{u}_1 = [1 \; 2]^T$, respectively. Moreover, we proved that $\mathbf{u}_1$ and $\mathbf{u}_2$ are orthogonal. Rescaling the vectors $\mathbf{u}_1$ and $\mathbf{u}_2$ by $1/\sqrt{5}$, they become orthonormal eigenvectors of the matrix $\mathbf{A}$. Consequently, the matrix

$$\mathbf{Q} = \frac{1}{\sqrt{5}} \begin{bmatrix} -2 & 1 \\ 1 & 2 \end{bmatrix}$$

yields a change of coordinates that transforms the given bilinear form into $-10y_1^2 + 5y_2^2$. Since multiplication by orthogonal matrices preserves lengths (Euclidean norms), the set of points where $x_1^2 + x_2^2 = 1$ is the same as the set where $y_1^2 + y_2^2 = 1$. It is not hard to see now that the maximum value of $-10y_1^2 + 5y_2^2 = 5 - 15y_1^2$ along the unit circle is 5, while the minimum value is -10, the largest and the smallest eigenvalues, respectively.

5. The symmetric matrix corresponding to the quadratic form is

$$\mathbf{A} = \begin{bmatrix} 1 & 4 & 3 \\ 4 & 1 & 0 \\ 3 & 0 & 1 \end{bmatrix}.$$

In Problem 3, Exercises 5.3, we found that $\mathbf{A}$ has eigenvalues $1, -4$ and 6, with the corresponding vectors $\mathbf{u}_1 = [0 \; 3 \; -4]^T$, $\mathbf{u}_2 = [5 \; -4 \; -3]^T$ and $\mathbf{u}_3 = [5 \; 4 \; 3]^T$, respectively. Moreover, we proved that $\mathbf{u}_1, \mathbf{u}_2$ and $\mathbf{u}_3$ are orthogonal. Consequently, the matrix

$$\mathbf{Q} = \frac{\sqrt{2}}{10} \begin{bmatrix} 0 & 5 & 5 \\ 3 & -4 & 4 \\ -4 & -3 & 3 \end{bmatrix}$$

provides a change of coordinates that transforms the given bilinear form into $y_1^2 - 4y_2^2 + 6y_3^2$. Since multiplication by orthogonal matrices preserves lengths (Euclidean norms), the set of points where $x_1^2 + x_2^2 + x_3^2 = 1$ is the same as the set where $y_1^2 + y_2^2 + y_3^2 = 1$. We see now that the maximum value of $y_1^2 - 4y_2^2 + 6y_3^2 = 1 - 5(y_2^2 - y_3^2)$ along the unit sphere is 6, while the minimum value is -4, the largest and the smallest eigenvalues, respectively.

7. The symmetric matrix corresponding to the quadratic form is

$$
\mathbf{A} = \begin{bmatrix} 2 & 1 & 1 \\ 1 & 2 & -1 \\ 1 & -1 & 2 \end{bmatrix}.
$$

In Problem 5, Exercises 5.3, we found that $\mathbf{A}$ has eigenvalues 0 and 3, with the and three orthogonal eigenvectors corresponding vectors $\mathbf{u_1} = [1\ -1\ -1]^T$, $\mathbf{u_2} = [1\ 1\ 0]^T$ and $\mathbf{u_3} = [1\ -1\ 1]^T$, respectively. Rescaling the vectors $\mathbf{u_1}$ by $1/\sqrt{3}$, $\mathbf{u_2}$ by $1/\sqrt{2}$ and $\mathbf{u_3}$ by $1/\sqrt{6}$, they become orthonormal eigenvectors of the matrix $\mathbf{A}$. Consequently, the matrix

$$
\mathbf{Q} = \begin{bmatrix} 1/\sqrt{3} & 1/\sqrt{2} & 1/\sqrt{6} \\ -1/\sqrt{3} & 1/\sqrt{2} & -1/\sqrt{6} \\ -1/\sqrt{3} & 0 & 2/\sqrt{6} \end{bmatrix}
$$

yields a change of coordinates that transforms the given bilinear form into $3y_2^2 + 3y_3^2$. Since multiplication by orthogonal matrices preserves lengths (Euclidean norms), the set of points where $x_1^2 + x_2^2 + x_3^2 = 1$ is the same as the set where $y_1^2 + y_2^2 + y_3^2 = 1$. We see now that the maximum value of $3y_2^2 + 3y_3^2 = 3 - 3y_1^2$ along the unit sphere is 3, while the minimum value is 0, the largest and the smallest eigenvalues, respectively.

9. (a) From the Principal Axes Theorem, we can find an orthogonal matrix $\mathbf{Q}$ such that

$$
\mathbf{v}^T \mathbf{A} \mathbf{v} = \mathbf{x}^T \mathbf{A} \mathbf{x} = r_1 x_1^2 + r_2 x_2^2 + \cdots + r_n x_n^2,
$$

where $\mathbf{v} = \mathbf{Q}\mathbf{x}$. Notice that the vectors $\mathbf{x}$ and $\mathbf{v}$ have the same length (=1), since the matrix $\mathbf{Q}$ is orthogonal. Moreover, the ith column of $\mathbf{Q}$ is an eigenvector of $\mathbf{A}$ corresponding to the eigenvalue r_i. We have

$$
r_1 x_1^2 + r_2 x_2^2 + \cdots + r_n x_n^2 \le r_1 (x_1^2 + x_2^2 + \cdots + x_n^2) = r_1,
$$

with equality if and only if $x_2 = x_3 = \cdots = x_n = 0$, in which case $x_1 = \pm 1$. Therefore, at a maximum point $\mathbf{x} = [\pm 1\ 0\ \ldots\ 0]^T$, we see that $\mathbf{v} = \mathbf{Q}\mathbf{x}$ is $\pm$ the first column of $\mathbf{Q}$, that is, the eigenvector corresponding to the eigenvalue r_1.

(b) As in part (a), we write

$$\mathbf{v}^T\mathbf{A}\mathbf{v} = \mathbf{x}^T\mathbf{A}\mathbf{x} = r_1x_1^2 + r_2x_2^2 + \cdots + r_nx_n^2,$$

where $\mathbf{v} = \mathbf{Q}\mathbf{x}$ is a unit vector orthogonal to the first column of $\mathbf{Q}$. That means $x_1 = 0$, where $\mathbf{x} = [x_1 \; x_2 \; \ldots \; x_n]^T$. Therefore

$$\mathbf{x}^T\mathbf{A}\mathbf{x} = r_2x_2^2 + \cdots + r_nx_n^2,$$

and we argue as above.

(c) The bilinear form $\mathbf{v}^T\mathbf{A}\mathbf{v}$ achieves its maximum value, over all unit vectors $\mathbf{v}$ orthogonal to the set of eigenvectors $\mathbf{u}_1, \mathbf{u}_2, \ldots, \mathbf{u}_{i-1}$, when $\mathbf{v}$ is an eigenvector $\mathbf{u}_i$ corresponding to the ith largest eigenvalue r_i; and that this maximum value is r_2. The proof is similar to the one in part (b) above, with minor modifications.

(d) We have

$$\frac{\mathbf{v}^T\mathbf{A}\mathbf{v}}{\mathbf{v}^T\mathbf{v}} = \frac{\mathbf{x}^T\mathbf{A}\mathbf{x}}{\mathbf{x}^T\mathbf{x}} = \frac{r_1x_1^2 + r_2x_2^2 + \cdots + r_nx_n^2}{x_1^2 + x_2^2 + \cdots + x_n^2}$$
$$\leq \frac{r_1(x_1^2 + x_2^2 + \cdots + x_n^2)}{x_1^2 + x_2^2 + \cdots + x_n^2} = r_1.$$

We continue the rest of the arguments as in parts (a), (b) and (c) above.

11. The symmetric matrix corresponding to the quadratic form is

$$\mathbf{A} = (0.5) \begin{bmatrix} 0 & 1 & \ldots & 1 \\ 1 & 0 & \ldots & 1 \\ \vdots & \vdots & \ddots & \vdots \\ 1 & 1 & \ldots & 0 \end{bmatrix}.$$

One can either compute inductively the characteristic equation of $\mathbf{A}$, which is $(r - n + 1)(r + 1)^{n-1} = 0$, to conclude that the eigenvalues of $\mathbf{A}$ are $r_1 = (n-1)$, of algebraic multiplicity 1, and $r_2 = -1$, with algebraic multiplicity $(n-1)$.

Alternatively, we can reach the same conclusion using the theory of circulant matrices from Problem 14, Exercises 5.1 in the following way. Let $f(x) = x + x^2 + \cdots x^{n-1}$ be the degree n polynomial associated to the circulant matrix (whose coefficients consist of the first row entries), and let $\omega_j = e^{\frac{2\pi i j}{n}}$, be the nth roots of unity (here $i^2 = -1$, and $j = 0, 1, \ldots, n$). Then, the eigenvalues of $\mathbf{A}$ are

$$r_j = f(\omega^j)/2.$$

However, since $\omega_j^n = 1$, then either $j = 0$, in which case $\omega_0 = 1$, or else $1 + \omega_j + \cdots + \omega_j^{n-1} = 0$. It follows immediately that the eigenvalues of $\mathbf{A}$ are $r_1 = (n-1)$, of algebraic multiplicity 1, and $r_2 = -1$, with algebraic multiplicity $(n-1)$.

Using now the theory of the Raleigh quotient from Problem 9, Exercises 6.2 above, it follows that the maximum of the given bilinear form, when $x_1^2 + x_2^2 + \cdots + x_n^2 = 1$, equals the largest eigenvalue, which is $(n - 1/2)$.

13. (a) Notice first from the fact that $\mathbf{A}$ is diagonalizable, and 1 is not one of its eigenvalues that $\det(\mathbf{I} - \mathbf{A}) \neq 0$. This shows that $(\mathbf{I} - \mathbf{A})^{-1}$ exists, and so $\mathbf{x_n} = (\mathbf{I} - \mathbf{A})^{-1}\mathbf{b}$ is well-defined. We have

$$\begin{aligned}
\mathbf{A}\mathbf{x_n} + \mathbf{b} &= \mathbf{A}(\mathbf{I} - \mathbf{A})^{-1}\mathbf{b} + \mathbf{b} \\
&= \mathbf{A}(\mathbf{I} - \mathbf{A})^{-1}\mathbf{b} - (\mathbf{I} - \mathbf{A})^{-1}\mathbf{b} + (\mathbf{I} - \mathbf{A})^{-1}\mathbf{b} + \mathbf{b} \\
&= (\mathbf{A} - \mathbf{I})(\mathbf{I} - \mathbf{A})^{-1}\mathbf{b} + (\mathbf{I} - \mathbf{A})^{-1}\mathbf{b} + \mathbf{b} \\
&= -\mathbf{b} + (\mathbf{I} - \mathbf{A})^{-1}\mathbf{b} + \mathbf{b} \\
&= (\mathbf{I} - \mathbf{A})^{-1}\mathbf{b} \\
&= \mathbf{x_n}.
\end{aligned}$$

Hence, $\mathbf{x_n}$ is a stationary solution.

(b) We argue by induction. Notice first that $\mathbf{x_1} = \mathbf{A}\mathbf{x_0} + \mathbf{b}$, and so

$$\mathbf{x_2} = \mathbf{A}\mathbf{x_1} + \mathbf{b} = \mathbf{A}(\mathbf{A}\mathbf{x_0} + \mathbf{b}) + \mathbf{b} = \mathbf{A}^2\mathbf{x_0} + (\mathbf{A} + \mathbf{I})\mathbf{b}.$$

Assume now that $x_n = A^n x_0 + (A^{n-1} + A^{n-2} + \cdots + A + I)b$. We have

$$x_{n+1} = Ax_n + b$$
$$= A\left(A^n x_0 + (A^{n-1} + A^{n-2} + \cdots + A + I)b\right) + b$$
$$= A^{n+1}x_0 + (A^n + A^{n-1} + \cdots + A + I)b.$$

(c) Notice that we have $x_n = A^n x_0 + (I - A^n)(I - A)^{-1}b$. Now, if D is a diagonal form of A, and its entries are all less than 1, then $\lim_{n \to \infty} D^n = 0$. Therefore, $\lim_{n \to \infty} A^n = 0$, and we have

$$\lim_{n \to \infty} x_n = \lim_{n \to \infty} \left(A^n x_0 + (I - A^n)(I - A)^{-1}b\right) = (I - A)^{-1}b.$$

(d) $B y_n = \begin{bmatrix} A & \vdots & b \\ 0 & \vdots & 1 \end{bmatrix} \begin{bmatrix} x_n \\ 1 \end{bmatrix} = \begin{bmatrix} Ax_n + b \\ 1 \end{bmatrix} \begin{bmatrix} x_{n+1} \\ 1 \end{bmatrix} = y_{n+1}.$

(e) Let $y_{n+1} = \begin{bmatrix} x_{n+1} \\ z_{n+1} \end{bmatrix}$ satisfying $y_{n+1} = B y_n$, where x_n is $m \times 1$ and z_n is a scalar. We have

$$\begin{bmatrix} x_{n+1} \\ z_{n+1} \end{bmatrix} = \begin{bmatrix} A & \vdots & b \\ 0 & \vdots & 1 \end{bmatrix} \begin{bmatrix} x_n \\ z_n \end{bmatrix} = \begin{bmatrix} Ax_n + bz_n \\ z_n \end{bmatrix},$$

and so $z_{n+1} = z_n$ for every $n \geq 0$. In particular $z_{n+1} = z_0 = c$. Assume $c \neq 0$. From $x_{n+1} = Ax_n + cb$, we see that $\dfrac{x_{n+1}}{c} = Ax_n c + b$, and so $x_n = (1/c)(\text{first } m \text{ entries of } y_n)$.

(f) Notice that $\det(B - rI) = \det(A - rI)(1 - r)$, and so the set of eigenvalues of B is the set of eigenvalues of A to which we added the eigenvalue 1.

(g) Let v_r be an eigenvector of A, corresponding to the eigenvalue r, and let $u_r = [v_r \ 0]^T$. Then

$$B u_r = \begin{bmatrix} A & \vdots & b \\ 0 & \vdots & 1 \end{bmatrix} \begin{bmatrix} v_r \\ 0 \end{bmatrix} = \begin{bmatrix} Av_r \\ 0 \end{bmatrix} = r \begin{bmatrix} v_r \\ 0 \end{bmatrix} = r u_r,$$

and so u_r is an eigenvector of B, corresponding to the eigenvalue r.

Let $\mathbf{u}_{m+1} = \begin{bmatrix} (\mathbf{I} - \mathbf{A})^{-1}\mathbf{b} \\ 1 \end{bmatrix}$. We have

$$\mathbf{B}\mathbf{u}_{m+1} = \begin{bmatrix} \mathbf{A} & \vdots & \mathbf{b} \\ \mathbf{0} & \vdots & 1 \end{bmatrix} \begin{bmatrix} (\mathbf{I} - \mathbf{A})^{-1}\mathbf{b} \\ 1 \end{bmatrix}$$

$$= \begin{bmatrix} \mathbf{A}(\mathbf{I} - \mathbf{A})^{-1}\mathbf{b} + \mathbf{b} \\ 1 \end{bmatrix}$$

$$= \begin{bmatrix} (\mathbf{I} - \mathbf{A})^{-1}\mathbf{b} \\ 1 \end{bmatrix} = \mathbf{u}_{m+1},$$

and so $\mathbf{u}_{m+1}$ is an eigenvector of $\mathbf{B}$ corresponding to the eigenvalue 1.

(h) Let $\mathbf{x}_0 = c_1\mathbf{v}_1 + c_2\mathbf{v}_2 + \cdots + c_n\mathbf{v}_m$, where $\mathbf{v}_1, \mathbf{v}_2, \ldots, \mathbf{v}_m$ are m linearly independent vectors of $\mathbf{A}$ corresponding to the eigenvalues r_i, and $c_1, c_2, \ldots, c_m$ are scalars. Let $\mathbf{u}_i = [\mathbf{v}_i \ 0]^T$, $i = 1, 2, \ldots, m$, and $\mathbf{u}_{m+1} = \begin{bmatrix} (\mathbf{I} - \mathbf{A})^{-1}\mathbf{b} \\ 1 \end{bmatrix}$. Notice now that the vector $\mathbf{y}_0 = [\mathbf{x}_0 \ 1]^T$ can be written as

$$\mathbf{y}_0 = \mathbf{u}_{m+1} + \begin{bmatrix} \mathbf{x}_0 - (\mathbf{I} - \mathbf{A})^{-1}\mathbf{b} \\ 0 \end{bmatrix}.$$

If the coefficients $d_1, d_2, \ldots, d_m$ are defined by

$$(\mathbf{I} - \mathbf{A})^{-1}\mathbf{b} = d_1\mathbf{v}_1 + d_2\mathbf{v}_2 + \cdots + d_m\mathbf{v}_m,$$

we can write now

$$\mathbf{y}_0 = (c_1 - d_1)\mathbf{u}_1 + (c_2 - d_2)\mathbf{u}_2 + \cdots + (c_m - d_m)\mathbf{u}_m + \mathbf{u}_{m+1}.$$

Multiplying by $\mathbf{B}$ repeatedly, this yields the following formula for $\mathbf{y}_n$:

$$\mathbf{y}_n = (c_1 - d_1)r_1^n\mathbf{u}_1 + (c_2 - d_2)r_2^n\mathbf{u}_2$$
$$+ \cdots + (c_m - d_m)r_m^n\mathbf{u}_m + \mathbf{u}_{m+1}.$$

We can recover now

$$\mathbf{x}_n = (c_1 - d_1)r_1^n\mathbf{v}_1 + (c_2 - d_2)r_2^n\mathbf{v}_2$$
$$+ \cdots + (c_m - d_m)r_m^n\mathbf{v}_m + (\mathbf{I} - \mathbf{A})^{-1}\mathbf{b}$$

$$= ((c_1 - d_1)r_1^n + d_1)\,\mathbf{v}_1 + ((c_2 - d_2)r_2^n + d_2)\,\mathbf{v}_2$$
$$+ \cdots + ((c_m - d_m)r_m^n + d_m)\,\mathbf{v}_m.$$

Moreover, if we write $\mathbf{b} = f_1\mathbf{v}_1 + f_2\mathbf{v}_2 + \cdots + f_m\mathbf{v}_m$, since the $\mathbf{v}_i$'s are also eigenvectors of $(\mathbf{I} - \mathbf{A})^{-1}$, corresponding to the eigenvalues $1/r_i$'s, then

$$(\mathbf{I} - \mathbf{A})^{-1}\mathbf{b} = \frac{f_1}{r_1}\mathbf{v}_1 + \frac{f_2}{r_2}\mathbf{v}_2 + \cdots + \frac{f_m}{r_m}\mathbf{v}_m.$$

This yields the formula for the coefficients d_i:

$$d_i = \frac{f_i}{r_i}, \quad i = 1, \ldots, m.$$

15. (a) Notice first that in the case of the sphere (denoted by B) $\rho(x_1, x_2, x_3) = 3M/4\pi r^3$. We compute first $\mathbf{I}_{ii}$, and we can assume without loss of generality $x_i = z$, and we use spherical coordinates (ρ, ϕ, θ):

$$\begin{aligned}
\mathbf{I}_{ii} &= \frac{3M}{4\pi r^3} \iiint_B z^2 \, dV \\
&= \frac{3M}{4\pi r^3} \int_0^{2\pi} \int_0^{\pi} \int_0^r (\rho\cos\phi)^2 \rho^2 \sin\phi \, d\rho \, d\phi \, d\theta \\
&= \frac{3M}{4\pi r^3} \int_0^{2\pi} d\theta \int_0^{\pi} \cos^2\phi \sin\phi \, d\phi \int_0^r \rho^4 \, d\rho \\
&= \frac{3M}{4\pi r^3} \cdot (2\pi) \cdot \left(-\frac{1}{3}\cos^2\phi \right)\Big|_0^{\pi} \cdot \frac{\rho^5}{5}\Big|_0^r \\
&= \frac{Mr^2}{5}.
\end{aligned}$$

To show $\mathbf{I}_{ij} = 0$ for $i \neq j$, we may assume without loss of generality $i = 1, j = 2$ and $x_1 = x, x_2 = y$. We have

$$\begin{aligned}
\mathbf{I}_{ij} &= \frac{3M}{4\pi r^3} \iiint_B xy \, dV \\
&= \frac{3M}{4\pi r^3} \int_0^{2\pi} \int_0^{\pi} \int_0^r (\rho\cos\theta\sin\phi) \\
&\quad \times (\rho\sin\theta\sin\phi)\rho^2 \sin\phi \, d\rho \, d\phi \, d\theta
\end{aligned}$$

$$= \frac{3M}{4\pi r^3} \int_0^{2\pi} \cos\theta \sin\theta d\theta \int_0^\pi \sin^3\phi\, d\phi \int_0^r \rho^4\, d\rho$$

$$= 0,$$

as the factor

$$\int_0^{2\pi} \cos\theta \sin\theta d\theta = \left(-\frac{\cos 2\theta}{4}\right)\Big|_0^{2\pi} = 0.$$

(b) Notice first that in the case of the cylinder (denoted by E) the density is given by $\rho(x_1, x_2, x_3) = M/\pi r^2 h$. For all of computations we use cylindrical coordinates. Furthermore, for symmetry reasons

$$\mathbf{I}_{13} = \mathbf{I}_{23} = \frac{M}{\pi r^2 h} \iiint_E xz\, dV$$

$$= \frac{M}{\pi r^2 h} \int_0^{2\pi} \int_0^r \int_{-h/2}^{h/2} (\rho\cos\theta)(z)\rho\, dz\, dr\, d\theta$$

$$= \frac{M}{\pi r^2 h} \int_0^{2\pi} \cos\theta \int_0^r \rho^2\, d\rho \int_{-h/2}^{h/2} z\, dz = 0,$$

as the first integral is zero. A similar computation which we will not reproduce here shows that $\mathbf{I}_{12} = 0$.

$$\mathbf{I}_{11} = \mathbf{I}_{22} = \frac{M}{\pi r^2 h} \iiint_E x^2\, dV$$

$$= \frac{M}{\pi r^2 h} \int_0^{2\pi} \int_0^r \int_{-h/2}^{h/2} (\rho\cos\theta)^2 \rho\, dz\, dr\, d\theta$$

$$= \frac{M}{\pi r^2 h} \int_0^{2\pi} \cos^2\theta \int_0^r \rho^3\, d\rho \int_{-h/2}^{h/2} dz$$

$$= \frac{M}{\pi r^2 h} \cdot \left(\frac{1}{2}\left(\theta + \frac{1}{2}\sin 2\theta\right)\right)\Big|_0^{2\pi} \cdot \left(\frac{1}{4}\rho^4\right)\Big|_0^r \cdot h$$

$$= \frac{Mr^2}{4}.$$

$$I_{33} = \frac{M}{\pi r^2 h} \iiint_E x^2 \, dV$$

$$= \frac{M}{\pi r^2 h} \int_0^{2\pi} \int_0^r \int_{-h/2}^{h/2} z^2 \rho \, dz \, dr \, d\theta$$

$$= \frac{M}{\pi r^2 h} \int_0^{2\pi} d\theta \int_0^r \rho \, d\rho \int_{-h/2}^{h/2} z^2 \, dz$$

$$= \frac{M}{\pi r^2 h} \cdot 2\pi \cdot \left(\frac{1}{2}\rho^2\right)\Big|_0^r \cdot h \cdot \left(\frac{z^3}{3}\right)\Big|_{-h/2}^{h/2}$$

$$= \frac{Mh^2}{12}.$$

(c) In this case, the density is $\rho = M/abc$. If E denotes the prism, we compute

$$I_{11} = \frac{M}{abc} \iiint_E x^2 \, dV = \frac{M}{abc} \int_{-c/2}^{c/2} \int_{-a/2}^{a/2} \int_{-b/2}^{b/2} x^2 \, dz \, dy \, dx$$

$$= \frac{M}{abc} ab \int_{-c/2}^{c/2} x^2 \, dx$$

$$= \frac{Mc^2}{12}$$

The same argument shows that $I_{22} = \dfrac{Ma^2}{12}$ and $I_{33} = \dfrac{Mb^2}{12}$.
A similar computation shows $I_{ij} = 0$ for $i \neq j$.

6.3   SCHUR DECOMPOSITION AND ITS IMPLICATIONS

1. The characteristic equation for the matrix $\mathbf{A} = \begin{bmatrix} 7 & 6 \\ -9 & -8 \end{bmatrix}$ is

$$|\mathbf{A} - r\mathbf{I}| = \begin{vmatrix} 7-r & 6 \\ -9 & -8-r \end{vmatrix} = 0$$

which is equivalent to $r^2 + r - 2 = (r+2)(r-1) = 0$. Hence the eigenvalues of $\mathbf{A}$ are $r_1 = 1$ and $r_2 = -2$. To find the eigenvectors corresponding to $r_1 = 1$, we must solve the equation $(\mathbf{A} - \mathbf{I})\mathbf{u} = \mathbf{0}$, which means

$$\begin{bmatrix} 6 & 6 \\ -9 & -9 \end{bmatrix} \begin{bmatrix} u_1 \\ u_2 \end{bmatrix} = \begin{bmatrix} 0 \\ 0 \end{bmatrix}.$$

Notice that this matrix equation is equivalent to the single scalar equation $6u_1 + 6u_2 = 0$, whose solutions are obtained by assigning an arbitrary value for u_1 (say $u_1 = s$) and setting $u_1 = -s$. A unit eigenvector is obtained for $s = 1/\sqrt{2}$

$$\mathbf{u_1} = \frac{1}{\sqrt{2}} \begin{bmatrix} 1 \\ -1 \end{bmatrix}.$$

We look now for a unit vector $\mathbf{u_2}$, orthogonal to $\mathbf{u_1}$. Such a vector can be found immediately to be

$$\mathbf{u_2} = \frac{1}{\sqrt{2}} \begin{bmatrix} 1 \\ 1 \end{bmatrix}.$$

Let now $\mathbf{Q}$ be the orthogonal matrix formed by the column vectors $\mathbf{u_1}$ and $\mathbf{u_2}$

$$\mathbf{Q} = \frac{1}{\sqrt{2}} \begin{bmatrix} 1 & 1 \\ -1 & 1 \end{bmatrix}.$$

We compute now the matrix $\mathbf{U}$, the upper triangular form of $\mathbf{A}$:

$$\mathbf{U} = \mathbf{Q}^H \mathbf{A} \mathbf{Q} = \frac{1}{2} \begin{bmatrix} 1 & 1 \\ -1 & 1 \end{bmatrix} \begin{bmatrix} 7 & 6 \\ -9 & -8 \end{bmatrix} \begin{bmatrix} 1 & -1 \\ 1 & 1 \end{bmatrix}$$

$$= \begin{bmatrix} 1 & 15 \\ 0 & -2 \end{bmatrix}.$$

3. The characteristic equation for the matrix $\mathbf{A} = \begin{bmatrix} 2 & 10 \\ -4 & -2 \end{bmatrix}$ is

$$|\mathbf{A} - r\mathbf{I}| = \begin{vmatrix} 2 - r & 10 \\ -4 & -2 - r \end{vmatrix} = 0$$

which is equivalent to $r^2 + 36 = 0$. Hence the eigenvalues of A are $r_1 = 6i$ and $r_2 = -6i$. To find the eigenvectors corresponding to $r_1 = -6i$, we must solve the equation $(A + 6iI)u = 0$, which means

$$\begin{bmatrix} 2+6i & 10 \\ -4 & -2+6i \end{bmatrix} \begin{bmatrix} u_1 \\ u_2 \end{bmatrix} = \begin{bmatrix} 0 \\ 0 \end{bmatrix}.$$

Notice that this matrix equation is equivalent to the single scalar equation $-4u_1 + 2(-1+3i)u_2 = 0$. A unit such eigenvector is

$$u_1 = \frac{1}{\sqrt{14}} \begin{bmatrix} 3i-1 \\ 2 \end{bmatrix}.$$

We look now for a unit vector $u_2 = [a \ b]^T$, orthogonal to u_1. This means the Hermitian product $u_1 \cdot u_2 = 0$, which is equivalent to $(-3i - 1)a + 2b = $ A particular unit vector solution u_2 is

$$u_2 = \frac{1}{\sqrt{14}} \begin{bmatrix} 2 \\ 3i+1 \end{bmatrix}.$$

Let now Q be the orthogonal matrix formed by the column vectors u_1 and u_2

$$Q = \frac{1}{\sqrt{14}} \begin{bmatrix} 3i-1 & 2 \\ 2 & 3i+1 \end{bmatrix}.$$

We compute now the matrix U, the upper triangular form of A:

$$U = Q^H A Q$$
$$= \frac{1}{\sqrt{14}} \begin{bmatrix} -3i-1 & 2 \\ 2 & -3i+1 \end{bmatrix} \begin{bmatrix} 2 & 10 \\ -4 & -2 \end{bmatrix} \begin{bmatrix} 3i-1 & 2 \\ 2 & 3i+1 \end{bmatrix}$$
$$= \begin{bmatrix} -6i & -6i+4 \\ 0 & 6i \end{bmatrix}.$$

5. The characteristic equation for the matrix $\mathbf{A} = \begin{bmatrix} 0 & -1 & 2 \\ -1 & 0 & 0 \\ -1 & 1 & -1 \end{bmatrix}$ is

$$|\mathbf{A} - r\mathbf{I}| = \begin{vmatrix} -r & -1 & 2 \\ -1 & -r & 0 \\ -1 & 1 & -1-r \end{vmatrix} = 0,$$

which is equivalent to $(r + 1)(r^2 + 1) = 0$. Hence the eigenvalues of $\mathbf{A}$ are $r_1 = -1, r_2 = i$ and $r_2 = -i$. We find first the eigenvectors corresponding to $r_1 = -1$, by solving the equation $(\mathbf{A} + \mathbf{I})\mathbf{u} = \mathbf{0}$, which means

$$\begin{bmatrix} 1 & -1 & 2 \\ -1 & 1 & 0 \\ -1 & 1 & 0 \end{bmatrix} \begin{bmatrix} u_1 \\ u_2 \\ u_3 \end{bmatrix} = \begin{bmatrix} 0 \\ 0 \\ 0 \end{bmatrix}.$$

By Gaussian elimination, this matrix equation is equivalent to $u_3 = 0$ and $u_1 - u_2 = 0$. A unit eigenvector is

$$\mathbf{u_1} = \frac{1}{\sqrt{2}} \begin{bmatrix} 1 \\ 1 \\ 0 \end{bmatrix}.$$

We look now for two orthogonal unit vectors $\mathbf{u_2}$ and $\mathbf{u_3}$ which are also orthogonal to $\mathbf{u_1}$. A possible choice can be seen to be

$$\mathbf{u_1} = \frac{1}{\sqrt{2}} \begin{bmatrix} 1 \\ -1 \\ 0 \end{bmatrix}, \quad \text{and} \quad \mathbf{u_2} = \begin{bmatrix} 0 \\ 0 \\ 1 \end{bmatrix}$$

Let now $\mathbf{Q_1}$ be the orthogonal matrix formed by the column vectors $\mathbf{u_1}, \mathbf{u_2}$ and $\mathbf{u_3}$

$$\mathbf{Q_1} = \frac{1}{\sqrt{2}} \begin{bmatrix} 1 & 1 & 0 \\ 1 & -1 & 0 \\ 0 & 0 & \sqrt{2} \end{bmatrix}.$$

We compute now the matrix:

$$Q_1^H A Q_1 = \frac{1}{2} \begin{bmatrix} 1 & 1 & 0 \\ 1 & -1 & 0 \\ 0 & 0 & \sqrt{2} \end{bmatrix} \begin{bmatrix} 0 & -1 & 2 \\ -1 & 0 & 0 \\ -1 & 1 & -1 \end{bmatrix}$$

$$\times \begin{bmatrix} 1 & 1 & 0 \\ 1 & -1 & 0 \\ 0 & 0 & \sqrt{2} \end{bmatrix}$$

$$= \begin{bmatrix} -1 & 0 & \sqrt{2} \\ 0 & 1 & \sqrt{2} \\ 0 & -\sqrt{2} & -1 \end{bmatrix}.$$

Consider now the companion 2×2 matrix $A_1 = \begin{bmatrix} 1 & \sqrt{2} \\ -\sqrt{2} & -1 \end{bmatrix}$. Its characteristic equation is

$$|A - rI| = \begin{bmatrix} 1-r & \sqrt{2} \\ -\sqrt{2} & -1-r \end{bmatrix} = 0$$

which is equivalent to $r^2 + 1 = 0$, and so its eigenvalues are $r_1 = -i$ and $r_2 = i$. To find an eigenvector corresponding to $r_1 = -i$, we must solve the equation $(A_1 + iI)v = 0$, which means

$$\begin{bmatrix} 1+i & \sqrt{2} \\ -\sqrt{2} & -1+i \end{bmatrix} \begin{bmatrix} v_1 \\ v_2 \end{bmatrix} = \begin{bmatrix} 0 \\ 0 \end{bmatrix}.$$

Notice that this matrix equation is equivalent to the single scalar equation $-\sqrt{2}v_1 + (i-1)v_2 = 0$. A unit such eigenvector is

$$v_2 = \frac{1}{2} \begin{bmatrix} i-1 \\ \sqrt{2} \end{bmatrix}.$$

We look now for a unit vector $v_3 = [a \; b]^T$, orthogonal to v_2. This means the Hermitian product $u_1 \cdot u_2 = 0$, which is equivalent to $(-i - 1)a + \sqrt{2}b = A$ particular unit vector solution u_2 is

$$v_2 = \frac{1}{2} \begin{bmatrix} \sqrt{2} \\ i+1 \end{bmatrix}.$$

We construct now the unitary matrix $\mathbf{Q}_2$ formed by the column vectors $[1\ 0\ 0]^T$, $\mathbf{v}_2$ and $\mathbf{v}_3$ (after padding $\mathbf{v}_2$ and $\mathbf{v}_3$ by 0 as a first entry)

$$\mathbf{Q}_2 = \begin{bmatrix} 1 & 0 & 0 \\ 0 & (i-1)/2 & \sqrt{2}/2 \\ 0 & \sqrt{2}/2 & (i+1)/2 \end{bmatrix}.$$

We compute now the matrix U, the upper triangular form of **A**:

$$\mathbf{U} = \mathbf{Q}_2^H(\mathbf{Q}_1^H\mathbf{A}\mathbf{Q}_1)\mathbf{Q}_2$$

$$= \frac{1}{4}\begin{bmatrix} 2 & 0 & 0 \\ 0 & (-i-1) & \sqrt{2} \\ 0 & \sqrt{2} & (-i+1) \end{bmatrix}\begin{bmatrix} -1 & 0 & \sqrt{2} \\ 0 & 1 & \sqrt{2} \\ 0 & -\sqrt{2} & -1 \end{bmatrix}$$

$$\times \begin{bmatrix} 2 & 0 & 0 \\ 0 & (i-1) & \sqrt{2} \\ 0 & \sqrt{2} & (i+1) \end{bmatrix}$$

$$= \begin{bmatrix} -1 & 1 & (i+1)/\sqrt{2} \\ 0 & -i & -\sqrt{2}(i+1) \\ 0 & 0 & i \end{bmatrix}.$$

7. Recall that the trace of a product of two $n \times n$ matrices **M** and **N** has the property $\text{Trace}(\mathbf{MN}) = \text{Trace}(\mathbf{NM})$.

Let now **A** be an arbitrary complex $n \times n$ matrix. From the Schur decomposition theorem we can write $\mathbf{U} = \mathbf{Q}^H\mathbf{AQ}$, where **Q** is a unitary matrix, and **U** is an upper triangular matrix. Then $\text{Trace}(\mathbf{U}) = \text{Trace}(\mathbf{Q}^H\mathbf{AQ}) = \text{Trace}\left((\mathbf{QQ}^H)\mathbf{A}\right) = \text{Trace}(\mathbf{IA}) = \text{Trace}(\mathbf{A})$. However, from Corollary 7, **U** has the eigenvalues of **A** on its diagonal, and so $\text{Trace}(\mathbf{U})(= \text{Trace}(\mathbf{A}))$ is the sum of the eigenvalues of **A**.

9. Take $\mathbf{A} = \begin{bmatrix} 1 & -1 \\ 1 & 1 \end{bmatrix}$ and $\mathbf{B} = \begin{bmatrix} 1 & 0 \\ 0 & -1 \end{bmatrix}$. By a direct computation, we can see that both **A** and **B** are normal. However, notice that for $\mathbf{A}+\mathbf{B} = \begin{bmatrix} 2 & -1 \\ 1 & 0 \end{bmatrix}$, we have

$$(\mathbf{A}+\mathbf{B})(\mathbf{A}+\mathbf{B})^H = \begin{bmatrix} 5 & 2 \\ 2 & 1 \end{bmatrix} \neq \begin{bmatrix} 5 & -2 \\ -2 & 1 \end{bmatrix} = (\mathbf{A}+\mathbf{B})^H(\mathbf{A}+\mathbf{B}).$$

Take now again $\mathbf{A} = \begin{bmatrix} 1 & -1 \\ 1 & 1 \end{bmatrix}$ and $\mathbf{B} = \begin{bmatrix} 1 & 1 \\ 1 & 1 \end{bmatrix}$. Clearly, both matrices are normal. However, notice that

$$\mathbf{AB} = \begin{bmatrix} 0 & 0 \\ 2 & 2 \end{bmatrix}$$

is a triangular, but not diagonal matrix. In particular, $\mathbf{AB}$ is not normal.

11. We compute:

$$\begin{aligned}
\|\mathbf{Av}\|^2 &= \left(\overline{\mathbf{Av}}\right)^T (\mathbf{Av}) = \left(\bar{\mathbf{A}}\bar{\mathbf{v}}\right)^T (\mathbf{Av}) = \bar{\mathbf{v}}^T \bar{\mathbf{A}}^T \mathbf{Av} = \bar{\mathbf{v}}^T \mathbf{A}^H \mathbf{Av} \\
&= \bar{\mathbf{v}}^T \mathbf{A}\mathbf{A}^H \mathbf{v} = \left(\mathbf{A}^T \bar{\mathbf{v}}\right)^T \left(\mathbf{A}^H \mathbf{v}\right) \\
&= \left(\overline{\bar{\mathbf{A}}^T \mathbf{v}}\right)^T \left(\mathbf{A}^H \mathbf{v}\right) = \left(\overline{\mathbf{A}^H \mathbf{v}}\right)^T \left(\mathbf{A}^H \mathbf{v}\right) \\
&= \|\mathbf{A}^H \mathbf{v}\|^2
\end{aligned}$$

13. Let $\mathbf{A} = \begin{bmatrix} 1 & i \\ i & 1 \end{bmatrix}$. We compute

$$\mathbf{A}\mathbf{A}^H = \begin{bmatrix} 1 & i \\ i & 1 \end{bmatrix} \begin{bmatrix} 1 & -i \\ -i & 1 \end{bmatrix} = \begin{bmatrix} 2 & 0 \\ 0 & 2 \end{bmatrix}.$$

Similarly,

$$\mathbf{A}^H \mathbf{A} = \begin{bmatrix} 1 & -i \\ -i & 1 \end{bmatrix} \begin{bmatrix} 1 & i \\ i & 1 \end{bmatrix} = \begin{bmatrix} 2 & 0 \\ 0 & 2 \end{bmatrix}.$$

Since $\mathbf{A}\mathbf{A}^H = \mathbf{A}^H \mathbf{A}$, and so $\mathbf{A}$ is normal.

The characteristic equation for the matrix $\mathbf{A}$ is is

$$|\mathbf{A} - r\mathbf{I}| = \begin{vmatrix} 1-r & i \\ i & 1-r \end{vmatrix} = 0$$

which is equivalent to $r^2 - 2r + 2 = (r+1)^2 + 1 = 0$. Hence the eigenvalues of $\mathbf{A}$ are $r_1 = i+1$ and $r_2 = -i+1$. To find the eigenvectors corresponding to $r_1 = i + 1$, we must solve the equation $(\mathbf{A} - (i + 1)\mathbf{I})\mathbf{u} = \mathbf{0}$, which means

$$\begin{bmatrix} -i & i \\ i & -i \end{bmatrix} \begin{bmatrix} u_1 \\ u_2 \end{bmatrix} = \begin{bmatrix} 0 \\ 0 \end{bmatrix}.$$

Notice that this matrix equation is equivalent to the single scalar equation $-iu_1 + iu_2 = 0$. A unit such is

$$\mathbf{u_1} = \frac{1}{\sqrt{2}} \begin{bmatrix} 1 \\ 1 \end{bmatrix}.$$

To find the eigenvectors corresponding to $r_1 = -i + 1$, we must solve the equation $(\mathbf{A} - (-i + 1)\mathbf{I})\mathbf{u} = \mathbf{0}$, which means

$$\begin{bmatrix} i & i \\ i & i \end{bmatrix} \begin{bmatrix} u_1 \\ u_2 \end{bmatrix} = \begin{bmatrix} 0 \\ 0 \end{bmatrix}.$$

Notice that this matrix equation is equivalent to the single scalar equation $iu_1 + iu_2 = 0$. A unit such is

$$\mathbf{u_2} = \frac{1}{\sqrt{2}} \begin{bmatrix} 1 \\ -1 \end{bmatrix}.$$

Hence, an orthonormal basis of eigenvectors for $\mathbf{A}$ is $\left\{ \begin{bmatrix} \dfrac{1}{\sqrt{2}} & \dfrac{1}{\sqrt{2}} \end{bmatrix}^T, \right.$
$\left. \begin{bmatrix} \dfrac{1}{\sqrt{2}} & -\dfrac{1}{\sqrt{2}} \end{bmatrix}^T \right\}.$

15. Notice that the matrix $\mathbf{A} = \begin{bmatrix} 0 & -2 & -2 \\ 2 & 0 & 1 \\ 2 & -1 & 0 \end{bmatrix}$ skew-symmetric, and so it is automatically normal $(\mathbf{A}\mathbf{A}^H = \mathbf{A}(-\mathbf{A}) = (-\mathbf{A})\mathbf{A} = \mathbf{A}^H\mathbf{A})$.
Its characteristic equation is

$$|\mathbf{A} - r\mathbf{I}| = \begin{vmatrix} -r & -2 & -2 \\ 2 & -r & 1 \\ 2 & -1 & -r \end{vmatrix} = 0,$$

which is equivalent to $r(r^2 + 9) = 0$. Hence the eigenvalues of $\mathbf{A}$ are $r_1 = 0, r_2 = 3i$ and $r_2 = -3i$. To find the eigenvectors corresponding to $r_1 = 0$, we must solve the equation $\mathbf{Au} = \mathbf{0}$, which means

$$
\begin{bmatrix} 0 & -2 & -2 \\ 2 & 0 & 1 \\ 2 & -1 & 0 \end{bmatrix} \begin{bmatrix} u_1 \\ u_2 \\ u_3 \end{bmatrix} = \begin{bmatrix} 0 \\ 0 \\ 0 \end{bmatrix}.
$$

By Gaussian elimination, this matrix equation is equivalent to

$$
\begin{aligned}
2u_1 \quad + \quad u_3 &= 0, \\
u_2 + \quad u_3 &= 0.
\end{aligned}
$$

A unit eigenvector is

$$
\mathbf{u_1} = \frac{1}{3} \begin{bmatrix} 1 \\ 2 \\ -2 \end{bmatrix}.
$$

We find next a unit eigenvector corresponding to the eigenvalue $r_2 = -3i$ by solving the equation $(\mathbf{A} + 3i\mathbf{I})\mathbf{u} = \mathbf{0}$, which means

$$
\begin{bmatrix} 3i & -2 & -2 \\ 2 & 3i & 1 \\ 2 & -1 & 3i \end{bmatrix} \begin{bmatrix} u_1 \\ u_2 \\ u_3 \end{bmatrix} = \begin{bmatrix} 0 \\ 0 \\ 0 \end{bmatrix}.
$$

By Gaussian elimination, this matrix equation is equivalent to

$$
\begin{aligned}
3iu_1 - \quad 2u_2 - \quad 2u_3 &= 0, \\
(3i + 1)u_2 - \quad (3i - 1)u_3 &= 0.
\end{aligned}
$$

A unit eigenvector is

$$
\mathbf{u_2} = \frac{1}{3\sqrt{10}} \begin{bmatrix} 2 - 6i \\ 4 + 3i \\ 5 \end{bmatrix}.
$$

Finally, we look for a unit eigenvector corresponding to the eigenvalue $r_2 = 3i$ by solving the equation $(\mathbf{A} - 3i\mathbf{I})\mathbf{u} = \mathbf{0}$, which means

$$\begin{bmatrix} -3i & -2 & -2 \\ 2 & -3i & 1 \\ 2 & -1 & -3i \end{bmatrix} \begin{bmatrix} u_1 \\ u_2 \\ u_3 \end{bmatrix} = \begin{bmatrix} 0 \\ 0 \\ 0 \end{bmatrix}.$$

By Gaussian elimination, this matrix equation is equivalent to

$$\begin{aligned} 3iu_1 + \quad 2u_2 + \quad\quad 2u_3 \quad\quad &= 0, \\ (3i-1)u_2 - \quad (3i+1)u_3 &= 0. \end{aligned}$$

A unit eigenvector is

$$\mathbf{u}_2 = \frac{1}{3\sqrt{10}} \begin{bmatrix} 2+6i \\ 4-3i \\ 5 \end{bmatrix}.$$

The set $\left\{ \begin{bmatrix} \dfrac{1}{3} & \dfrac{2}{3} & -\dfrac{2}{3} \end{bmatrix}^T, \begin{bmatrix} \dfrac{2-6i}{3\sqrt{10}} & \dfrac{4+3i}{3\sqrt{10}} & 5 \end{bmatrix}^T, \begin{bmatrix} \dfrac{2+6i}{3\sqrt{10}} & \dfrac{4-3i}{3\sqrt{10}} & 5 \end{bmatrix}^T \right\}$ is a basis of orthonormal vectors.

17. Let

$$\mathbf{A} = [\mathbf{u}_1\ \mathbf{u}_2\ \cdots\ \mathbf{u}_n] = [\mathbf{v}_1^{\mathbf{T}}\ \mathbf{v}_2^{\mathbf{T}}\ \cdots\ \mathbf{v}_n^{\mathbf{T}}]$$

be a normal $n \times n$ matrix, and let $\mathbf{e}_i$ denote the standard vector $[0\cdots0\,1\,0\cdots0]^T$, with the non-zero entry on the ith row. We have $\|\mathbf{Ae}_i\| = \|\mathbf{u}_i\|$, while $\|\mathbf{A}^H\mathbf{e}_i\| = \|\bar{\mathbf{v}}_i\|$. From Problem 11, Exercises 6.3 above, we know that $\mathbf{A}$ normal implies $\|\mathbf{Ae}_i\| = \|\mathbf{A}^H\mathbf{e}_i\|$, and so $\|\mathbf{u}_i\| = \|\bar{\mathbf{v}}_i\| = \|\mathbf{v}_i\|$.

19. A straightforward but laborious procedure for this problem would be to observe that if r_A and r_B are eigenvalues of $\mathbf{A}$ and $\mathbf{B}$ respectively, then the 3 columns of the unitary matrix generating the similarity transformation must each be a nonzero solution to one of the following 9 systems:

$$\begin{bmatrix} \mathbf{A} - r_A\mathbf{I} \\ \mathbf{B} - r_B\mathbf{I} \end{bmatrix} \mathbf{u} = \mathbf{0}.$$

However the following is enlightening. We start by first diagonalizing the matrix

$$A = \begin{bmatrix} 0 & 1 & 1 \\ 1 & 0 & 1 \\ 1 & 1 & 0 \end{bmatrix},$$

by determining its diagonal form and then we study the action of the similarity transformation matrix U_1. We study next the effect of the action of U_1 on the matrix

$$B = \begin{bmatrix} 0 & -1 & 1 \\ 1 & 0 & -1 \\ -1 & 1 & 0 \end{bmatrix},$$

by computing the matrix $B = U_1^H B U_1$. We diagonalize next the matrix B_1 by a similarity transformation matrix U_2. Finally, we need to check that the matrix $(U_1 U_2)^H A (U_1 U_2)$ is diagonal.

The characteristic equation of A is

$$|A - rI| = \begin{vmatrix} -r & 1 & 1 \\ 1 & -r & 1 \\ 1 & 1 & -r \end{vmatrix} = 0,$$

which is equivalent to $(r + 1)^2 (r - 2) = 0$. Hence the eigenvalues of A are $r_1 = 2$ with algebraic multiplicity 1 and $r_2 = -1$ with algebraic multiplicity 2. To find the eigenvectors corresponding to $r_1 = 2$, we must solve the equation $(A - 2I)u = 0$, which means

$$\begin{bmatrix} -2 & 1 & 1 \\ 1 & -2 & 1 \\ 1 & 1 & -2 \end{bmatrix} \begin{bmatrix} u_1 \\ u_2 \\ u_3 \end{bmatrix} = \begin{bmatrix} 0 \\ 0 \\ 0 \end{bmatrix}.$$

This is equivalent to solving the reduced system of equations

$$\begin{aligned} -2u_1 + u_2 + u_3 &= 0, \\ u_2 - u_3 &= 0. \end{aligned}$$

A unit eigenvector is

$$\mathbf{u_1} = \frac{1}{\sqrt{3}} \begin{bmatrix} 1 \\ 1 \\ 1 \end{bmatrix}.$$

To find the eigenvectors corresponding to $r_1 = -1$, we must solve the equation $(\mathbf{A} + \mathbf{I})\mathbf{u} = \mathbf{0}$, which means

$$\begin{bmatrix} 1 & 1 & 1 \\ 1 & 1 & 1 \\ 1 & 1 & 1 \end{bmatrix} \begin{bmatrix} u_1 \\ u_2 \\ u_3 \end{bmatrix} = \begin{bmatrix} 0 \\ 0 \\ 0 \end{bmatrix}.$$

This is equivalent to solving the scalar equation $u_1 + u_2 + u_3 = 0$. A pair of orthonormal eigenvectors corresponding to the eigenvalue $r_2 = -1$ are

$$\mathbf{u_2} = \frac{1}{\sqrt{2}} \begin{bmatrix} 1 \\ -1 \\ 0 \end{bmatrix} \quad \text{and} \quad \mathbf{u_3} = \frac{1}{\sqrt{6}} \begin{bmatrix} -1 \\ -1 \\ 2 \end{bmatrix}.$$

Therefore, the matrix

$$\mathbf{U_1} = \begin{bmatrix} 1/\sqrt{3} & 1/\sqrt{2} & -1/\sqrt{6} \\ 1/\sqrt{3} & -1/\sqrt{2} & -1/\sqrt{6} \\ 1/\sqrt{3} & 0 & 2/\sqrt{6} \end{bmatrix}$$

diagonalizes $\mathbf{A}$ and

$$\mathbf{U_1^H A U_1} = \begin{bmatrix} 2 & 0 & 0 \\ 0 & -1 & 0 \\ 0 & 0 & -1 \end{bmatrix}.$$

We compute next

$$\mathbf{B}_1 = \mathbf{U}_1^H \mathbf{B} \mathbf{U}_1$$

$$= \begin{bmatrix} 1/\sqrt{3} & 1/\sqrt{3} & 1/\sqrt{3} \\ 1/\sqrt{2} & -1/\sqrt{2} & 0 \\ -1/\sqrt{6} & -1/\sqrt{6} & 2/\sqrt{6} \end{bmatrix} \begin{bmatrix} 0 & -1 & 1 \\ 1 & 0 & -1 \\ -1 & 1 & 0 \end{bmatrix}$$

$$\times \begin{bmatrix} 1/\sqrt{3} & 1/\sqrt{2} & -1/\sqrt{6} \\ 1/\sqrt{3} & -1/\sqrt{2} & -1/\sqrt{6} \\ 1/\sqrt{3} & 0 & 2/\sqrt{6} \end{bmatrix}$$

$$= \begin{bmatrix} 0 & 0 & 0 \\ 0 & 0 & \sqrt{3} \\ 0 & -\sqrt{3} & 0 \end{bmatrix}$$

The characteristic equation of $\mathbf{B}_1$ is $r(r^2 + 3) = 0$, and so its eigenvalues are $r_1 = 0$, $r_2 = i\sqrt{3}$ and $r_3 = -i\sqrt{3}$. An eigenvector corresponding to the eigenvalue $r_1 = 0$ is

$$\mathbf{v}_1 = \begin{bmatrix} 1 \\ 0 \\ 0 \end{bmatrix}$$

To find an eigenvector corresponding to the eigenvalue $r_2 = i\sqrt{3}$, we must solve the equation $(\mathbf{A} + i\sqrt{3}\mathbf{I})\mathbf{v} = \mathbf{0}$, which is equivalent to solving

$$\begin{bmatrix} -i\sqrt{3} & 0 & 0 \\ 0 & -i\sqrt{3} & \sqrt{3} \\ 0 & -\sqrt{3} & -i\sqrt{3} \end{bmatrix} \begin{bmatrix} v_1 \\ v_2 \\ v_3 \end{bmatrix} = \begin{bmatrix} 0 \\ 0 \\ 0 \end{bmatrix}.$$

This is equivalent to $v_1 = 0$ and $-iv_2 + v_3 = 0$. A unit eigenvector corresponding to $r_2 = \sqrt{3}$ is

$$\mathbf{v}_2 = \frac{1}{\sqrt{2}} \begin{bmatrix} 0 \\ 1 \\ i \end{bmatrix}.$$

A similar computation yields the unit eigenvector corresponding to $r_2 = -i\sqrt{3}$ is

$$\mathbf{v}_3 = \frac{1}{\sqrt{2}} \begin{bmatrix} 0 \\ i \\ 1 \end{bmatrix}.$$

Let

$$\mathbf{U}_2 = \begin{bmatrix} 1 & 0 & 0 \\ 0 & 1/\sqrt{2} & i/\sqrt{2} \\ 0 & i/\sqrt{2} & 1/\sqrt{2} \end{bmatrix}.$$

From our construction, it is clear that $\mathbf{U}_2\mathbf{U}_1$ diagonalizes the matrix $\mathbf{B}$:

$$(\mathbf{U}_1\mathbf{U}_2)^{\mathbf{H}}\mathbf{B}(\mathbf{U}_1\mathbf{U}_2) = \mathbf{U}_2^{\mathbf{H}}(\mathbf{U}_1^{\mathbf{H}}\mathbf{B}\mathbf{U}_1)\mathbf{U}_2 = \mathbf{U}_2^{\mathbf{H}}\mathbf{B}_1\mathbf{U}_2$$

$$= \begin{bmatrix} 0 & 0 & 0 \\ 0 & i\sqrt{3} & 0 \\ 0 & 0 & -i\sqrt{3} \end{bmatrix}.$$

It remains to check that $\mathbf{U}_2$ still diagonalizes $\mathbf{A}$ (remember that $\mathbf{U}_1$ diagonalizes $\mathbf{A}$):

$$(\mathbf{U}_1\mathbf{U}_2)^{\mathbf{H}}\mathbf{A}(\mathbf{U}_1\mathbf{U}_2) = \mathbf{U}_2^{\mathbf{H}}(\mathbf{U}_1^{\mathbf{H}}\mathbf{A}\mathbf{U}_1)\mathbf{U}_2 = \mathbf{U}_2^{\mathbf{H}}\begin{bmatrix} 2 & 0 & 0 \\ 0 & -1 & 0 \\ 0 & 0 & -1 \end{bmatrix}\mathbf{U}_2$$

$$= \begin{bmatrix} 1 & 0 & 0 \\ 0 & 1/\sqrt{2} & -i/\sqrt{2} \\ 0 & -i/\sqrt{2} & 1/\sqrt{2} \end{bmatrix}\begin{bmatrix} 2 & 0 & 0 \\ 0 & -1 & 0 \\ 0 & 0 & -1 \end{bmatrix}$$

$$\times \begin{bmatrix} 1 & 0 & 0 \\ 0 & 1/\sqrt{2} & i/\sqrt{2} \\ 0 & i/\sqrt{2} & 1/\sqrt{2} \end{bmatrix}$$

$$= \begin{bmatrix} 2 & 0 & 0 \\ 0 & -1 & 0 \\ 0 & 0 & -1 \end{bmatrix}.$$

It is a two steps procedure, and we have $\mathbf{U} = \mathbf{U}_1\mathbf{U}_2$, where

$$\mathbf{U}_1 = \begin{bmatrix} 1/\sqrt{3} & 1/\sqrt{2} & -1/\sqrt{6} \\ 1/\sqrt{3} & 1-/\sqrt{2} & -1/\sqrt{6} \\ 1/\sqrt{3} & 0 & 2/\sqrt{6} \end{bmatrix} \quad \text{and}$$

$$\mathbf{U}_2 = \begin{bmatrix} 1 & 0 & 0 \\ 0 & 1/\sqrt{2} & i/\sqrt{2} \\ 0 & i/\sqrt{2} & 1/\sqrt{2} \end{bmatrix}.$$

21. From the Cayley-Hamilton Theorem we know that

$$(-\mathbf{A})^n + c_{n-1}\mathbf{A}^{n-1} + \cdots + c_1\mathbf{A} + c_0\mathbf{I} = 0.$$

Multiplying by $\mathbf{A}^{-1}$, we find

$$-(-\mathbf{A})^{n-1} + c_{n-1}\mathbf{A}^{n-2} + \cdots + c_1\mathbf{I} + c_0\mathbf{A}^{-1} = 0,$$

and so

$$\mathbf{A}^{-1} = \frac{1}{c_0}\left((-\mathbf{A})^{n-1} - c_{n-1}\mathbf{A}^{n-2} - \cdots - c_1\mathbf{I}\right).$$

6.4   THE SINGULAR VALUE DECOMPOSITION

1. Recall that $\|\mathbf{A}\| = \sigma_1$, the largest singular value of $\mathbf{A}$. Furthermore, $\mu(\mathbf{A}) = \sigma_{\max}/\sigma_{\min}$. It remains to use Theorem 6 of Section 6.4 to compute the two singular values of $\mathbf{A}$ as square roots of eigenvalues of

$$\mathbf{A}^T\mathbf{A} = \begin{bmatrix} 2 & 1 \\ 1 & 1 \end{bmatrix}.$$

Thus $\|\mathbf{A}\| = \sqrt{\frac{1}{2}(\sqrt{5}+3)} \approx 1.618$ and $\mu(\mathbf{A}) = \dfrac{\sqrt{\frac{1}{2}(\sqrt{5}+3)}}{\sqrt{\frac{1}{2}(3-\sqrt{5})}} = \frac{1}{2}(\sqrt{5}+3) \approx 2.618$.

3. Norm of the given matrix: $\|\mathbf{A}\| \approx 16.848$, its condition number $\mu(\mathbf{A}) = \sigma_{\max}/\sigma_{\min} = \|\mathbf{A}\|/0 = \infty$.

5. Norm of the given matrix: $\|\mathbf{A}\| = \sigma_1 = 4$ (the absolute value of the largest eigenvalue of $\mathbf{A}$), its condition number $\mu(\mathbf{A}) = \sigma_{\max}/\sigma_{\min} = 4/1 = 4$.

7. Recall that by definition of the largest singular value, $\|\mathbf{A}\| = \|\mathbf{A}\mathbf{v}_1\| = \sigma_1$, and because $\mathbf{v}_1$ is a hypersphere vector, $\|\mathbf{v}_1\| = 1$, so the first of the desired equalities is satisfied for $\mathbf{w}_1 = \mathbf{v}_1$. Also, recall that $\|\mathbf{A}^{-1}\| = \sigma_{\min}^{-1}$. The vector $\mathbf{w}_2$ must therefore satisfy

$$\|\mathbf{A}^{-1}\mathbf{w}_2\| = \|\mathbf{w}_2\|/\sigma_{\min}.$$

We now observe that for $\mathbf{w}_2 = \mathbf{u}_r = \mathbf{A}\mathbf{v}_r$ the above equality holds, as $\|\mathbf{u}_r\| = \sigma_r = \sigma_{\min}$ and $\|\mathbf{v}_r\| = 1$.

9. To verify that the vector (19) minimizes the norm in (20), we find the minimum of the norm of (20), squared:

$$\|\mathbf{v}(t)\|^2 = (-2t + 1)^2 + t^2 + t^2,$$

and the quadratic function on the right attains its minimum when $d\|\mathbf{v}(t)\|^2/dt = 0 = -4(-2t+1) + 2t + 2t = 12t - 4$, whence $t = 1/3$.

11. From the definition of Euclidean norm in equation (12), one has

$$\|\mathbf{A}\mathbf{B}\| = \max_{\mathbf{v}\neq 0} \frac{\|\mathbf{A}\mathbf{B}\,\mathbf{v}\|}{\|\mathbf{v}\|} \leq \max_{\mathbf{v}\neq 0} \frac{\|\mathbf{A}\|\|\mathbf{B}\,\mathbf{v}\|}{\|\mathbf{v}\|} = \|\mathbf{A}\|\|\mathbf{B}\|,$$

where we used that $\|\mathbf{A}\mathbf{x}\| \leq \|\mathbf{A}\|\|\mathbf{x}\|$ for any vector $\mathbf{x}$.

13. Let λ_1 be the largest-by-magnitude eigenvalue of $\mathbf{A}$, then λ_1^n is such eigenvalue for $\mathbf{A}^n$. By Corollary 10 of Section 6.4, singular values of a symmetric matrix equal the magnitudes of its eigenvalues, so that, using equation (14), $\|\mathbf{A}\| = \sigma_1 = |\lambda_1|$. Since $\mathbf{A}^n$ is also a symmetric matrix, applying Corollary 10 to it implies $|\lambda_1^n|$ is its largest singular value, so $\|\mathbf{A}^n\| = |\lambda_1^n| = \|\mathbf{A}\|^n$.

15. Because transpose of an orthogonal matrix is again an orthogonal one, if $\mathbf{A} = \mathbf{U}\boldsymbol{\Sigma}\mathbf{V}^T$ is a singular value decomposition, then also $\mathbf{A}^T = \mathbf{V}\boldsymbol{\Sigma}^T\mathbf{U}^T$ is an SVD. Since for any SVD, the singular values are uniquely defined up to a permutation, this means $\mathbf{A}$ and $\mathbf{A}^T$ have the same singular values. Hence, from Theorem 6 of Section 6.4 follows that eigenvalues of $\mathbf{A}^T\mathbf{A}$ and $\mathbf{A}\mathbf{A}^T$ coincide, and so the former can be replaced by the latter in the formulation of Theorem 6.

17. The SVD representation of the given matrix is

$$
\mathbf{A} = \begin{bmatrix} 1 & 1 \\ 1 & -1 \end{bmatrix} = \begin{bmatrix} 0.707107 & 0.707107 \\ -0.707107 & 0.707107 \end{bmatrix}
$$
$$
\begin{bmatrix} 1.41421 & 0. \\ 0. & 1.41421 \end{bmatrix} \begin{bmatrix} 0. & 1. \\ 1. & 0. \end{bmatrix},
$$

so to get a rank one approximation one could use either

$$
\mathbf{\Sigma}_1 = \begin{bmatrix} 1.41421 & 0. \\ 0. & 0. \end{bmatrix} \quad \text{or} \quad \mathbf{\Sigma}_1' = \begin{bmatrix} 0. & 0. \\ 0. & 1.41421 \end{bmatrix}.
$$

The resulting approximations are

$$
\mathbf{A}_1 = \mathbf{U}\mathbf{\Sigma}_1\mathbf{V}^T = \begin{bmatrix} 0 & 1 \\ 0 & -1 \end{bmatrix} \quad \text{and} \quad \mathbf{A}_1' = \mathbf{U}\mathbf{\Sigma}_1'\mathbf{V}^T = \begin{bmatrix} 1 & 0 \\ 1 & 0 \end{bmatrix}
$$

respectively. They are, in fact, equally good approximations:

$$
\|\mathbf{A} - \mathbf{A}_1\|_{\text{Frob}} = \|\mathbf{A} - \mathbf{A}_1'\|_{\text{Frob}} = \sqrt{2}.
$$

19. The SVD of the first matrix is

$$
\mathbf{A} = \mathbf{U}\mathbf{\Sigma}\mathbf{V}^T
$$
$$
= \begin{bmatrix} -0.67691 & -0.20483 & 0.49983 & 0.5 \\ -0.67662 & -0.20499 & -0.50017 & -0.5 \\ -0.20506 & 0.67684 & 0.49983 & -0.5 \\ -0.20476 & 0.67669 & -0.50017 & 0.5 \end{bmatrix}
$$
$$
\times \begin{bmatrix} 5.51563 & 0. & 0. \\ 0. & 1.2561 & 0. \\ 0. & 0. & 0.00115 \\ 0. & 0. & 0. \end{bmatrix}
$$
$$
\times \begin{bmatrix} -0.3197 & 0.7513 & -0.57735 \\ -0.4908 & -0.65252 & -0.57735 \\ -0.8105 & 0.09878 & 0.57735 \end{bmatrix}^T,
$$

whence we obtain the rank two approximation

$$\mathbf{A}_2 = \mathbf{U}\boldsymbol{\Sigma}_2\mathbf{V}^T = \mathbf{U} \begin{bmatrix} 5.51563 & 0. & 0. \\ 0. & 1.2561 & 0. \\ 0. & 0. & 0. \\ 0. & 0. & 0. \end{bmatrix} \mathbf{V}^T$$

$$= \begin{bmatrix} 1.0003 & 2.0003 & 3.0007 \\ 0.9997 & 1.9997 & 2.9993 \\ 1.0003 & 0.0003 & 1.0007 \\ 0.9997 & -0.0003 & 0.9993 \end{bmatrix}.$$

Let us write $\mathbf{B}$ for the second given matrix. Then distances from $\mathbf{A}$ to $\mathbf{B}$ and $\mathbf{A}_2$ in the Frobenius norm are respectively

$$\|\mathbf{A} - \mathbf{B}\|_{\text{Frob}} = 0.002, \quad \text{and} \quad \|\mathbf{A} - \mathbf{A}_2\|_{\text{Frob}} = 0.0012.$$

21. Recall the formula (3) for SVD:

$$\big[\mathbf{A}\big]_{m\text{-by-}n} = \big[\mathbf{U}\big]_{m\text{-by-}m}\big[\boldsymbol{\Sigma}\big]_{m\text{-by-}n}\big[\mathbf{V}^T\big]_{n\text{-by-}n}.$$

Let us write a_{kl}, u_{ki}, σ_{ij}, v'_{jl} for the entries of $\mathbf{A}$, $\mathbf{U}$, $\boldsymbol{\Sigma}$, $\mathbf{V}^T$ respectively, where $1 \le k, i \le m$, $1 \le j, l \le n$. Using the definition of matrix multiplication, we have

$$a_{kl} = \sum_{i=1}^{m}\sum_{j=1}^{n} u_{ki}\sigma_{ij}v'_{jl}.$$

Let, on the other hand, b_{kl} be the corresponding entry of $\mathbf{U}_r\boldsymbol{\Sigma}_r\mathbf{V}_r^T$, then similarly to the above,

$$b_{kl} = \sum_{i=1}^{r}\sum_{j=1}^{r} u_{ki}\sigma_{ij}v'_{jl}.$$

Now since $\sigma_{ij} = 0$ as soon as either $i > r$, or $j > r$, the sums for a_{kl} and b_{kl} coincide. This proves $\mathbf{A} = \mathbf{U}\boldsymbol{\Sigma}\mathbf{V}^T = \mathbf{U}_r\boldsymbol{\Sigma}_r\mathbf{V}_r^T$.

23. Since det $\mathbf{A}$ is defined, $\mathbf{A}$ is a square matrix, and therefore so are $\mathbf{U}$, $\mathbf{V}$. We have

$$\det \mathbf{A} = \det \mathbf{U} \cdot \det \mathbf{\Sigma} \cdot \det \mathbf{V}.$$

Recall that $\mathbf{U}$, $\mathbf{V}$ are orthogonal matrices, so $\det \mathbf{U} = \pm 1$ and $\det \mathbf{V} = \pm 1$, whence follows

$$\det \mathbf{A} = \pm \det \mathbf{\Sigma}$$

as desired. Finally, since singular values are real *nonnegative* numbers, it is easy to construct examples for each sign in the above equality. The simplest such pair of examples is 1-by-1 matrices $\begin{bmatrix} 1 \end{bmatrix}$ and $\begin{bmatrix} -1 \end{bmatrix}$: the corresponding matrices of singular values are $\begin{bmatrix} 1 \end{bmatrix}$ for both of them.

25. (a) Let $\mathbf{x}$ and $\mathbf{y}$ be two fixed vectors. The dot product between $\mathbf{A}\mathbf{x}$ and $\mathbf{y}$ is $(\mathbf{A}\mathbf{x})^T \mathbf{y} = \mathbf{x}^T \mathbf{A}^T \mathbf{y}$. On the other hand, dot product between $\mathbf{x}$ and $\mathbf{A}^T \mathbf{y}$ also is $\mathbf{x}^T \mathbf{A}^T \mathbf{y}$. This shows that vectors $\mathbf{A}\mathbf{x}$ and $\mathbf{y}$ are orthogonal if and only if so are vectors $\mathbf{x}$ and $\mathbf{A}^T \mathbf{y}$.

Now put $\mathbf{x} = \mathbf{v}$, $\mathbf{y} = \mathbf{A}\mathbf{v}_0$. The above argument then shows that orthogonality of $\mathbf{A}\mathbf{v}$ and $\mathbf{A}\mathbf{v}_0$ is equivalent to that of $\mathbf{v}$ and $\mathbf{A}^T \mathbf{A}\mathbf{v}_0$.

(b) It is known from vector calculus that the directional derivative of function $f(\mathbf{x})$ along a vector $\mathbf{v}$ equals

$$f'_{\mathbf{v}}(\mathbf{x}) = \mathbf{grad} f(\mathbf{x}) \cdot \mathbf{v}.$$

We have noted that moving in any of the directions tangent to the sphere at the North Pole $\mathbf{v}_0$ does not increase the function $f(\mathbf{x})$. It means, along every such direction $f(\mathbf{x})$ has a local minimum at $\mathbf{v}_0$, and so the corresponding directional derivative is zero. Equivalently, for every $\mathbf{v}$ tangent to the sphere at $\mathbf{v}_0$ (that is, for $\mathbf{v} \perp \mathbf{v}_0$),

$$\mathbf{grad} f(\mathbf{v}_0) \cdot \mathbf{v} = 0.$$

This proves the desired statement.

(c) Observe that $A^T A$ is symmetric, so Lemma 3 of Section 6.2 is indeed applicable. Taking $b = 0$ yields

$$\mathbf{grad}(x^T A^T A x) = 2A^T A x.$$

We need to evaluate the gradient at the North Pole, i.e., $x = v_0$. The above formula then gives

$$2A^T A v_0.$$

(d) Let v be a vector in the equatorial plane. As we have shown, v is orthogonal to the gradient at the North Pole, that is, to $2A^T A v_0$. By (a), it is equivalent to Av being orthogonal to Av_0, so the statement of Lemma 4 follows.

27. (a) Put $W = R_{col}^n$.

    (b) Put W to be the subspace of vectors orthogonal to u_1.

    (c) Put W to be the subspace of vectors orthogonal to $u_1, u_2, \ldots,$ u_{k-1}.

6.5 THE POWER METHOD AND THE QR ALGORITHM

1. With this matrix and initial vector $[1 \ \ 1 \ \ 1]^T$, the first three decimals stabilize after the 5th iteration. With $[1 \ \ 0 \ \ 0]^T$ as initial vector, stabilization also occurs after 5 iterations. In both cases, the eigenvalue obtained in this way is 4, corresponding eigenvector is $[0 \ \ 1 \ \ 0]^T$.

3. Let $|\lambda_1| \leq \ldots \leq |\lambda_n|$ be the eigenvalues of A ordered by increasing magnitude. Then $|1/\lambda_n| \leq \ldots \leq |1/\lambda_1|$ are the eigenvalues of A^{-1} by increasing magnitude. The inverse power method may be seen as the direct method applied to A^{-1}. Its speed of convergence is therefore governed by the ratio of the largest-in-magnitude to the second largest eigenvalue of A^{-1}:

$$\frac{1/\lambda_1}{1/\lambda_2} = \frac{\lambda_2}{\lambda_1},$$

which is the ratio of the second smallest-in-magnitude eigenvalue of $\mathbf{A}$ to the smallest one. The larger the ratio, the faster is convergence of the inverse power method. This explains its efficiency for nearly-singular matrices: for those, λ_1 is small in magnitude, therefore the above ratio may be large.

5. Iterations with the initial vector $[1 \quad 0]^T$ stabilize to the third decimal after the 6th one, initial vector $[1 \quad 1]^T$ – after the 5th. Resulting eigenvector is $[-0.4472 \quad 0.8944]^T$ with the eigenvalue 5.

   Iterations with the matrix $\mathbf{A}^{-1}$ and the same initial vectors stabilize after the 4th one (for both vectors). The smallest eigenvalue obtained in this way is 1, a corresponding eigenvector: $[0.83205 \quad -0.5547]^T$.

7. Using the same pair of initial vectors as above leads to an interesting effect here: $[1 \quad 1]^T$ is an eigenvector for the smaller eigenvalue, so accidentally taking it as a starting iteration may be misleading.

   Using the basis vectors $[1 \quad 0]^T$ and $[0 \quad 1]^T$ as initial points of power method results in a series of iterations terminating after the 11th and 9th step respectively. The largest eigenvalue obtained in this way is 2, its eigenvector: $[0.3713 \quad 0.9284]^T$.

   Inverse power method with the initial basis vectors stabilizes to the third decimal after 11 and 13 iterations respectively for the first and second basis vector. The smallest eigenvalue is -1, corresponding eigenvector $[0.707 \quad 0.707]^T$.

9. Recalling the notation of Example 1, express the initial vector as

$$\mathbf{v} = c_1\mathbf{u}_1 + c_2\mathbf{u}_2 + \cdots + c_n\mathbf{u}_n,$$

where $\mathbf{u}_1, \ldots, \mathbf{u}_n$ are the eigenvectors of $\mathbf{A}$, and the corresponding eigenvalues are ordered so that $|\lambda_1| \leq \ldots \leq |\lambda_{n-2}| \leq |\lambda_{n-1}| = |\lambda_n|$ and $\lambda_{n-1} = -\lambda_n$. Then for the p-th iteration of the power method,

$$\frac{\mathbf{A}^p\mathbf{v}}{\lambda_n^p} = c_1\left(\frac{\lambda_1}{\lambda_n}\right)^p\mathbf{u}_1 + \cdots + c_{n-2}\left(\frac{\lambda_{n-2}}{\lambda_n}\right)^p\mathbf{u}_{n-2}$$
$$+ c_{n-1}(-1)^p\mathbf{u}_{n-1} + c_n\mathbf{u}_n,$$

so if $|\lambda_{n-2}| < |\lambda_{n-1}|$, the sequence of such iterations has *two* limit points: $c_n\mathbf{u}_n + c_{n-1}\mathbf{u}_{n-1}$ and $c_n\mathbf{u}_n - c_{n-1}\mathbf{u}_{n-1}$ for even and odd sequence elements respectively. This implies, sums and differences of consecutive iterations $A^{p+1}v + A^p v$ and $A^{2p}v - A^{2p-1}v$ converge to the two eigenvectors of dominant eigenvalues: $2c_n u_n$ and $2c_{n-1}u_{n-1}$.

11. (a) One needs 13 iterations for eigenvector entries to stabilize up to three decimals. The largest in magnitude eigenvalue is ≈ 3.000 (after 20 iterations).

    (b) The iterative process stabilizes up to three decimals after 10 steps, the resulting smallest eigenvalue is 1 (after 20 iterations, before roundoff: 0.99999).

    (c) Only 4 iterations are required to stabilize up to three decimals in this case. The largest eigenvalue of $M = (A - 1.9I)^{-1}$ is 10 (after 20 iterations; with precision by far larger than 3 decimals), which means that the smallest eigenvalue of $A - 1.9I$ is 0.1, thus A has the eigenvalue 2.

    (d) By the above, A has eigenvalues 1, 2, 3. According to Example 1, the speed of convergence is defined by the ratio of the largest-in-magnitude to the second largest eigenvalue (the larger the ratio, the faster is the convergence). It remains to observe that for the matrices A, A^{-1}, $(A - 1.9I)^{-1}$ this ratio is 1.5, 2, 9 respectively.

13. Iterations with the given matrix and initial vector will oscillate between vectors proportional to $[1 \ 1 \ 1]^T$ and $[1 \ 1 \ 5]^T$. The reason is – eigenvalues of the given matrix are $r_1 = -3$ and $r_2 = 3$ of multiplicity 2. Having two eigenvalues of the largest magnitude means that the convergence estimates in Example 1 need not hold, which has in fact been observed.

15. Recall that according to Theorem 4 of Section 4.3, the vector providing a least square approximation for an inconsistent linear system is the one satisfying normal equations (equation (6) of Section 4.3). In our case the inconsistent system is

$$vr = b = Av,$$

where r is the unknown, v is the coefficient matrix, and vector b is defined by the second equality. Hence the normal equation takes the form

$$v^T vr = v^T b = v^T Av.$$

Solving with respect to r, we obtain $r = v^T Av / v^T v$ as desired.

17. The derivation consists of multiplying equation

$$\mathbf{v} = c_1\mathbf{u}_1 + \cdots + c_n\mathbf{u}_n$$

by matrix $\mathbf{A}^p$ on the left and using linearity and associativity of matrix multiplication.

7

LINEAR SYSTEMS OF DIFFERENTIAL EQUATIONS

7.1 FIRST-ORDER LINEAR SYSTEMS

1. Differentiating each entry of the given matrix,
$$\mathbf{X}(t) = \begin{bmatrix} e^{5t} & 3e^{5t} \\ -2e^{5t} & -e^{5t} \end{bmatrix}, \quad \text{gives} \quad \frac{d\mathbf{X}(t)}{dt} = \begin{bmatrix} 5e^{5t} & 15e^{5t} \\ -10e^{5t} & -5e^{5t} \end{bmatrix}.$$

3. (a) To calculate $\int \mathbf{A}(t)\, dt$, we integrate each entry of $\mathbf{A}(t)$ to obtain

$$\int \mathbf{A}(t)\, dt = \begin{bmatrix} \int t\, dt & \int e^t\, dt \\ \int 1\, dt & \int e^t\, dt \end{bmatrix} = \begin{bmatrix} t^2/2 + c_1 & e^t + c_2 \\ t + c_3 & e^t + c_4 \end{bmatrix}.$$

(b) Taking the definite integral of each entry in $\mathbf{B}(t)$ yields

$$\int\limits_0^1 \mathbf{B}(t)\, dt = \begin{bmatrix} \int_0^1 \cos t\, dt & -\int_0^1 \sin t\, dt \\ \int_0^1 \sin t\, dt & \int_0^1 \cos t\, dt \end{bmatrix}$$

$$= \begin{bmatrix} \sin t \,\big|_0^1 & \cos t \,\big|_0^1 \\ -\cos t \,\big|_0^1 & \sin t \,\big|_0^1 \end{bmatrix} = \begin{bmatrix} \sin 1 & \cos 1 - 1 \\ 1 - \cos 1 & \sin 1 \end{bmatrix}.$$

(c) By the product rule, we see that

$$\frac{d}{dt}\, [\mathbf{A}(t)\mathbf{B}(t)] = \mathbf{A}(t)\mathbf{B}'(t) + \mathbf{A}'(t)\mathbf{B}(t).$$

Solutions Manual to Accompany Fundamentals of Matrix Analysis with Applications,
First Edition. Edward Barry Saff and Arthur David Snider.
© 2016 John Wiley & Sons, Inc. Published 2016 by John Wiley & Sons, Inc.

Therefore, we first calculate $\mathbf{A}'(t)$ and $\mathbf{B}'(t)$ by differentiating each entry of $\mathbf{A}(t)$ and $\mathbf{B}(t)$, respectively, to obtain

$$\mathbf{A}'(t) = \begin{bmatrix} 1 & e^t \\ 0 & e^t \end{bmatrix} \quad \text{and} \quad \mathbf{B}'(t) = \begin{bmatrix} -\sin t & -\cos t \\ \cos t & -\sin t \end{bmatrix}.$$

Hence, by matrix multiplication we have

$$\frac{d}{dt}[\mathbf{A}(t)\mathbf{B}(t)] = \mathbf{A}(t)\mathbf{B}'(t) + \mathbf{A}'(t)\mathbf{B}(t)$$

$$= \begin{bmatrix} t & e^t \\ 1 & e^t \end{bmatrix} \begin{bmatrix} -\sin t & -\cos t \\ \cos t & -\sin t \end{bmatrix} + \begin{bmatrix} 1 & e^t \\ 0 & e^t \end{bmatrix} \begin{bmatrix} \cos t & -\sin t \\ \sin t & \cos t \end{bmatrix}$$

$$= \begin{bmatrix} e^t \cos t - t \sin t & -t \cos t - e^t \sin t \\ e^t \cos t - \sin t & -\cos t - e^t \sin t \end{bmatrix}$$

$$+ \begin{bmatrix} \cos t + e^t \sin t & e^t \cos t - \sin t \\ e^t \sin t & e^t \cos t \end{bmatrix}$$

$$= \begin{bmatrix} (1+e^t)\cos t + (e^t - t)\sin t & (e^t - t)\cos t - (e^t + 1)\sin t \\ e^t \cos t + (e^t - 1)\sin t & (e^t - 1)\cos t - e^t \sin t \end{bmatrix}.$$

Thus,

$$\frac{d}{dt}[\mathbf{A}(t)\mathbf{B}(t)]$$

$$= \begin{bmatrix} (1+e^t)\cos t + (e^t - t)\sin t & (e^t - t)\cos t - (e^t + 1)\sin t \\ e^t \cos t + (e^t - 1)\sin t & (e^t - 1)\cos t - e^t \sin t \end{bmatrix}.$$

One can also get this answer by differentiating term-wise the product

$$\mathbf{A}(t)\mathbf{B}(t) = \begin{bmatrix} t \cos t + e^t \sin t & e^t \cos t - t \sin t \\ \cos t + e^t \sin t & e^t \cos t - \sin t \end{bmatrix}.$$

5. We simply compare the two following expressions:

$$\mathbf{x}'(t) = \begin{bmatrix} 3e^{3t} \\ 6e^{3t} \end{bmatrix}$$

and

$$\begin{bmatrix} 1 & 1 \\ -2 & 4 \end{bmatrix} \mathbf{x}(t) = \begin{bmatrix} 1 & 1 \\ -2 & 4 \end{bmatrix} \begin{bmatrix} e^{3t} \\ 2e^{3t} \end{bmatrix} = \begin{bmatrix} 3e^{3t} \\ 6e^{3t} \end{bmatrix},$$

which proves the desired statement.

7. We first calculate $\mathbf{X}'(t)$ by differentiating each entry of $\mathbf{X}(t)$. Thus, we obtain

$$\mathbf{X}'(t) = \begin{bmatrix} 2e^{2t} & 3e^{3t} \\ -2e^{2t} & -6e^{3t} \end{bmatrix}.$$

Substituting the matrix $\mathbf{X}(t)$ into the given differential equation and performing matrix multiplication yields

$$\begin{bmatrix} 1 & -1 \\ 2 & 4 \end{bmatrix} \mathbf{X} = \begin{bmatrix} 1 & -1 \\ 2 & 4 \end{bmatrix} \begin{bmatrix} e^{2t} & e^{3t} \\ -e^{2t} & -2e^{3t} \end{bmatrix}$$

$$= \begin{bmatrix} e^{2t} + e^{2t} & e^{3t} + 2e^{3t} \\ 2e^{2t} - 4e^{2t} & 2e^{3t} - 8e^{3t} \end{bmatrix} = \begin{bmatrix} 2e^{2t} & 3e^{3t} \\ -2e^{2t} & -6e^{3t} \end{bmatrix} = \mathbf{X}'.$$

So, we see that $\mathbf{X}(t)$ does satisfy the given differential equation.

9. We first check that $\mathbf{x}_1$, $\mathbf{x}_2$ satisfy $\mathbf{x}' = \mathbf{A}\mathbf{x}$:

$\mathbf{x}_1'(t) = \begin{bmatrix} e^t \\ e^t \end{bmatrix}$, on the other hand, $\mathbf{A}\mathbf{x}_1(t) = \begin{bmatrix} 2 & -1 \\ 3 & -2 \end{bmatrix} \begin{bmatrix} e^t \\ e^t \end{bmatrix} = \begin{bmatrix} e^t \\ e^t \end{bmatrix}.$

Similarly for $\mathbf{x}_2$:

$\mathbf{x}_2'(t) = \begin{bmatrix} -e^{-t} \\ -3e^{-t} \end{bmatrix}$, and $\mathbf{A}\mathbf{x}_2(t) = \begin{bmatrix} 2 & -1 \\ 3 & -2 \end{bmatrix} \begin{bmatrix} e^{-t} \\ 3e^{-t} \end{bmatrix} = \begin{bmatrix} -e^{-t} \\ -3e^{-t} \end{bmatrix}.$

To verify that $\mathbf{x}_p$ is a solution of the given nonhomogeneous system, we compute

$$\mathbf{x}_p' = \frac{3}{2} \begin{bmatrix} e^t + te^t \\ e^t + te^t \end{bmatrix} - \frac{1}{4} \begin{bmatrix} e^t \\ 3e^t \end{bmatrix} + \begin{bmatrix} 1 \\ 2 \end{bmatrix} = \begin{bmatrix} \frac{5}{4}e^t + \frac{3}{2}te^t + 1 \\ \frac{3}{4}e^t + \frac{3}{2}te^t + 2 \end{bmatrix}.$$

On the other hand,

$$\mathbf{A}\mathbf{x}_p + \mathbf{f}(t) = \begin{bmatrix} 2 & -1 \\ 3 & -2 \end{bmatrix} \begin{bmatrix} \frac{3}{2}te^t - \frac{1}{4}e^t + t \\ \frac{3}{2}te^t - \frac{3}{4}e^t + 2t - 1 \end{bmatrix} + \begin{bmatrix} e^t \\ t \end{bmatrix}$$

$$= \begin{bmatrix} \frac{3}{2}te^t + \frac{5}{4}e^t + 1 \\ \frac{3}{2}te^t + \frac{3}{4}e^t + 2 \end{bmatrix},$$

where the last expression is precisely $\mathbf{x}_p'$.

11. The right-hand expressions are:

$$7x + 2y = [7 \ 2] \cdot [x \ y],$$

so the given system takes form

$$\begin{bmatrix} x \\ y \end{bmatrix}' = \begin{bmatrix} 7 & 2 \\ 3 & -2 \end{bmatrix} \begin{bmatrix} x \\ y \end{bmatrix}.$$

13. We start by expressing right-hand sides of all equations as dot products.

$$x + y + z = [1 \ 1 \ 1] \cdot [x \ y \ z], \quad 2z - x = [-1 \ 0 \ 2] \cdot [x \ y \ z],$$
$$4y = [0 \ 4 \ 0] \cdot [x \ y \ z].$$

Thus, by the definition of the product of a matrix and a column vector, the matrix form of the given system is

$$\begin{bmatrix} x \\ y \\ z \end{bmatrix}' = \begin{bmatrix} 1 & 1 & 1 \\ -1 & 0 & 2 \\ 0 & 4 & 0 \end{bmatrix} \begin{bmatrix} x \\ y \\ z \end{bmatrix}.$$

15. To rewrite this system in matrix form, we define the vectors $\mathbf{x}(t) = [x(t) \ y(t)]^T$ (which gives $\mathbf{x}'(t) = [x'(t) \ y'(t)]^T$), $\mathbf{f}(t) = [t^2 \ e^t]^T$, and the matrix

$$\mathbf{A}(t) = \begin{bmatrix} 3 & -1 \\ -1 & 2 \end{bmatrix}.$$

Thus, this system becomes a matrix differential equation

$$\begin{bmatrix} x(t) \\ y(t) \end{bmatrix}' = \begin{bmatrix} 3 & -1 \\ -1 & 2 \end{bmatrix} \begin{bmatrix} x(t) \\ y(t) \end{bmatrix} + \begin{bmatrix} t^2 \\ e^t \end{bmatrix}.$$

We can verify that this equation is equivalent to the original system by performing matrix multiplication and addition to obtain

$$\begin{bmatrix} x'(t) \\ y'(t) \end{bmatrix} = \begin{bmatrix} 3x(t) - y(t) \\ -x(t) + 2y(t) \end{bmatrix} + \begin{bmatrix} t^2 \\ e^t \end{bmatrix} = \begin{bmatrix} 3x(t) - y(t) + t^2 \\ -x(t) + 2y(t) + e^t \end{bmatrix}.$$

Since two vectors are equal if and only if their corresponding components are equal, we see that this vector equation is equivalent to

$$x'(t) = 3x(t) - y(t) + t^2,$$
$$y'(t) = -x(t) + 2y(t) + e^t,$$

which is the original system.

17. Arguing along the lines of Exercise 15, we introduce vectors $x(t) = [x(t) \ y(t) \ z(t)]^T$, $\mathbf{f}(t) = [3 \ \sin t \ 0]^T$ and the matrix

$$\mathbf{A} = \begin{bmatrix} 1 & 1 & 1 \\ 2 & -1 & 3 \\ 1 & 0 & 5 \end{bmatrix}.$$

The given system then becomes

$$\mathbf{x}'(t) = \begin{bmatrix} x \\ y \\ z \end{bmatrix}' = \begin{bmatrix} 1 & 1 & 1 \\ 2 & -1 & 3 \\ 1 & 0 & 5 \end{bmatrix} \begin{bmatrix} x \\ y \\ z \end{bmatrix} + \begin{bmatrix} 3 \\ \sin t \\ 0 \end{bmatrix} = \mathbf{A}\mathbf{x}(t) + \mathbf{f}(t).$$

19. This exercise is similar to Example 5 in Section 7.1. First of all, observe that the given system contains second derivatives of both x and y. To express it as a first-order system, we will introduce the vector $[x_1(t) \ x_2(t) \ x_3(t) \ x_4(t)]^T$, where $x_1(t) = x(t)$, $x_2(t) = x'(t)$, $x_3(t) = y(t)$ and $x_4(t) = y'(t)$. The original system can now be rewritten as

$$x_2' = -3x_2 + x_4 - 2x_3$$
$$x_4' = -x_2 - 3x_4 - x_3.$$

Adding the equations $x_2(t) = x_1'(t)$, $x_4(t) = x_3'(t)$, we obtain

$$x_1' = x_2$$
$$x_2' = -3x_2 + x_4 - 2x_3$$
$$x_3' = x_4$$
$$x_4' = -x_2 - 3x_4 - x_3,$$

and in the matrix form:

$$
\begin{bmatrix} x_1 \\ x_2 \\ x_3 \\ x_4 \end{bmatrix}' = \begin{bmatrix} 0 & 1 & 0 & 0 \\ 0 & -3 & -2 & 1 \\ 0 & 0 & 0 & 1 \\ 0 & -1 & -1 & -3 \end{bmatrix} \begin{bmatrix} x_1 \\ x_2 \\ x_3 \\ x_4 \end{bmatrix}.
$$

21. This equation can be written as a first order system in normal form by using the substitutions $x_1(t) = y(t)$ and $x_2(t) = y'(t)$. With these substitutions, we have

$$
x_1'(t) = 0 \cdot x_1(t) + x_2(t),
$$
$$
x_2'(t) = 10x_1(t) + 3x_2(t) + \sin t.
$$

Let $\mathbf{x}(t) = [x_1(t) \ \ x_2(t)]^T$ (which means that $\mathbf{x}'(t) = [x_1'(t) \ \ x_2'(t)]^T$). We now write the system as a matrix differential equation using the vector $\mathbf{f}(t) = [0 \ \ \sin t]^T$ and the matrix

$$
\mathbf{A} = \begin{bmatrix} 0 & 1 \\ 10 & 3 \end{bmatrix}.
$$

Hence, the system in normal form yields a matrix differential equation

$$
\begin{bmatrix} x_1'(t) \\ x_2'(t) \end{bmatrix} = \begin{bmatrix} 0 & 1 \\ 10 & 3 \end{bmatrix} \begin{bmatrix} x_1(t) \\ x_2(t) \end{bmatrix} + \begin{bmatrix} 0 \\ \sin t \end{bmatrix}.
$$

23. This equation can be written as a first order system in normal form by using the substitutions $x_1(t) = w(t)$, $x_2(t) = w'(t)$, $x_3(t) = w''(t)$, and $x_4(t) = w'''(t)$. These substitutions give

$$
x_1'(t) = 0 \cdot x_1(t) + x_2(t) + 0 \cdot x_3(t) + 0 \cdot x_4(t),
$$
$$
x_2'(t) = 0 \cdot x_1(t) + 0 \cdot x_2(t) + x_3(t) + 0 \cdot x_4(t),
$$
$$
x_3'(t) = 0 \cdot x_1(t) + 0 \cdot x_2(t) + 0 \cdot x_3(t) + x_4(t),
$$
$$
x_4'(t) = -x_1(t) + 0 \cdot x_2(t) + 0 \cdot x_3(t) + 0 \cdot x_4(t) + t^2.
$$

We now rewrite this system as a matrix differential equation, i.e., $\mathbf{x}' = \mathbf{A}\mathbf{x}$, by defining vectors $\mathbf{x}(t) = [x_1(t) \ \ x_2(t) \ \ x_3(t) \ \ x_4(t)]^T$ (so that $\mathbf{x}'(t) = [x_1'(t) \ \ x_2'(t) \ \ x_3'(t) \ \ x_4'(t)]^T$) and $\mathbf{f}(t) = [0 \ \ 0 \ \ 0 \ \ t^2]^T$, and the matrix

$$A = \begin{bmatrix} 0 & 1 & 0 & 0 \\ 0 & 0 & 1 & 0 \\ 0 & 0 & 0 & 1 \\ -1 & 0 & 0 & 0 \end{bmatrix}.$$

Thus, the given fourth order differential equation is equivalent to the matrix system

$$\begin{bmatrix} x_1'(t) \\ x_2'(t) \\ x_3'(t) \\ x_4'(t) \end{bmatrix} = \begin{bmatrix} 0 & 1 & 0 & 0 \\ 0 & 0 & 1 & 0 \\ 0 & 0 & 0 & 1 \\ -1 & 0 & 0 & 0 \end{bmatrix} \begin{bmatrix} x_1(t) \\ x_2(t) \\ x_3(t) \\ x_4(t) \end{bmatrix} + \begin{bmatrix} 0 \\ 0 \\ 0 \\ t^2 \end{bmatrix}.$$

25. The assumption of the liquid in tanks being well stirred means that the concentration of salt in solution is $x_1(t)/75$ and $x_2(t)/75$ kg/L in the first and the second tank respectively. Multiplying this concentration by the rate of flows from both tanks gives us the mass of salt that exits each tank every minute (in particular, the flows between tanks and out of the system). Taking into account also the mass of salt flowing into both tanks, we have

$$x_1'(t) = 0.8 + \frac{x_2(t)}{75} - 5\frac{x_1(t)}{75}$$
$$x_2'(t) = 0.1 + 4\frac{x_1(t)}{75} - 6\frac{x_2(t)}{75}.$$

The units of the above equations are kg/min. Writing them in the matrix form yields

$$\begin{bmatrix} x_1 \\ x_2 \end{bmatrix}' = \begin{bmatrix} \frac{-1}{15} & \frac{1}{75} \\ \frac{4}{75} & \frac{-2}{25} \end{bmatrix} \begin{bmatrix} x_1 \\ x_2 \end{bmatrix} + \begin{bmatrix} \frac{4}{5} \\ \frac{1}{10} \end{bmatrix}.$$

27. Recall that the equation of heat transfer is

$$\frac{dx}{dt} = -\frac{x - x_\infty}{\tau},$$

where $x(t)$ denotes temperature of the body, x_∞ is the temperature of the environment, τ is the time constant, t – time. If the transfer occurs between two bodies, x, x_∞ should be replaced with their temperatures (and thus both can be functions of t). Applying the heat transfer equation to zone A gives

$$x_A'(t) = 80 \times 1/4 - \frac{x_A(t)}{4} - \frac{x_A(t) - x_B(t)}{3},$$

where the first term accounts for the heat generated by the furnace, the second – for the heat transfer between A and the outside, the third – between A and B. Units of this equation are °F/h. Note that here the temperature of the environment is $0°$ Similarly, for area B:

$$x_B'(t) = -\frac{x_B(t)}{5} - \frac{x_B(t) - x_A(t)}{3}.$$

Writing the above equations in matrix form, one obtains

$$\begin{bmatrix} x_A \\ x_B \end{bmatrix}' = \begin{bmatrix} -7/12 & 1/3 \\ 1/3 & -8/15 \end{bmatrix} \begin{bmatrix} x_A \\ x_B \end{bmatrix} + \begin{bmatrix} 20 \\ 0 \end{bmatrix}.$$

29. Applying the Newton's law to the leftmost mass, we conclude that mx'' must equal the sum of all forces acting on it. There are only two such forces: created on the left and on the right by two adjacent springs. Hence,

$$mx'' = -kx + k(y - x).$$

The equation for the rightmost mass is identical to the above with x replaced by z. Computing the deformations of springs adjacent to the middle mass in terms of displacements x, y, z, one has

$$my'' = k(x - y) + k(z - y).$$

In order to write the three above equations as a first-order matrix system, we introduce the new set of unknowns, x_i, $1 \le i \le 6$, such that $x_1(t) = x(t)$, $x_2(t) = x'(t)$, $x_3(t) = y(t)$, $x_4(t) = y'(t)$, $x_5(t) = z(t)$, $x_6(t) = z'(t)$. Taking these defining equations into account, the resulting matrix equation becomes

$$
\begin{bmatrix} x_1 \\ x_2 \\ x_3 \\ x_4 \\ x_5 \\ x_6 \end{bmatrix}' =
\begin{bmatrix}
0 & 1 & 0 & 0 & 0 & 0 \\
-\frac{2k}{m} & 0 & \frac{k}{m} & 0 & 0 & 0 \\
0 & 0 & 0 & 1 & 0 & 0 \\
\frac{k}{m} & 0 & -\frac{2k}{m} & 0 & \frac{k}{m} & 0 \\
0 & 0 & 0 & 0 & 0 & 1 \\
0 & 0 & \frac{k}{m} & 0 & -\frac{2k}{m} & 0
\end{bmatrix}
\begin{bmatrix} x_1 \\ x_2 \\ x_3 \\ x_4 \\ x_5 \\ x_6 \end{bmatrix}.
$$

31. Arguing similarly to the Exercise 29, we obtain the following expressions for Newton's law for the two masses:

$$
m_1 x'' = -k_1 x + b(y' - x')
$$
$$
m_2 y'' = -k_2 y + b(x' - y').
$$

Substituting $x_1(t) = x(t)$, $x_2(t) = x'(t)$, $x_3(t) = y(t)$, $x_4(t) = y'(t)$, we write the previous system as a matrix differential equation:

$$
\begin{bmatrix} x_1 \\ x_2 \\ x_3 \\ x_4 \end{bmatrix}' =
\begin{bmatrix}
0 & 1 & 0 & 0 \\
\frac{-k_1}{m_1} & \frac{-b}{m_1} & 0 & \frac{b}{m_1} \\
0 & 0 & 1 & 0 \\
0 & \frac{b}{m_2} & \frac{-k_2}{m_2} & \frac{-b}{m_2}
\end{bmatrix}
\begin{bmatrix} x_1 \\ x_2 \\ x_3 \\ x_4 \end{bmatrix}.
$$

33. Since we want I_1 and I_2 to be dummy variables, we list them last and write the final three equations as

$$
\begin{bmatrix}
0 & -30 & 60 & 0 & 0 \\
-1 & 0 & 0 & 1 & -1 \\
0 & -1 & -1 & 0 & 1
\end{bmatrix}
\begin{bmatrix} I_2 \\ I_4 \\ I_5 \\ I_1 \\ I_3 \end{bmatrix} =
\begin{bmatrix} 0 \\ 0 \\ 0 \end{bmatrix}.
$$

Gauss elimination yields $I_5 = I_3/2$, $I_4 = 2I_3/3$, $I_2 = I_1 - I_3$.
Substitution into the first two equations yields

$$
2I_1' + 90(I_1 - I_3) = 9, \quad I_3' + 30(\tfrac{2}{3}I_3) - 90(I_1 - I_3) = 0, \text{ or}
$$
$$
\begin{bmatrix} I_1 \\ I_3 \end{bmatrix}' =
\begin{bmatrix} -45 & 45 \\ 90 & -110 \end{bmatrix}
\begin{bmatrix} I_1 \\ I_3 \end{bmatrix} +
\begin{bmatrix} 9/2 \\ 0 \end{bmatrix}.
$$

35. Kirchhoff's current Law at the bottom node shows $I_1 = I_3 + I_2$.
The Voltage Law, around the left loop, says $-10 + 10I_1 + 0.02I_2' = 0$.
The Voltage Law, around the right loop, says $0.025I_3' - 0.02I_2$.
Substituting $I_1 = I_3 + I_2$ gives the matrix format

$$
\begin{bmatrix} 0.02 & 0 \\ -0.02 & 0.025 \end{bmatrix}
\begin{bmatrix} I_2 \\ I_3 \end{bmatrix}' =
\begin{bmatrix} -10 & = 10 \\ 0 & 0 \end{bmatrix}
\begin{bmatrix} I_1 \\ I_3 \end{bmatrix} +
\begin{bmatrix} 10 \\ 0 \end{bmatrix}
$$

and multiplying by the inverse, $\begin{bmatrix} 50 & 0 \\ 40 & 40 \end{bmatrix}$, gives

$$\begin{bmatrix} I_2 \\ I_3 \end{bmatrix}' = \begin{bmatrix} -500 & -500 \\ -400 & -400 \end{bmatrix} \begin{bmatrix} I_2 \\ I_3 \end{bmatrix} + \begin{bmatrix} 500 \\ 400 \end{bmatrix}.$$

37. Kirchhoff's Current Law, applied at the top right node, tells us $I_3 = I_1 - I_2$.

Kirchhoff's Voltage Law, applied around the left loop, tells us $50I_1' + 80I_2 - 160 = 0$, or $I_1' = -\frac{8}{5}I_2 + \frac{16}{5}$.

The Voltage Law, applied around the right loop, says $q_3/\frac{1}{800} - 80I_2 = 0$, where $I_3 = q_3'$.

Differentiating the latter and substituting for I_3 gives $800I_3 - 80I_2' = 0 = 800(I_1 - I_2) - 80I_2'$, or $I_2' = 10I_1 - 10I_2$.

The matrix form is

$$\begin{bmatrix} I_1 \\ I_2 \end{bmatrix}' = \begin{bmatrix} 0 & -8/5 \\ 10 & -10 \end{bmatrix} \begin{bmatrix} I_1 \\ I_2 \end{bmatrix} + \begin{bmatrix} 16/5 \\ 0 \end{bmatrix}.$$

39. Kirchhoff's Current Law, applied at the bottom node, says $I_3 = I_1 - I_2$.

Kirchhoff's Voltage Law for the left loop says $0.05I_{11}' + (1)I_2 = 0$.

The Voltage Law for the right loop says $-I_2 + q_3/0.5 + \cos 3t = 0$, where $I_3 = q_3'$.

Differentiating the latter and substituting for I_3 gives $-I_2' + 2I_3 - 3\sin 3t = 0 = -I_2' + 2(I_1 - I_2) - 3\sin 3t$.

The matrix form is

$$\begin{bmatrix} I_1 \\ I_2 \end{bmatrix}' = \begin{bmatrix} 0 & -2 \\ 2 & -2 \end{bmatrix} \begin{bmatrix} I_1 \\ I_2 \end{bmatrix} + \begin{bmatrix} 0 \\ -3\sin 37 \end{bmatrix}.$$

7.2 THE MATRIX EXPONENTIAL FUNCTION

1. Recall that for a diagonalizable 2-by-2 matrix $\mathbf{A} = \mathbf{P}^{-1}\mathbf{DP}$ it holds

$$e^{\mathbf{A}t} = \mathbf{P} \begin{bmatrix} e^{r_1 t} & 0 \\ 0 & e^{r_2 t} \end{bmatrix} \mathbf{P}^{-1},$$

where r_1, r_2 are the eigenvalues, $\mathbf{P}$ is the matrix of eigenvectors, $\mathbf{D} = \text{diag}(r_1, r_2)$ is the diagonal matrix of eigenvalues. Computing this expression in our case gives

$$e^{\mathbf{A}t} = \begin{bmatrix} 1 & 1 \\ -\frac{1}{3} & 2 \end{bmatrix} \begin{bmatrix} e^{10t} & 0 \\ 0 & e^{3t} \end{bmatrix} \begin{bmatrix} 1 & 1 \\ -\frac{1}{3} & 2 \end{bmatrix}^{-1} = \begin{bmatrix} \frac{e^{3t}}{7} + \frac{6e^{10t}}{7} & \frac{3e^{3t}}{7} - \frac{3e^{10t}}{7} \\ \frac{2e^{3t}}{7} - \frac{2e^{10t}}{7} & \frac{6e^{3t}}{7} + \frac{e^{10t}}{7} \end{bmatrix}.$$

In Exercises 3–11 we follow the same argument as outlined above.

3.

$$e^{\mathbf{A}t} = \begin{bmatrix} 1 & 1 \\ -1 & 1 \end{bmatrix} \begin{bmatrix} e^{3t} & 0 \\ 0 & e^{-t} \end{bmatrix} \begin{bmatrix} 1 & 1 \\ -1 & 1 \end{bmatrix}^{-1} = \begin{bmatrix} \frac{e^{-t}}{2} + \frac{e^{3t}}{2} & \frac{e^{-t}}{2} - \frac{e^{3t}}{2} \\ \frac{e^{-t}}{2} - \frac{e^{3t}}{2} & \frac{e^{-t}}{2} + \frac{e^{3t}}{2} \end{bmatrix}.$$

5.

$$e^{\mathbf{A}t} = \begin{bmatrix} 1 & 1 \\ 1 & -1 \end{bmatrix} \begin{bmatrix} e^{2t} & 0 \\ 0 & 1 \end{bmatrix} \begin{bmatrix} 1 & 1 \\ 1 & -1 \end{bmatrix}^{-1} = \begin{bmatrix} \frac{1}{2} + \frac{e^{2t}}{2} & -\frac{1}{2} + \frac{e^{2t}}{2} \\ -\frac{1}{2} + \frac{e^{2t}}{2} & \frac{1}{2} + \frac{e^{2t}}{2} \end{bmatrix}.$$

7.

$$e^{\mathbf{A}t} = \begin{bmatrix} 1 & 1 & 1 \\ 1 & 0 & 1 \\ 0 & 1 & 1 \end{bmatrix} \begin{bmatrix} e^{3t} & 0 & 0 \\ 0 & e^{2t} & 0 \\ 0 & 0 & e^{t} \end{bmatrix} \begin{bmatrix} 1 & 1 & 1 \\ 1 & 0 & 1 \\ 0 & 1 & 1 \end{bmatrix}^{-1}$$

$$= \begin{bmatrix} -e^{t} + e^{2t} + e^{3t} & e^{t} - e^{2t} & e^{t} - e^{3t} \\ -e^{t} + e^{3t} & e^{t} & e^{t} - e^{3t} \\ -e^{t} + e^{2t} & e^{t} - e^{2t} & e^{t} \end{bmatrix}.$$

9.

$$e^{\mathbf{A}t} = \begin{bmatrix} 1 & 1 \\ i & -i \end{bmatrix} \begin{bmatrix} e^{-it} & 0 \\ 0 & e^{it} \end{bmatrix} \begin{bmatrix} 1 & 1 \\ i & -i \end{bmatrix}^{-1} = \begin{bmatrix} \cos(t) & -\sin(t) \\ \sin(t) & \cos(t) \end{bmatrix}.$$

11.

$$e^{\mathbf{A}t} = \begin{bmatrix} 1 & 1 \\ \frac{1}{2} & -i \end{bmatrix} \begin{bmatrix} (1+i)t & 0 \\ 0 & 3t \end{bmatrix} \begin{bmatrix} 1 & 1 \\ \frac{1}{2} & -i \end{bmatrix}^{-1}$$

$$= \begin{bmatrix} e^{(1+i)t}\left(\frac{4}{5} + \frac{2i}{5}\right) + \left(\frac{1}{5} - \frac{2i}{5}\right)e^{3t} & \left(-\frac{2}{5} + \frac{4i}{5}\right)e^{t}\left(-e^{it} + e^{2t}\right) \\ \left(\frac{2}{5} + \frac{i}{5}\right)e^{t}\left(e^{it} - e^{2t}\right) & e^{3t}\left(\frac{4}{5} + \frac{2i}{5}\right) + \left(\frac{1}{5} - \frac{2i}{5}\right)e^{(1+i)t} \end{bmatrix}.$$

13. For the given matrix

$$\mathbf{A} = \begin{bmatrix} 1 & -1 \\ 1 & 3 \end{bmatrix},$$

the unique eigenvalue is $r_1 = 2$, and therefore $(\mathbf{A}t - tr_1\mathbf{I})^2 = 0$. Following the strategy described before the Exercise 13, we obtain

$$e^{\mathbf{A}t} = e^{tr_1}e^{(\mathbf{A}t - tr_1\mathbf{I})}.$$

Since in the above equation

$$e^{(\mathbf{A}t - tr_1\mathbf{I})} = \mathbf{I} + (\mathbf{A}t - tr_1\mathbf{I}),$$

we have

$$e^{\mathbf{A}t} = e^{2t}(\mathbf{I} + (\mathbf{A}t - 2t\mathbf{I})) = \begin{bmatrix} (1-t)e^{2t} & -te^{2t} \\ te^{2t} & (1+t)e^{2t} \end{bmatrix}.$$

15. For the 3-by-3 matrix we'll follow the same strategy, that is, using expression

$$e^{\mathbf{A}t} = e^{tr_1}e^{(\mathbf{A}t - tr_1\mathbf{I})},$$

where

$$e^{(\mathbf{A}t - tr_1\mathbf{I})} = \mathbf{I} + (\mathbf{A}t - tr_1\mathbf{I}) + \frac{(\mathbf{A}t - tr_1\mathbf{I})^2}{2!}.$$

The given matrix has the unique eigenvalue $r_1 = -1$, whence the two previous equations give

$$e^{\mathbf{A}t} = e^{-t}\left(\mathbf{I} + (\mathbf{A}t + t\mathbf{I}) + \frac{(\mathbf{A}t + t\mathbf{I})^2}{2!}\right)$$

$$= \begin{bmatrix} e^{-t}\left(1 - \frac{3}{2}(t-2)t\right) & e^{-t}t & \frac{1}{2}e^{-t}(t-2)t \\ -3e^{-t}t & e^{-t} & e^{-t}t \\ \frac{-9}{2}e^{-t}(t-2)t & 3e^{-t}t & \frac{1}{2}e^{-t}(3(t-2)t+2) \end{bmatrix}.$$

17. Arguing in the same way as in Exercise 15, since the unique eigenvalue is $r_1 = -1$,

$$e^{\mathbf{A}t} = e^{-t}\left(\mathbf{I} + (\mathbf{A}t + t\mathbf{I}) + \frac{(\mathbf{A}t + t\mathbf{I})^2}{2!}\right)$$

$$= \begin{bmatrix} e^{-t}\left(\frac{t^2}{2} + t + 1\right) & e^{-t}t(t+1) & \frac{1}{2}e^{-t}t^2 \\ -\frac{1}{2}e^{-t}t^2 & e^{-t}\left(-t^2 + t + 1\right) & -\frac{1}{2}e^{-t}(t-2)t \\ \frac{1}{2}e^{-t}(t-2)t & e^{-t}(t-3)t & \frac{1}{2}e^{-t}((t-4)t+2) \end{bmatrix}.$$

19. Let us first evaluate the second term in the right-hand side of formula (4) taking $t_2 = t$, $t_1 = 0$. Using Exercise 15, we have

$$e^{At} \int_0^t [e^{-As}\mathbf{f}(s)]ds =$$

$$= \begin{bmatrix} e^{-t}\left(1 - \frac{3}{2}(t-2)t\right) & e^{-t}t & \frac{1}{2}e^{-t}(t-2)t \\ -3e^{-t}t & e^{-t} & e^{-t}t \\ \frac{1}{2}(-9)e^{-t}(t-2)t & 3e^{-t}t & \frac{1}{2}e^{-t}(3(t-2)t+2) \end{bmatrix} \int_0^t \begin{bmatrix} -e^s s^2 \\ e^s s \\ -3e^s s^2 \end{bmatrix} ds$$

$$= \begin{bmatrix} t + e^{-t}(t+2) - 2 \\ t + e^{-t} - 1 \\ 3\left(t + e^{-t}(t+2) - 2\right) \end{bmatrix}.$$

On the other hand, the first term in formula (4) is evaluated as follows:

$$e^{At}\mathbf{x}(0) = \begin{bmatrix} e^{-t}\left(1 - \frac{3}{2}(t-2)t\right) & e^{-t}t & \frac{1}{2}e^{-t}(t-2)t \\ -3e^{-t}t & e^{-t} & e^{-t}t \\ \frac{1}{2}(-9)e^{-t}(t-2)t & 3e^{-t}t & \frac{1}{2}e^{-t}(3(t-2)t+2) \end{bmatrix} \begin{bmatrix} 0 \\ 3 \\ 0 \end{bmatrix}$$

$$= \begin{bmatrix} 3e^{-t}t \\ 3e^{-t} \\ 9e^{-t}t \end{bmatrix}.$$

To summarize, the solution of the given initial-value problem is

$$\mathbf{x}(t) = \begin{bmatrix} t + e^{-t}(4t+2) - 2 \\ t + 4e^{-t} - 1 \\ 3\left(t + e^{-t}(4t+2) - 2\right) \end{bmatrix}.$$

21. Because of the initial condition $\mathbf{x}(0) = \mathbf{0}$, we obtain from formula (4):

$$\mathbf{x}(t) = e^{At} \int_0^t [e^{-As}\mathbf{f}(s)]ds =$$

$$= \begin{bmatrix} e^{-t}\left(\frac{t^2}{2} + t + 1\right) & e^{-t}t(t+1) & \frac{1}{2}e^{-t}t^2 \\ -\frac{1}{2}e^{-t}t^2 & e^{-t}\left(-t^2 + t + 1\right) & -\frac{1}{2}e^{-t}(t-2)t \\ \frac{1}{2}e^{-t}(t-2)t & e^{-t}(t-3)t & \frac{1}{2}e^{-t}((t-4)t+2) \end{bmatrix}$$

$$\times \int_0^t \begin{bmatrix} \frac{1}{2}e^s((s-1)s(s+6) + 6) \\ -\frac{1}{2}e^s(s+1)(s(s+6) - 2) \\ \frac{1}{2}e^s s(s(s+9) + 16) \end{bmatrix} ds$$

$$= \begin{bmatrix} e^{-t}\left(-2t^2 - 7t + 2e^t(t+4) - 8\right) \\ -e^{-t}\left(-2t + e^t - 1\right)(t+1) \\ e^{-t}\left(-2t^2 + t + e^t(t-2) + 2\right) \end{bmatrix}.$$

23. Using diagonalization as in Exercises 1–11, we compute the following expression for e^{At}:

$$e^{At} = \begin{bmatrix} 0 & 1 \\ 1 & -\frac{2}{3} \end{bmatrix} \begin{bmatrix} e^{2t} & 0 \\ 0 & e^{-t} \end{bmatrix} \begin{bmatrix} 0 & 1 \\ 1 & -\frac{2}{3} \end{bmatrix}^{-1}.$$

This allows us to obtain the solution of the given system:

$$
\begin{aligned}
\mathbf{x}(t) &= e^{At} \int_0^t [e^{-As}\mathbf{f}(s)]ds \\
&= \begin{bmatrix} e^{-t} & 0 \\ \frac{2}{3}e^{-t}\left(-1 + e^{3t}\right) & e^{2t} \end{bmatrix} \\
&\quad \times \int_0^t \begin{bmatrix} e^s s^2 \\ \frac{1}{3}e^{-2s}\left(-2\left(-1 + e^{3s}\right)s^2 + 3s + 3\right) \end{bmatrix} ds \\
&= \begin{bmatrix} (t-2)t - 2e^{-t} + 2 \\ \frac{1}{12}\left(-12t^2 + 6t + 16e^{-t} + 11e^{2t} - 27\right) \end{bmatrix}.
\end{aligned}
$$

25. Computing e^{At} gives:

$$e^{At} = \begin{bmatrix} 1 & 1 \\ -i & i \end{bmatrix} \begin{bmatrix} e^{-it} & 0 \\ 0 & e^{it} \end{bmatrix} \begin{bmatrix} 1 & 1 \\ -i & i \end{bmatrix}^{-1},$$

and accordingly

$$
\begin{aligned}
\mathbf{x}(t) &= e^{At} \int_0^t [e^{-As}\mathbf{f}(s)]ds \\
&= \begin{bmatrix} \cos(t) & \sin(t) \\ -\sin(t) & \cos(t) \end{bmatrix} \int_0^t \begin{bmatrix} \cos(s) \\ \sin(s) \end{bmatrix} ds \\
&= \begin{bmatrix} \sin(t) \\ \cos(t) - 1 \end{bmatrix}.
\end{aligned}
$$

31. (a) For the two given matrices we compute:

$$AB = \begin{bmatrix} 2 & 3 \\ -2 & 2 \end{bmatrix} \neq \begin{bmatrix} 1 & 7 \\ -1 & 3 \end{bmatrix} = BA.$$

(b) Using diagonalizations of **A** and **B** one obtains

$$
e^{At} = \begin{bmatrix} 2 & 2 \\ 1-i & 1+i \end{bmatrix} \begin{bmatrix} e^{t(2-i)} & 0 \\ 0 & e^{t(2+i)} \end{bmatrix} \begin{bmatrix} 2 & 2 \\ 1-i & 1+i \end{bmatrix}^{-1}
$$

$$
= e^{2t} \begin{bmatrix} \cos(t) - \sin(t) & 2\sin(t) \\ -\sin(t) & \cos(t) + \sin(t) \end{bmatrix}.
$$

and

$$
e^{Bt} = \begin{bmatrix} 1 & -1 \\ 0 & 1 \end{bmatrix} \begin{bmatrix} e^{2t} & 0 \\ 0 & e^t \end{bmatrix} \begin{bmatrix} 1 & -1 \\ 0 & 1 \end{bmatrix}^{-1} = e^t \begin{bmatrix} e^t & -1+e^t \\ 0 & 1 \end{bmatrix}.
$$

Furthermore,

$$
e^{(A+B)t} = \begin{bmatrix} \frac{1}{2}\left(1-i\sqrt{11}\right) & \frac{1}{2}\left(i\sqrt{11}+1\right) \\ 1 & 1 \end{bmatrix} \begin{bmatrix} e^{\frac{t}{2}(i\sqrt{11}+7)} & 0 \\ 0 & e^{\frac{t}{2}(7-i\sqrt{11})} \end{bmatrix}
$$

$$
\begin{bmatrix} \frac{1}{2}\left(1-i\sqrt{11}\right) & \frac{1}{2}\left(i\sqrt{11}+1\right) \\ 1 & 1 \end{bmatrix}^{-1}
$$

$$
= \frac{e^{\frac{7t}{2}}}{\sqrt{11}} \begin{bmatrix} -\sin\frac{\sqrt{11}t}{2} + \sqrt{11}\cos\frac{\sqrt{11}t}{2} & 6\sin\frac{\sqrt{11}t}{2} \\ -2\sin\frac{\sqrt{11}t}{2} & \sqrt{11}\cos\frac{\sqrt{11}t}{2} + \sin\frac{\sqrt{11}t}{2} \end{bmatrix}.
$$

It remains to observe that

$$
e^{At}e^{Bt} =
$$

$$
= e^{3t} \begin{bmatrix} e^t(\cos(t) - \sin(t))(-1 + e^t)\cos(t) - (-3 + e^t)\sin(t) \\ -e^t\sin(t) \qquad \cos(t) - (-2 + e^t)\sin(t) \end{bmatrix}
$$

$$
\neq e^{(A+B)t}.
$$

33.

$$
e^{At} = \begin{bmatrix} e^t & 0 & 0 & 0 & 0 \\ 0 & e^{-t}(t+1) & e^{-t}t & 0 & 0 \\ 0 & -e^{-t}t & e^{-t}(1-t) & 0 & 0 \\ 0 & 0 & 0 & \cos(t) & \sin(t) \\ 0 & 0 & 0 & -\sin(t) & \cos(t) \end{bmatrix}.
$$

35.

$$e^{At} = \begin{bmatrix} e^{-t} & 0 & 0 & 0 & 0 \\ 0 & e^{-t}(t+1) & e^{-t}t & 0 & 0 \\ 0 & -e^{-t}t & e^{-t}(1-t) & 0 & 0 \\ 0 & 0 & 0 & e^{-2t}(2t+1) & e^{-2t}t \\ 0 & 0 & 0 & -4e^{-2t}t & e^{-2t}(1-2t) \end{bmatrix}.$$

7.3   THE JORDAN NORMAL FORM

1. The relation $\mathbf{J} = \mathbf{P}^{-1}\mathbf{AP}$ implies that there is a 1-to-1 correspondence between the eigenvectors of $\mathbf{J}$ and those of $\mathbf{A}$. In fact, for $\mathbf{x}$ – an eigenvector of $\mathbf{A}$, the vector $\mathbf{P}^{-1}\mathbf{x}$ is an eigenvector of $\mathbf{J}$ (and all eigenvectors of $\mathbf{J}$ are of this form). Recalling that columns of $\mathbf{P}$ contain the eigenvectors of $\mathbf{A}$ as a subset, we conclude that $\mathbf{P}^{-1}\mathbf{x}$ is a vector of the form $[0 \ 0 \ \ldots \ 0 \ 1 \ 0 \ \ldots \ 0]^T$, where 1 is in the j-th row if the j-th column of $\mathbf{P}$ is an eigenvector of $\mathbf{A}$.

   Equivalently, if $\mathbf{J}$ consists of q Jordan blocks, their upper left corners occupying positions (i_p, j_p), $1 \le p \le q$, then the eigenvectors of $\mathbf{J}$ are column vectors with 1 in the row i_p, zeros elsewhere, and the corresponding eigenvalue r_p.

3. A nonsingular 3-by-3 Jordan block is

$$\mathbf{J}_{3\times3} = \begin{bmatrix} r & 1 & 0 \\ 0 & r & 1 \\ 0 & 0 & r \end{bmatrix}$$

with $r \ne 0$. It is easy to verify that

$$\mathbf{J}_{3\times3}^{-1} = \begin{bmatrix} r^{-1} & -r^{-2} & r^{-3} \\ 0 & r^{-1} & -r^{-2} \\ 0 & 0 & r^{-1} \end{bmatrix}$$

is its inverse.

5. Depending on how many linearly independent eigenvectors such a matrix has, its possible Jordan forms are

$$
\begin{bmatrix} r & 1 & 0 & 0 \\ 0 & r & 1 & 0 \\ 0 & 0 & r & 1 \\ 0 & 0 & 0 & r \end{bmatrix}, \quad \text{one eigenvector,}
$$

$$
\begin{bmatrix} r & 1 & 0 & 0 \\ 0 & r & 0 & 0 \\ 0 & 0 & r & 1 \\ 0 & 0 & 0 & r \end{bmatrix} \quad \text{and} \quad \begin{bmatrix} r & 1 & 0 & 0 \\ 0 & r & 1 & 0 \\ 0 & 0 & r & 0 \\ 0 & 0 & 0 & r \end{bmatrix}, \quad \text{two eigenvectors,}
$$

$$
\begin{bmatrix} r & 1 & 0 & 0 \\ 0 & r & 0 & 0 \\ 0 & 0 & r & 0 \\ 0 & 0 & 0 & r \end{bmatrix} \quad \text{and} \quad \begin{bmatrix} r & 0 & 0 & 0 \\ 0 & r & 0 & 0 \\ 0 & 0 & r & 0 \\ 0 & 0 & 0 & r \end{bmatrix}, \quad \text{three and four eigenvectors}
$$

respectively.

Observe that the number of independent eigenvectors is the same as the number of Jordan blocks.

To determine the Jordan normal forms in Exercises 7–13 it is enough to find eigenvalues of the given matrix and determine the number of respective eigenvectors (equivalently, Jordan blocks). Recall that the eigenvectors can be found as null vectors of $\mathbf{A} - r\mathbf{I}$. Finding the similarity transform, however, will require applying the method described in Section 7.3.

7. The single eigenvalue of multiplicity 2 is $r_1 = 4$. Since the given matrix has only one eigenvector with this eigenvalue, it must be that the Jordan form contains only one block, that is,

$$
\mathbf{J} = \begin{bmatrix} 4 & 1 \\ 0 & 4 \end{bmatrix}.
$$

9. The given matrix $\mathbf{A}$ has a single eigenvalue of multiplicity 3, $r_1 = 1$. Furthermore, the matrix $\mathbf{A} - \mathbf{I}$ has one null vector: $[0 \ 0 \ 1]^T$. The Jordan form therefore consists of s single block:

$$
\mathbf{J} = \begin{bmatrix} 1 & 1 & 0 \\ 0 & 1 & 1 \\ 0 & 0 & 1 \end{bmatrix}.
$$

11. The matrix $\mathbf{A}$ has two eigenvalues: $r_1 = 3$ and $r_2 = 5$, the latter of multiplicity 2. Solving the homogeneous system $(\mathbf{A} - 5\mathbf{I})\mathbf{x} = \mathbf{0}$ gives a single null vector $[-1 \ 0 \ 1]^T$, so the Jordan form of $\mathbf{A}$ has a single block for r_2. To summarize,

$$\mathbf{J} = \begin{bmatrix} 3 & 0 & 0 \\ 0 & 5 & 1 \\ 0 & 0 & 5 \end{bmatrix}.$$

13. Eigenvalues of matrix $\mathbf{A}$ are $r_1 = 1$ of multiplicity 3 and $r_2 = 2$. The matrix $\mathbf{A} - \mathbf{I}$ has a single null vector $[0 \ 1 \ 0 \ 2]^T$, and so the Jordan form of $\mathbf{A}$ contains a single block for r_1:

$$\mathbf{J} = \begin{bmatrix} 1 & 1 & 0 & 0 \\ 0 & 1 & 1 & 0 \\ 0 & 0 & 1 & 0 \\ 0 & 0 & 0 & 2 \end{bmatrix}.$$

15. Characteristic polynomial of the given matrix is r^4, that is, $r_1 = 0$ is the eigenvalue of multiplicity 4. Furthermore, r_1 has two eigenvectors: $[0 \ -1 \ 0 \ 1]^T$ and $[1 \ 1 \ 1 \ 0]^T$ To identify the number of Jordan blocks in $\mathbf{J}$, we will argue as follows. Observe that for a Jordan block $\mathbf{J}_b$ of size at least 2×2 with eigenvalue 0, $\text{rank}(\mathbf{J}_b^2) = \text{rank}(\mathbf{J}_b) - 1$. It follows that $\text{rank}(\mathbf{J}^2) = \text{rank}(\mathbf{J}) - q$, where q is the number of such Jordan blocks in $\mathbf{J}$. Since ranks of similar matrices coincide, $\text{rank}(\mathbf{A}^2) = \text{rank}(\mathbf{A}) - q$. As $\mathbf{A}^2 = 0$ (check it!), $\mathbf{J}$ consists of two 2×2 Jordan blocks:

$$\mathbf{J} = \begin{bmatrix} 0 & 1 & 0 & 0 \\ 0 & 0 & 0 & 0 \\ 0 & 0 & 0 & 1 \\ 0 & 0 & 0 & 0 \end{bmatrix}.$$

17. Eigenvalues of $\mathbf{A}$ are $r_1 = 0$ of multiplicity 1 and $r_2 = 2$ of multiplicity 5. To find out the number of blocks for r_2, we use the method from Exercise 15. Namely, the number of blocks for r_2 of size at least 2×2 in the Jordan form of $\mathbf{A}$ is $\text{rank}(\mathbf{A} - 2\mathbf{I}) - \text{rank}((\mathbf{A} - 2\mathbf{I})^2) = 4 - 2 = 2$. Similarly, the number of r_2-blocks of size at least 3×3 in $\mathbf{J}$ is $\text{rank}((\mathbf{A} - 2\mathbf{I})^2) - \text{rank}((\mathbf{A} - 2\mathbf{I})^3) = 2 - 1 = 1$. Indeed, to prove this equality observe that

$$\begin{bmatrix} 0 & 1 \\ 0 & 0 \end{bmatrix}^2 = \mathbf{0},$$

and, since block-diagonal matrices (the Jordan form in particular) are multiplied blockwise, the blocks of size at most 2×2 do not contribute to the difference between ranks of $(\mathbf{A} - 2\mathbf{I})^2$ and $(\mathbf{A} - 2\mathbf{I})^3$. Since the sum of dimensions of r_2-blocks is at most 5, this completely describes $\mathbf{J}$. By this method we can, of course, compute the number of blocks of size at least $k \times k$ for any $k = 1, 2 \ldots$ and any eigenvalue of a given matrix.

Summarizing the above argument, matrix $\mathbf{A}$ has the Jordan form

$$\mathbf{J} = \begin{bmatrix} 0 & 0 & 0 & 0 & 0 & 0 \\ 0 & 2 & 1 & 0 & 0 & 0 \\ 0 & 0 & 2 & 0 & 0 & 0 \\ 0 & 0 & 0 & 2 & 1 & 0 \\ 0 & 0 & 0 & 0 & 2 & 1 \\ 0 & 0 & 0 & 0 & 0 & 2 \end{bmatrix}.$$

19. The given matrix has a single eigenvalue $r_1 = 3$ of multiplicity 2. Accordingly, there are either two eigenvectors for r_1, or an eigenvector and a generalized eigenvector. Arguing as in Example 2 of Section 7.3, we write down two systems for these two cases:

$$\begin{bmatrix} \mathbf{Au}^{(1)} - 3\mathbf{u}^{(1)} \\ \mathbf{Au}^{(2)} - 3\mathbf{u}^{(2)} \end{bmatrix} = \begin{bmatrix} \mathbf{0} \\ \mathbf{0} \end{bmatrix} \quad \text{and} \quad \begin{bmatrix} \mathbf{Au}^{(1)} - 3\mathbf{u}^{(1)} \\ \mathbf{Au}^{(2)} - 3\mathbf{u}^{(2)} - \mathbf{u}^{(1)} \end{bmatrix} = \begin{bmatrix} \mathbf{0} \\ \mathbf{0} \end{bmatrix}.$$

It is not hard to see that the second case holds (for example, arguing as we did in Exercises 15–17; alternatively, verify that the first system gives linearly dependent vectors after partitioning the general solution), so we will write it in terms of vector components.

$$\begin{bmatrix} 0 & -2 & 0 & 0 & \vdots & 0 \\ 0 & 0 & 0 & 0 & \vdots & 0 \\ -1 & 0 & 0 & -2 & \vdots & 0 \\ 0 & -1 & 0 & 0 & \vdots & 0 \end{bmatrix}.$$

A general solution of this system is

$$\mathbf{u} = \begin{bmatrix} -2t_1 & 0 & t_2 & t_1 \end{bmatrix}^T,$$

whence

$$\mathbf{u}_1 = t_1[-2 \; 0]^T, \qquad \mathbf{u}_2 = [t_2 \; t_1]^T.$$

Choosing $t_1 = t_2 = 1$ gives a system of generalized eigenvectors $\mathbf{u}_1$, $\mathbf{u}_2$, and the similarity transform to Jordan form is

$$\mathbf{A} = \begin{bmatrix} -2 & 1 \\ 0 & 1 \end{bmatrix} \begin{bmatrix} 3 & 1 \\ 0 & 3 \end{bmatrix} \begin{bmatrix} -2 & 1 \\ 0 & 1 \end{bmatrix}^{-1}.$$

Finally

$$e^{\mathbf{A}t} = \begin{bmatrix} -2 & 1 \\ 0 & 1 \end{bmatrix} \begin{bmatrix} e^{3t} & te^{3t} \\ 0 & e^{3t} \end{bmatrix} \begin{bmatrix} -2 & 1 \\ 0 & 1 \end{bmatrix}^{-1} = \begin{bmatrix} e^{3t} & -2e^{3t}t \\ 0 & e^{3t} \end{bmatrix}.$$

21. The given matrix has an eigenvalue $r_1 = 1$ of multiplicity 2 and a simple eigenvalue $r_2 = 3$. We therefore need to consider the two homogeneous systems:

$$\begin{bmatrix} \mathbf{A}\mathbf{u}^{(1)} - \mathbf{u}^{(1)} \\ \mathbf{A}\mathbf{u}^{(2)} - \mathbf{u}^{(2)} \\ \mathbf{A}\mathbf{u}^{(3)} - 3\mathbf{u}^{(3)} \end{bmatrix} = \begin{bmatrix} \mathbf{0} \\ \mathbf{0} \\ \mathbf{0} \end{bmatrix} \quad \text{and} \quad \begin{bmatrix} \mathbf{A}\mathbf{u}^{(1)} - \mathbf{u}^{(1)} \\ \mathbf{A}\mathbf{u}^{(2)} - \mathbf{u}^{(2)} - \mathbf{u}^{(1)} \\ \mathbf{A}\mathbf{u}^{(3)} - 3\mathbf{u}^{(3)} \end{bmatrix} = \begin{bmatrix} \mathbf{0} \\ \mathbf{0} \\ \mathbf{0} \end{bmatrix}.$$

Rewriting them in terms of vector components gives respectively

$$\left[\begin{array}{ccc|ccc|ccc:c}
0 & 0 & 0 & 0 & 0 & 0 & 0 & 0 & 0 & 0 \\
1 & 2 & 0 & 0 & 0 & 0 & 0 & 0 & 0 & 0 \\
0 & 1 & 0 & 0 & 0 & 0 & 0 & 0 & 0 & 0 \\
0 & 0 & 0 & 0 & 0 & 0 & 0 & 0 & 0 & 0 \\
0 & 0 & 0 & 1 & 2 & 0 & 0 & 0 & 0 & 0 \\
0 & 0 & 0 & 0 & 1 & 0 & 0 & 0 & 0 & 0 \\
0 & 0 & 0 & 0 & 0 & 0 & -2 & 0 & 0 & 0 \\
0 & 0 & 0 & 0 & 0 & 0 & 1 & 0 & 0 & 0 \\
0 & 0 & 0 & 0 & 0 & 0 & 0 & 1 & -2 & 0
\end{array}\right]$$

and

$$
\left[\begin{array}{ccc|ccc|ccc:c}
0 & 0 & 0 & 0 & 0 & 0 & 0 & 0 & 0 & 0 \\
1 & 2 & 0 & 0 & 0 & 0 & 0 & 0 & 0 & 0 \\
0 & 1 & 0 & 0 & 0 & 0 & 0 & 0 & 0 & 0 \\
-1 & 0 & 0 & 0 & 0 & 0 & 0 & 0 & 0 & 0 \\
0 & -1 & 0 & 1 & 2 & 0 & 0 & 0 & 0 & 0 \\
0 & 0 & -1 & 0 & 1 & 0 & 0 & 0 & 0 & 0 \\
0 & 0 & 0 & 0 & 0 & 0 & -2 & 0 & 0 & 0 \\
0 & 0 & 0 & 0 & 0 & 0 & 1 & 0 & 0 & 0 \\
0 & 0 & 0 & 0 & 0 & 0 & 0 & 1 & -2 & 0
\end{array}\right].
$$

The general solution of the first system is

$$
\mathbf{u} = [0 \ 0 \ t_1 \ 0 \ 0 \ t_2 \ 0 \ 2t_3 \ t_3]^T,
$$

and of the second

$$
\mathbf{u} = [0 \ 0 \ t_1 \ -2t_1 \ t_1 \ t_2 \ 0 \ 2t_3 \ t_3]^T.
$$

Clearly, partitioning the first one into vectors of length 3 gives a linearly dependent triple. Taking $t_1 = t_2 = t_3 = 1$ in the second solution we obtain the system of linearly independent generalized eigenvectors:

$$
[\mathbf{u}_1 \ \mathbf{u}_2 \ \mathbf{u}_3] = \begin{bmatrix} 0 & -2 & 0 \\ 0 & 1 & 2 \\ 1 & 1 & 1 \end{bmatrix}.
$$

To summarize, the matrix $\mathbf{A}$ is expressed through its Jordan form as

$$
\mathbf{A} = \begin{bmatrix} 0 & -2 & 0 \\ 0 & 1 & 2 \\ 1 & 1 & 1 \end{bmatrix} \begin{bmatrix} 1 & 1 & 0 \\ 0 & 1 & 0 \\ 0 & 0 & 3 \end{bmatrix} \begin{bmatrix} 0 & -2 & 0 \\ 0 & 1 & 2 \\ 1 & 1 & 1 \end{bmatrix}^{-1}.
$$

Accordingly,

$$
e^{At} = \begin{bmatrix} 0 & -2 & 0 \\ 0 & 1 & 2 \\ 1 & 1 & 1 \end{bmatrix} \begin{bmatrix} e^t & te^t & 0 \\ 0 & e^t & 0 \\ 0 & 0 & e^{3t} \end{bmatrix} \begin{bmatrix} 0 & -2 & 0 \\ 0 & 1 & 2 \\ 1 & 1 & 1 \end{bmatrix}^{-1}
$$

$$
= e^t \begin{bmatrix} 1 & 0 & 0 \\ \frac{1}{2}\left(-1 + e^{2t}\right) & e^{2t} & 0 \\ \frac{1}{4}\left(-2t + e^{2t} - 1\right) & \frac{1}{2}\left(-1 + e^{2t}\right) & 1 \end{bmatrix}.
$$

23. The given matrix A has an eigenvalue $r_1 = 1$ of multiplicity 3, and a simple eigenvalue $r_2 = 2$. There are 1, 2 or 3 Jordan blocks corresponding to r_1. The homogeneous systems to be solved in these three cases are

$$
\begin{bmatrix} A u^{(1)} - u^{(1)} \\ A u^{(2)} - u^{(2)} - u^{(1)} \\ A u^{(3)} - u^{(3)} - u^{(2)} \\ A u^{(4)} - 2u^{(4)} \end{bmatrix} = \begin{bmatrix} 0 \\ 0 \\ 0 \\ 0 \end{bmatrix}, \quad \begin{bmatrix} A u^{(1)} - u^{(1)} \\ A u^{(2)} - u^{(2)} \\ A u^{(3)} - u^{(3)} - u^{(2)} \\ A u^{(4)} - 2u^{(4)} \end{bmatrix} = \begin{bmatrix} 0 \\ 0 \\ 0 \\ 0 \end{bmatrix},
$$

$$
\begin{bmatrix} A u^{(1)} - u^{(1)} \\ A u^{(2)} - u^{(2)} \\ A u^{(3)} - u^{(3)} \\ A u^{(4)} - 2u^{(4)} \end{bmatrix} = \begin{bmatrix} 0 \\ 0 \\ 0 \\ 0 \end{bmatrix}
$$

respectively. Only the first one yields linearly independent vectors after partitioning the general solution into columns of length 4 (alternatively, one could use the approach of Exercises 15–17 and observe that $\operatorname{rank}(A - I) - \operatorname{rank}((A - I)^2) = 3 - 2 = 1$, so there is only one Jordan block for r_1). Namely, the general solution is

$$
u = [0 \ -2t_1 \ 0 \ 0 \ -2t_1 \ -t_1 - t_2 \ 0 \ 0 \ -t_2 \ -t_3
$$
$$
0 \ -t_1 \ -3t_4 \ -6t_4 \ -t_4 \ -t_4]^T,
$$

which after setting $t_i = 1$, $1 \le i \le 4$, gives the generalized eigenvectors

$$
[u_1 \ u_2 \ u_3 \ u_4] = \begin{bmatrix} 0 & -2 & -1 & -3 \\ -2 & -2 & -1 & -6 \\ 0 & 0 & 0 & -1 \\ 0 & 0 & -1 & -1 \end{bmatrix}.
$$

Hence **A** is expressed through its Jordan form as

$$\mathbf{A} = \begin{bmatrix} 0 & -2 & -1 & -3 \\ -2 & -2 & -1 & -6 \\ 0 & 0 & 0 & -1 \\ 0 & 0 & -1 & -1 \end{bmatrix} \begin{bmatrix} 1 & 1 & 0 & 0 \\ 0 & 1 & 1 & 0 \\ 0 & 0 & 1 & 0 \\ 0 & 0 & 0 & 2 \end{bmatrix} \begin{bmatrix} 0 & -2 & -1 & -3 \\ -2 & -2 & -1 & -6 \\ 0 & 0 & 0 & -1 \\ 0 & 0 & -1 & -1 \end{bmatrix}^{-1}$$

and

$$e^{\mathbf{A}t} = \begin{bmatrix} 0 & -2 & -1 & -3 \\ -2 & -2 & -1 & -6 \\ 0 & 0 & 0 & -1 \\ 0 & 0 & -1 & -1 \end{bmatrix} \begin{bmatrix} e^t & te^t & \frac{t^2}{2}e^t & 0 \\ 0 & e^t & te^t & 0 \\ 0 & 0 & e^t & 0 \\ 0 & 0 & 0 & e^{2t} \end{bmatrix} \begin{bmatrix} 0 & -2 & -1 & -3 \\ -2 & -2 & -1 & -6 \\ 0 & 0 & 0 & -1 \\ 0 & 0 & -1 & -1 \end{bmatrix}^{-1}$$

$$= e^t \begin{bmatrix} 1 & 0 & -2t + 3e^t - 3 & 2t \\ t & 1 & -t^2 - 4t + 6e^t - 6 & t(t+1) \\ 0 & 0 & e^t & 0 \\ 0 & 0 & -1 + e^t & 1 \end{bmatrix}.$$

25. Eigenvalues of **A** are: the simple $r_1 = 2$, and $r_2 = 3$ of multiplicity 4. Depending on the number of Jordan blocks for r_2, generalized eigenvectors for **A** are the solution of one of the following homogeneous systems (cf. Exercise 5):

$$\begin{bmatrix} \mathbf{Au}^{(1)} - 2\mathbf{u}^{(1)} \\ \mathbf{Au}^{(2)} - 3\mathbf{u}^{(2)} \\ \mathbf{Au}^{(3)} - 3\mathbf{u}^{(3)} - \mathbf{u}^{(2)} \\ \mathbf{Au}^{(4)} - 3\mathbf{u}^{(4)} - \mathbf{u}^{(3)} \\ \mathbf{Au}^{(5)} - 3\mathbf{u}^{(5)} - \mathbf{u}^{(4)} \end{bmatrix} = \begin{bmatrix} 0 \\ 0 \\ 0 \\ 0 \\ 0 \end{bmatrix},$$

$$\begin{bmatrix} \mathbf{Au}^{(1)} - 2\mathbf{u}^{(1)} \\ \mathbf{Au}^{(2)} - 3\mathbf{u}^{(2)} \\ \mathbf{Au}^{(3)} - 3\mathbf{u}^{(3)} \\ \mathbf{Au}^{(4)} - 3\mathbf{u}^{(4)} - \mathbf{u}^{(3)} \\ \mathbf{Au}^{(5)} - 3\mathbf{u}^{(5)} - \mathbf{u}^{(4)} \end{bmatrix} = \begin{bmatrix} 0 \\ 0 \\ 0 \\ 0 \\ 0 \end{bmatrix}, \quad \begin{bmatrix} \mathbf{Au}^{(1)} - 2\mathbf{u}^{(1)} \\ \mathbf{Au}^{(2)} - 3\mathbf{u}^{(2)} \\ \mathbf{Au}^{(3)} - 3\mathbf{u}^{(3)} - \mathbf{u}^{(2)} \\ \mathbf{Au}^{(4)} - 3\mathbf{u}^{(4)} \\ \mathbf{Au}^{(5)} - 3\mathbf{u}^{(5)} - \mathbf{u}^{(4)} \end{bmatrix} = \begin{bmatrix} 0 \\ 0 \\ 0 \\ 0 \\ 0 \end{bmatrix},$$

$$\begin{bmatrix} \mathbf{Au}^{(1)} - 2\mathbf{u}^{(1)} \\ \mathbf{Au}^{(2)} - 3\mathbf{u}^{(2)} \\ \mathbf{Au}^{(3)} - 3\mathbf{u}^{(3)} \\ \mathbf{Au}^{(4)} - 3\mathbf{u}^{(4)} \\ \mathbf{Au}^{(5)} - 3\mathbf{u}^{(5)} - 3\mathbf{u}^{(4)} \end{bmatrix} = \begin{bmatrix} 0 \\ 0 \\ 0 \\ 0 \\ 0 \end{bmatrix}, \quad \begin{bmatrix} \mathbf{Au}^{(1)} - 2\mathbf{u}^{(1)} \\ \mathbf{Au}^{(2)} - 3\mathbf{u}^{(2)} \\ \mathbf{Au}^{(3)} - 3\mathbf{u}^{(3)} \\ \mathbf{Au}^{(4)} - 3\mathbf{u}^{(4)} \\ \mathbf{Au}^{(5)} - 3\mathbf{u}^{(5)} \end{bmatrix} = \begin{bmatrix} 0 \\ 0 \\ 0 \\ 0 \\ 0 \end{bmatrix},$$

with 1, 2 (2 matrices), 3 and 4 r_2-blocks respectively. Only the first one has a general solution that yields a linearly independent collection after partitioning into vectors of length 5 (see the solutions of Exercise 23 and Exercises 15–17 on using ranks to avoid solving all the five systems above). Indeed, after partitioning the general solution

$$[t_1 \quad -t_1 \ t_1 \quad -t_1 \ t_1 \ 0 \ 0 \ 0 \ 0 \ t_2 \ 0 \ 0 \ 0 \ t_2 \ t_3 \ 0 \ 0 \ t_2 \ t_3 \ t_4 \ 0$$

$$t_2 \ t_3 \ t_4 \ t_5]$$

with $t_i = 1$, $1 \le i \le 5$, we obtain a collection of generalized eigenvectors

$$[\mathbf{u}_1 \ \mathbf{u}_2 \ \mathbf{u}_3 \ \mathbf{u}_4 \ \mathbf{u}_5] = \begin{bmatrix} 1 & 0 & 0 & 0 & 0 \\ -1 & 0 & 0 & 0 & 1 \\ 1 & 0 & 0 & 1 & 1 \\ -1 & 0 & 1 & 1 & 1 \\ 1 & 1 & 1 & 1 & 1 \end{bmatrix}.$$

Using the above matrix as a similarity transform, we have

$$\mathbf{A} = \begin{bmatrix} 1 & 0 & 0 & 0 & 0 \\ -1 & 0 & 0 & 0 & 1 \\ 1 & 0 & 0 & 1 & 1 \\ -1 & 0 & 1 & 1 & 1 \\ 1 & 1 & 1 & 1 & 1 \end{bmatrix} \begin{bmatrix} 2 & 0 & 0 & 0 & 0 \\ 0 & 3 & 1 & 0 & 0 \\ 0 & 0 & 3 & 1 & 0 \\ 0 & 0 & 0 & 3 & 1 \\ 0 & 0 & 0 & 0 & 3 \end{bmatrix} \begin{bmatrix} 1 & 0 & 0 & 0 & 0 \\ -1 & 0 & 0 & 0 & 1 \\ 1 & 0 & 0 & 1 & 1 \\ -1 & 0 & 1 & 1 & 1 \\ 1 & 1 & 1 & 1 & 1 \end{bmatrix}^{-1}$$

and

$$e^{\mathbf{A}t} = \begin{bmatrix} 1 & 0 & 0 & 0 & 0 \\ -1 & 0 & 0 & 0 & 1 \\ 1 & 0 & 0 & 1 & 1 \\ -1 & 0 & 1 & 1 & 1 \\ 1 & 1 & 1 & 1 & 1 \end{bmatrix} \begin{bmatrix} e^{2t} & 0 & 0 & 0 & 0 \\ 0 & e^{3t} & e^{3t}t & \frac{1}{2}e^{3t}t^2 & \frac{1}{6}e^{3t}t^3 \\ 0 & 0 & e^{3t} & e^{3t}t & \frac{1}{2}e^{3t}t^2 \\ 0 & 0 & 0 & e^{3t} & e^{3t}t \\ 0 & 0 & 0 & 0 & e^{3t} \end{bmatrix}$$

$$\begin{bmatrix} 1 & 0 & 0 & 0 & 0 \\ -1 & 0 & 0 & 0 & 1 \\ 1 & 0 & 0 & 1 & 1 \\ -1 & 0 & 1 & 1 & 1 \\ 1 & 1 & 1 & 1 & 1 \end{bmatrix}^{-1}$$

$$= e^{2t} \begin{bmatrix} 1 & 0 & 0 & 0 & 0 \\ -1 + e^t & e^t & 0 & 0 & 0 \\ e^t(t-1) + 1 & e^t t & e^t & 0 & 0 \\ \frac{1}{2}\left(e^t\left(t^2 - 2t + 2\right) - 2\right) & \frac{e^t t^2}{2} & e^t t & e^t & 0 \\ \frac{1}{6}\left(e^t\left(t^3 - 3t^2 + 6t - 6\right) + 6\right) & \frac{e^t t^3}{6} & \frac{e^t t^2}{2} & e^t t & e^t \end{bmatrix}.$$

31. Observe that the described matrix **A** has a single eigenvalue r_1 of the multiplicity equal to the number of columns in **A**. Let us first assume that all the entries on superdiagonal are nonzero. With this assumption, the desired statement is equivalent to the Jordan form of **A** being a single r_1-block. And indeed, using our assumption, rank$(\mathbf{A} - r_1\mathbf{I}) = n - 1$, the number of superdiagonal entries, so the null space of $\mathbf{A} - r_1\mathbf{I}$ is one-dimensional, and **A** in fact has only one eigenvector.

In the general case, **A** can be seen as a block-diagonal matrix, each of the blocks with only nonzero entries on superdiagonal. Since the Jordan form of a block-diagonal matrix is itself block-diagonal, consisting of Jordan forms of respective blocks of **A**, the general statement now follows from the above argument for a single block.

33. To rewrite the given equation as a first-order system, we introduce the vector $\mathbf{x}(t) = [x_1(t)\ \ x_2(t)]^T$ with $y = x_1$, $x_1' = x_2$. The resulting system is

$$\mathbf{x}'(t) = \begin{bmatrix} x_1 \\ x_2 \end{bmatrix}' = \begin{bmatrix} 0 & 1 \\ 0 & 0 \end{bmatrix} \begin{bmatrix} x_1 \\ x_2 \end{bmatrix} = \mathbf{Ax},$$

and according to formula (4) of Section 7.2 its solution is

$$\begin{bmatrix} x_1 \\ x_2 \end{bmatrix} = e^{\mathbf{A}t}\mathbf{x}(0) = \begin{bmatrix} 1 & t \\ 0 & 1 \end{bmatrix} \begin{bmatrix} C_1 \\ C_2 \end{bmatrix} = \begin{bmatrix} C_1 + C_2 t \\ C_2 \end{bmatrix},$$

so we have indeed obtained $y(t) = C_1 + C_2 t$. The matrix **A** above is defective: its only eigenvector is $[1\ \ 0]^T$ with the eigenvalue 0.

35. Using the formula (4) of Section 7.2, we have

$$\mathbf{x} = e^{\mathbf{A}t}\mathbf{x}(0) = e^t \begin{bmatrix} 1 & 0 & 0 \\ \frac{1}{2}(-1 + e^{2t}) & e^{2t} & 0 \\ \frac{1}{4}(-2t + e^{2t} - 1) & \frac{1}{2}(-1 + e^{2t}) & 1 \end{bmatrix} \begin{bmatrix} 1 \\ 0 \\ 1 \end{bmatrix}$$

$$= \begin{bmatrix} e^t \\ \frac{1}{2}e^t(-1 + e^{2t}) \\ \frac{1}{4}e^t(-2t + e^{2t} + 3) \end{bmatrix},$$

where we took the second equality from Exercise 21.

7.4 MATRIX EXPONENTIATION VIA GENERALIZED EIGENVECTORS

1. The given matrix has the eigenvalue $r_1 = 4$ of multiplicity 2. Furthermore, as $(A - 4I)^2 = 0$, the set of linearly independent generalized eigenvectors is $\mathbf{u}_1 = [1\ 0]^T$, $\mathbf{u}_2 = [0\ 1]^T$. We have

$$e^{At}\mathbf{u}_1 = e^{4tI}e^{t(A-4I)}\mathbf{u}_1 = e^{4t}\left[I + t(A - 4I)\right]\begin{bmatrix}1\\0\end{bmatrix}$$

$$= e^{4t}\begin{bmatrix}1 & -2t\\0 & 1\end{bmatrix}\begin{bmatrix}1\\0\end{bmatrix} = \begin{bmatrix}e^{4t}\\0\end{bmatrix},$$

and

$$e^{At}\mathbf{u}_2 = e^{4t}\begin{bmatrix}1 & -2t\\0 & 1\end{bmatrix}\begin{bmatrix}0\\1\end{bmatrix} = \begin{bmatrix}-2te^{4t}\\e^{4t}\end{bmatrix}.$$

Thus

$$e^{At} = (e^{At}[\mathbf{u}_1\ \mathbf{u}_2])[\mathbf{u}_1\ \mathbf{u}_2]^{-1} = \begin{bmatrix}e^{4t} & -2te^{4t}\\0 & e^{4t}\end{bmatrix}.$$

3. The single eigenvalue of the given matrix is $r_1 = 1$ of multiplicity 3. It holds $(A - I)^3 = 0$, so the canonical basis vectors can be used as generalized eigenvectors:

$$[\mathbf{u}_1\ \mathbf{u}_2\ \mathbf{u}_3] = \begin{bmatrix}1 & 0 & 0\\0 & 1 & 0\\0 & 0 & 1\end{bmatrix}.$$

Computing as above,

$$e^{At}\mathbf{u}_1 = e^{tI}e^{t(A-I)}\mathbf{u}_1 = e^{t}\left[I + t(A - I) + \frac{t^2}{2}(A - I)^2\right]\begin{bmatrix}1\\0\\0\end{bmatrix}$$

$$= e^{t}\begin{bmatrix}1 & 0 & 0\\t & 1 & 0\\\frac{t^2}{2}+t & t & 1\end{bmatrix}\begin{bmatrix}1\\0\\0\end{bmatrix} = \begin{bmatrix}e^{t}\\te^{t}\\(\frac{t^2}{2}+t)e^{t}\end{bmatrix},$$

$$e^{At}\mathbf{u}_2 = e^t \begin{bmatrix} 1 & 0 & 0 \\ t & 1 & 0 \\ \frac{t^2}{2}+t & t & 1 \end{bmatrix} \begin{bmatrix} 0 \\ 1 \\ 0 \end{bmatrix} = \begin{bmatrix} 0 \\ e^t \\ te^t \end{bmatrix},$$

$$e^{At}\mathbf{u}_3 = e^t \begin{bmatrix} 1 & 0 & 0 \\ t & 1 & 0 \\ \frac{t^2}{2}+t & t & 1 \end{bmatrix} \begin{bmatrix} 0 \\ 0 \\ 1 \end{bmatrix} = \begin{bmatrix} 0 \\ 0 \\ e^t \end{bmatrix}.$$

It remains to summarize

$$e^{At} = (e^{At}[\mathbf{u}_1 \ \mathbf{u}_2 \ \mathbf{u}_3])[\mathbf{u}_1 \ \mathbf{u}_2 \ \mathbf{u}_3]^{-1} = \begin{bmatrix} e^t & 0 & 0 \\ te^t & e^t & 0 \\ (\frac{t^2}{2}+t)e^t & te^t & e^t \end{bmatrix}.$$

Of course, we could have written the above equality even earlier – when we saw that the matrix of generalized eigenvectors is in fact the identity matrix.

5. Eigenvalues of the given matrix: a simple $r_1 = 3$ and $r_2 = 5$ of multiplicity 2. The eigenvector corresponding to r_1 is $[0 \ -1 \ 1]$. The generalized eigenvectors for r_2 are found from the system

$$(\mathbf{A} - 5\mathbf{I})^2 = \mathbf{0} \qquad \Longleftrightarrow \qquad \begin{bmatrix} 4 & 0 & 4 & \vdots & 0 \end{bmatrix},$$

whence $\mathbf{u}_2 = [-1 \ 0 \ 1]$ and $\mathbf{u}_3 = [0 \ 1 \ 0]$. This allows us to obtain

$$e^{At}\mathbf{u}_1 = e^{3t\mathbf{I}}e^{t(\mathbf{A}-3\mathbf{I})}\mathbf{u}_1 = e^{3t}\mathbf{I}\mathbf{u}_1 = \begin{bmatrix} 0 \\ -e^{3t} \\ e^{3t} \end{bmatrix},$$

$$e^{At}\mathbf{u}_2 = e^{5t\mathbf{I}}e^{t(\mathbf{A}-5\mathbf{I})}\mathbf{u}_2 = e^{5t}\left[\mathbf{I} + t(\mathbf{A} - 5\mathbf{I})\right] \begin{bmatrix} -1 \\ 0 \\ 1 \end{bmatrix}$$

$$= e^{5t} \begin{bmatrix} 1-t & -t & -t \\ 2t & 1 & 2t \\ -t & t & 1-t \end{bmatrix} \begin{bmatrix} -1 \\ 0 \\ 1 \end{bmatrix} = \begin{bmatrix} -e^{5t} \\ 0 \\ e^{5t} \end{bmatrix},$$

$$e^{At}\mathbf{u}_3 = e^{5t} \begin{bmatrix} 1-t & -t & -t \\ 2t & 1 & 2t \\ -t & t & 1-t \end{bmatrix} \begin{bmatrix} 0 \\ 1 \\ 0 \end{bmatrix} = \begin{bmatrix} -te^{5t} \\ e^{5t} \\ te^{5t} \end{bmatrix}.$$

Summarizing,

$$e^{At} = (e^{At}[\mathbf{u}_1 \ \mathbf{u}_2 \ \mathbf{u}_3])[\mathbf{u}_1 \ \mathbf{u}_2 \ \mathbf{u}_3]^{-1}$$

$$= \begin{bmatrix} 0 & -e^{5t} & -te^{5t} \\ -e^{3t} & 0 & e^{5t} \\ e^{3t} & e^{5t} & te^{5t} \end{bmatrix} \begin{bmatrix} 0 & -1 & 0 \\ -1 & 0 & 1 \\ 1 & 1 & 0 \end{bmatrix}^{-1}$$

$$= \begin{bmatrix} -e^{5t}(t-1) & -e^{5t}t & -e^{5t}t \\ e^{3t}(-1+e^{2t}) & e^{5t} & e^{3t}(-1+e^{2t}) \\ e^{5t}(t-1)+e^{3t} & e^{5t}t & e^{5t}t+e^{3t} \end{bmatrix}.$$

7. Eigenvalues of matrix $\mathbf{A}$ are: $r_1 = 1$ of multiplicity 3, $r_2 = 2$. First, we use the above methods to find generalized eigenvectors:

$$[\mathbf{u}_1 \ \mathbf{u}_2 \ \mathbf{u}_3 \ \mathbf{u}_4] = \begin{bmatrix} 0 & 0 & 1 & 3 \\ 1 & 0 & 0 & 4 \\ 0 & 1 & 0 & -3 \\ 2 & 0 & 0 & 9 \end{bmatrix},$$

where the first three vectors are the solutions of the homogeneous system $(\mathbf{A} - \mathbf{I})^3 \mathbf{u} = \mathbf{0}$, that is,

$$\begin{bmatrix} 0 & -2 & 0 & 1 & \vdots & 0 \end{bmatrix},$$

and $\mathbf{u}_4$ is the eigenvector for r_2. Furthermore,

$$e^{At}\mathbf{u}_1 = e^{tI}e^{t(\mathbf{A}-\mathbf{I})}\mathbf{u}_1 = e^{t}\mathbf{I}\mathbf{u}_1 = \begin{bmatrix} 0 \\ e^t \\ 0 \\ 2e^t \end{bmatrix},$$

where we used that $\mathbf{u}_1$ is an eigenvector for r_1,

$$e^{At}\mathbf{u}_2 = e^{tI}e^{t(\mathbf{A}-\mathbf{I})}\mathbf{u}_2 = e^t \left[\mathbf{I} + t(\mathbf{A}-\mathbf{I}) + \frac{t^2}{2}(\mathbf{A}-\mathbf{I})^2 \right] \begin{bmatrix} 0 \\ 0 \\ 1 \\ 0 \end{bmatrix}$$

$$= e^t \begin{bmatrix} 2t+1 & -3t^2 & t & \frac{3t^2}{2} \\ t(t+1) & -t^2-2t+1 & \frac{t^2}{2} & \frac{1}{2}t(t+2) \\ -4t & 3(t-2)t & 1-2t & \frac{1}{2}(-3)(t-2)t \\ 2t(t+1) & -3t(t+2) & t^2 & \frac{3t^2}{2}+3t+1 \end{bmatrix} \begin{bmatrix} 0 \\ 0 \\ 1 \\ 0 \end{bmatrix}$$

$$= \begin{bmatrix} te^t \\ \frac{t^2}{2}e^t \\ (1-2t)e^t \\ t^2 e^t \end{bmatrix},$$

$$e^{At}\mathbf{u}_3 = e^t \begin{bmatrix} 2t+1 & -3t^2 & t & \frac{3t^2}{2} \\ t(t+1) & -t^2-2t+1 & \frac{t^2}{2} & \frac{1}{2}t(t+2) \\ -4t & 3(t-2)t & 1-2t & \frac{1}{2}(-3)(t-2)t \\ 2t(t+1) & -3t(t+2) & t^2 & \frac{3t^2}{2}+3t+1 \end{bmatrix} \begin{bmatrix} 1 \\ 0 \\ 0 \\ 0 \end{bmatrix}$$

$$= \begin{bmatrix} (2t+1)e^t \\ t(t+1)e^t \\ -4te^t \\ 2t(t+1)e^t \end{bmatrix},$$

$$e^{At}\mathbf{u}_4 = e^{2tI}e^{t(A-2I)}\mathbf{u}_4 = e^{2t}I\mathbf{u}_4 = \begin{bmatrix} 3e^{2t} \\ 4e^{2t} \\ -3e^{2t} \\ 9e^{2t} \end{bmatrix}.$$

From the above we have

$$e^{At} = (e^{At}[\mathbf{u}_1\ \mathbf{u}_2\ \mathbf{u}_3\ \mathbf{u}_4])[\mathbf{u}_1\ \mathbf{u}_2\ \mathbf{u}_3\ \mathbf{u}_4]^{-1}$$

$$= \begin{bmatrix} 0 & te^t & (2t+1)e^t & 3e^{2t} \\ e^t & \frac{t^2}{2}e^t & t(t+1)e^t & 4e^{2t} \\ 0 & (1-2t)e^t & -4te^t & -3e^{2t} \\ 2e^t & t^2e^t & 2t(t+1)e^t & 9e^{2t} \end{bmatrix} \begin{bmatrix} 0 & 9 & 0 & -4 \\ 0 & -6 & 1 & 3 \\ 1 & 6 & 0 & -3 \\ 0 & -2 & 0 & 1 \end{bmatrix}$$

$$= e^t \begin{bmatrix} 2t+1 & 6t-6e^t+6 \\ t(t+1) & 3(t(t+2)+3)-8e^t \\ -4t & 6(-2t+e^t-1) \\ 2t(t+1) & 6(t(t+2)-3e^t+3) \end{bmatrix}$$

$$\begin{bmatrix} t & 3(-t+e^t-1) \\ \frac{t^2}{2} & -\frac{1}{3}3t(t+2)+4e^t-4 \\ 1-2t & 6t-3e^t+3 \\ t^2 & -3t(t+2)+9e^t-8 \end{bmatrix},$$

where e^t has been factored out from each of the matrix entries.

9. The given matrix has a single eigenvalue $r_1 = 0$ of multiplicity 4. In fact, $\mathbf{A}^4 = \mathbf{0}$, so the canonical basis vectors may serve as the generalized eigenvectors. Recall that we have already encountered such a situation in exercises 1 and 3 (and of course, what follows is essentially the application of method described in Exercises 7.2: computing the

exponent of the matrix with a single eigenvalue). Taking $\mathbf{u}_i$, $1 \leq i \leq 4$ to be the canonical basis vectors, we have, as in Exercise 3,

$$e^{\mathbf{A}t} = (e^{\mathbf{A}t}[\mathbf{u}_1 \ \mathbf{u}_2 \ \mathbf{u}_3 \ \mathbf{u}_4])[\mathbf{u}_1 \ \mathbf{u}_2 \ \mathbf{u}_3 \ \mathbf{u}_4]^{-1} = \mathbf{I} + t\mathbf{A} + \frac{t^2}{2}\mathbf{A}^2 + \frac{t^3}{6}\mathbf{A}^3$$

$$= \begin{bmatrix} 1-t & t & 0 & t \\ -2t & 3t+1 & -t & 3t \\ -t & t & 1 & t \\ t & -2t & t & 1-2t \end{bmatrix}.$$

The fact that the right-hand side is linear in t may look surprising, and the explanation is that $\mathbf{A}^2 = 0$. We also want to note that we skipped the process of constructing matrix $e^{\mathbf{A}t}$ from vectors $e^{\mathbf{A}t}\mathbf{u}_i$, $1 \leq i \leq 4$, because it is particularly simple for $\mathbf{u}_i$ being the basis vectors, see Exercises 1, 3 above.

11. The given matrix $\mathbf{A}$ has a simple eigenvalue $r_1 = 0$, and $r_2 = 2$ of multiplicity 5. Its generalized eigenvectors are

$$[\mathbf{u}_1 \ \mathbf{u}_2 \ \mathbf{u}_3 \ \mathbf{u}_4 \ \mathbf{u}_5 \ \mathbf{u}_6] = \begin{bmatrix} 0 & 0 & 0 & 0 & 0 & 1 \\ 0 & 0 & 0 & 0 & 1 & 0 \\ 0 & 0 & 0 & 1 & 0 & 0 \\ 0 & 0 & 1 & 0 & 0 & 0 \\ -1 & 1 & 0 & 0 & 0 & 0 \\ 1 & 1 & 0 & 0 & 0 & 0 \end{bmatrix},$$

where $\mathbf{u}_1$ is the eigenvector for r_1, and $\mathbf{u}_i$, $2 \leq i \leq 6$ are solutions of $(\mathbf{A} - 2\mathbf{I})^5 = \mathbf{0}$. As above, we first find the vectors $e^{\mathbf{A}t}\mathbf{u}_i$:

$$e^{\mathbf{A}t}\mathbf{u}_1 = e^{0t\mathbf{I}}e^{t\mathbf{A}}\mathbf{u}_1 = \mathbf{I}\mathbf{u}_1 = \begin{bmatrix} 0 \\ 0 \\ 0 \\ 0 \\ -1 \\ 1 \end{bmatrix},$$

$$e^{\mathbf{A}t}\mathbf{u}_2 = e^{2t\mathbf{I}}e^{t(\mathbf{A}-2\mathbf{I})}\mathbf{u}_2$$

$$= e^{2t}\left[\mathbf{I} + t(\mathbf{A} - 2\mathbf{I}) + \frac{t^2}{2!}(\mathbf{A} - 2\mathbf{I})^2 + \frac{t^3}{3!}(\mathbf{A} - 2\mathbf{I})^3\right.$$

$$\left. + \frac{t^4}{4!}(\mathbf{A} - 2\mathbf{I})^4\right]\mathbf{u}_2$$

$$= e^{2t} \begin{bmatrix} t+1 & -t & t(t+1) & t(t+1) \\ t & 1-t & (t-1)t & (t-1)t \\ 0 & 0 & 1 & 0 \\ 0 & 0 & 0 & 1 \\ 0 & 0 & 0 & 0 \\ 0 & 0 & 0 & 0 \end{bmatrix}$$

$$\begin{bmatrix} 0 & 0 \\ 0 & 0 \\ t & t \\ -t & -t \\ \frac{t^4-2t^3+3t^2-3t+3}{3} & -t\frac{t^3-2t^2+3t-3}{3} \\ -t\frac{t^3-2t^2+3t-3}{3} & \frac{t^4-2t^3+3t^2-3t+3}{3} \end{bmatrix} \begin{bmatrix} 0 \\ 0 \\ 0 \\ 0 \\ 1 \\ 1 \end{bmatrix}$$

$$= \begin{bmatrix} 0 \\ 0 \\ 2e^{2t}t \\ -2e^{2t}t \\ e^{2t} \\ e^{2t} \end{bmatrix},$$

$$e^{At}\mathbf{u}_3 = e^{2tI}e^{t(A-2I)}\mathbf{u}_3$$

$$= e^{2t} \begin{bmatrix} t+1 & -t & t(t+1) & t(t+1) \\ t & 1-t & (t-1)t & (t-1)t \\ 0 & 0 & 1 & 0 \\ 0 & 0 & 0 & 1 \\ 0 & 0 & 0 & 0 \\ 0 & 0 & 0 & 0 \end{bmatrix}$$

$$\begin{bmatrix} 0 & 0 \\ 0 & 0 \\ t & t \\ -t & -t \\ \frac{t^4-2t^3+3t^2-3t+3}{3} & -t\frac{t^3-2t^2+3t-3}{3} \\ -t\frac{t^3-2t^2+3t-3}{3} & \frac{t^4-2t^3+3t^2-3t+3}{3} \end{bmatrix} \begin{bmatrix} 0 \\ 0 \\ 0 \\ 1 \\ 0 \\ 0 \end{bmatrix}$$

$$= \begin{bmatrix} e^{2t}t(t+1) \\ e^{2t}(t-1)t \\ 0 \\ e^{2t} \\ 0 \\ 0 \end{bmatrix}.$$

Similarly to $e^{At}\mathbf{u}_3$ above,

$$
e^{At}\mathbf{u}_4 = \begin{bmatrix} e^{2t}t(t+1) \\ e^{2t}(t-1)t \\ e^{2t} \\ 0 \\ 0 \\ 0 \end{bmatrix}, \quad
e^{At}\mathbf{u}_5 = \begin{bmatrix} -e^{2t}t \\ e^{2t}(1-t) \\ 0 \\ 0 \\ 0 \\ 0 \end{bmatrix}, \quad
e^{At}\mathbf{u}_6 = \begin{bmatrix} e^{2t}(t+1) \\ e^{2t}t \\ 0 \\ 0 \\ 0 \\ 0 \end{bmatrix}.
$$

We conclude:

$$
e^{At} = (e^{At}[\mathbf{u}_1 \ \mathbf{u}_2 \ \mathbf{u}_3 \ \mathbf{u}_4 \ \mathbf{u}_5 \ \mathbf{u}_6])[\mathbf{u}_1 \ \mathbf{u}_2 \ \mathbf{u}_3 \ \mathbf{u}_4 \ \mathbf{u}_5 \ \mathbf{u}_6]^{-1}
$$

$$
= \begin{bmatrix}
0 & 0 & e^{2t}t(t+1) & e^{2t}t(t+1) & -e^{2t}t & e^{2t}(t+1) \\
0 & 0 & e^{2t}(t-1)t & e^{2t}(t-1)t & e^{2t}(1-t) & e^{2t}t \\
0 & 2e^{2t}t & 0 & e^{2t} & 0 & 0 \\
0 & -2e^{2t}t & e^{2t} & 0 & 0 & 0 \\
-1 & e^{2t} & 0 & 0 & 0 & 0 \\
1 & e^{2t} & 0 & 0 & 0 & 0
\end{bmatrix}
$$

$$
\times \begin{bmatrix}
0 & 0 & 0 & 0 & -\frac{1}{2} & \frac{1}{2} \\
0 & 0 & 0 & 0 & \frac{1}{2} & \frac{1}{2} \\
0 & 0 & 0 & 1 & 0 & 0 \\
0 & 0 & 1 & 0 & 0 & 0 \\
0 & 1 & 0 & 0 & 0 & 0 \\
1 & 0 & 0 & 0 & 0 & 0
\end{bmatrix}
$$

$$
= \begin{bmatrix}
e^{2t}(t+1) & -e^{2t}t & e^{2t}t(t+1) & e^{2t}t(t+1) \\
e^{2t}t & -e^{2t}(t-1) & e^{2t}(t-1)t & e^{2t}(t-1)t \\
0 & 0 & e^{2t} & 0 \\
0 & 0 & 0 & e^{2t} \\
0 & 0 & 0 & 0 \\
0 & 0 & 0 & 0
\end{bmatrix}
$$

$$
\begin{matrix}
0 & 0 \\
0 & 0 \\
e^{2t}t & e^{2t}t \\
-e^{2t}t & -e^{2t}t \\
\frac{1}{2}\left(1+e^{2t}\right) & \frac{1}{2}\left(-1+e^{2t}\right) \\
\frac{1}{2}\left(-1+e^{2t}\right) & \frac{1}{2}\left(1+e^{2t}\right)
\end{matrix} \; .
$$

13. To see why a nilpotent matrix $\mathbf{A}$ has the only eigenvalue $r_1 = 0$, observe that otherwise, if, say, $r_2 \neq 0$ is also an eigenvalue and $\mathbf{u}_2 \neq \mathbf{0}$ is a corresponding eigenvector, we have

$$r_2^p \mathbf{u} = \mathbf{A}^p \mathbf{u} = 0\mathbf{u} = \mathbf{0},$$

a contradiction. This shows also that the characteristic polynomial of $\mathbf{A}$ is $p(r) = r^d$ for d – dimension of $\mathbf{A}$. By Cayley-Hamilton theorem then $\mathbf{A}^d = \mathbf{0}$. Since generalized eigenvectors of A are precisely the null vectors of $\mathbf{A}^d$, they may be in particular chosen to equal columns of the identity matrix (or any other basis of the space $\mathbf{R}^d_{\text{col}}$).

15. In order to solve this initial value problem, we need to find $e^{\mathbf{A}t}$ for

$$\mathbf{A} = \begin{bmatrix} 3 & -2 \\ 0 & 3 \end{bmatrix}.$$

The only eigenvalue is clearly $r_1 = 3$. Since $(\mathbf{A} - 3\mathbf{I})^2 = \mathbf{0}$, we take the generalized eigenvectors $\mathbf{u}_1 = \begin{bmatrix} 1 & 0 \end{bmatrix}$, $\mathbf{u}_2 = \begin{bmatrix} 0 & 1 \end{bmatrix}$. Then

$$e^{\mathbf{A}t}\mathbf{u}_1 = e^{3t\mathbf{I}}e^{t(\mathbf{A}-3\mathbf{I})}\mathbf{u}_1 = e^{3t}\left[\mathbf{I} + t(\mathbf{A} - 3\mathbf{I})\right]\begin{bmatrix} 1 \\ 0 \end{bmatrix}$$

$$= e^{3t}\begin{bmatrix} 1 & -2t \\ 0 & 1 \end{bmatrix}\begin{bmatrix} 1 \\ 0 \end{bmatrix} = \begin{bmatrix} e^{3t} \\ 0 \end{bmatrix},$$

and

$$e^{\mathbf{A}t}\mathbf{u}_2 = e^{3t}\begin{bmatrix} 1 & -2t \\ 0 & 1 \end{bmatrix}\begin{bmatrix} 0 \\ 1 \end{bmatrix} = \begin{bmatrix} -2te^{3t} \\ e^{3t} \end{bmatrix}.$$

Thus

$$e^{\mathbf{A}t} = (e^{\mathbf{A}t}[\mathbf{u}_1 \ \mathbf{u}_2])[\mathbf{u}_1 \ \mathbf{u}_2]^{-1} = \begin{bmatrix} e^{3t} & -2te^{3t} \\ 0 & e^{3t} \end{bmatrix}.$$

Solution of the system $\mathbf{x}' = \mathbf{A}\mathbf{x}$ with $\mathbf{x}(0) = \begin{bmatrix} -1 & 2 \end{bmatrix}^T$ is then

$$\mathbf{x}(t) = \begin{bmatrix} e^{3t} & -2te^{3t} \\ 0 & e^{3t} \end{bmatrix}\begin{bmatrix} -1 \\ 2 \end{bmatrix} = \begin{bmatrix} -e^{3t} - 4te^{3t} \\ 2e^{3t} \end{bmatrix}.$$

17. As in Exercise 15, we need to compute e^{At}. The matrix

$$A = \begin{bmatrix} 1 & 0 & 1 & 2 \\ 1 & 1 & 2 & 1 \\ 0 & 0 & 2 & 0 \\ 0 & 0 & 1 & 1 \end{bmatrix}$$

has two eigenvalues: $r_1 = 1$ of multiplicity 3, and $r_2 = 2$. The generalized eigenvectors are

$$[\mathbf{u}_1 \ \mathbf{u}_2 \ \mathbf{u}_3 \ \mathbf{u}_4] = \begin{bmatrix} 0 & 0 & 1 & 3 \\ 1 & 0 & 0 & 6 \\ 0 & 0 & 0 & 1 \\ 0 & 1 & 0 & 1 \end{bmatrix},$$

where $\mathbf{u}_1$, $\mathbf{u}_2$, $\mathbf{u}_3$ are solutions of $(A - I)^3 \mathbf{u} = 0$, $\mathbf{u}_4$ is the eigenvector for r_2. Now observe that the initial value supplied by our problem is $\mathbf{x}(0) = [1 \ 0 \ 0 \ 0]^T = \mathbf{u}_3$, and since the solution of system $\mathbf{x}' = A\mathbf{x}$ is $\mathbf{x}(t) = e^{At}\mathbf{x}(0)$, it suffices to compute

$$e^{At}\mathbf{x}(0) = e^{At}\mathbf{u}_3 = e^{tI}e^{t(A-I)}\mathbf{u}_3 = e^t \left[I + t(A - I) + \frac{t^2}{2!}(A - I)^2 \right] \mathbf{u}_3$$

$$= e^t \begin{bmatrix} 1 & 0 & \frac{3t^2}{2} + t & 2t \\ t & 1 & 2t(t+1) & t(t+1) \\ 0 & 0 & \frac{t^2}{2} + t + 1 & 0 \\ 0 & 0 & \frac{1}{2}t(t+2) & 1 \end{bmatrix} \begin{bmatrix} 1 \\ 0 \\ 0 \\ 0 \end{bmatrix} = \begin{bmatrix} e^t \\ te^t \\ 0 \\ 0 \end{bmatrix}.$$

PART III

REVIEW PROBLEMS FOR PART III

1. If a matrix $\mathbf{A}$ is similar to the zero matrix $\mathbf{O}$, we have $\mathbf{A} = \mathbf{P}^{-1}\mathbf{OP} = \mathbf{O}$. On the other hand, if $\mathbf{A}$ is similar to the identity matrix $\mathbf{E}$, it holds $\mathbf{A} = \mathbf{P}^{-1}\mathbf{EP} = \mathbf{P}^{-1}\mathbf{P} = \mathbf{E}$.

3. Adding a (non-zero) constant multiple of the j-th row to the i-th one corresponds to multiplying by the matrix $\mathbf{E}_1(\alpha)$ on the left:

$$\mathbf{E}_1(\alpha) = \begin{bmatrix} 1 & & & \cdots & & & 0 \\ & \ddots & & & & \cdot{}^{\cdot{}^{\cdot}} & \\ & & 1 & & 0 & & \\ \vdots & & & \ddots & & & \vdots \\ & & \alpha & & 1 & & \\ & \cdot{}^{\cdot{}^{\cdot}} & & & & \ddots & \\ 0 & & & \cdots & & & 1 \end{bmatrix} \quad \text{\} } i\text{-th row}$$

$\underbrace{}_{j\text{-th column}}$

Characteristic polynomial of this matrix is $(1 - r)^n$. On the other hand, it is immediate from the definition that the set of eigenvectors with eigenvalue 1 consists of all vectors with 0 in the j-th coordinate, i.e. does not contain a full basis. Thus $\mathbf{E}_1(\alpha)$ is defective.

Solutions Manual to Accompany Fundamentals of Matrix Analysis with Applications,
First Edition. Edward Barry Saff and Arthur David Snider.
© 2016 John Wiley & Sons, Inc. Published 2016 by John Wiley & Sons, Inc.

Multiplying the i-th row by factor α corresponds to multiplying on the left by the matrix

$$\mathbf{E}_2(\alpha) = \begin{bmatrix} 1 & \cdots & 0 & \cdots & 0 \\ \vdots & \ddots & & 0 & \vdots \\ 0 & & \alpha & & 0 \\ \vdots & 0 & & \ddots & \vdots \\ 0 & \cdots & 0 & \cdots & 1 \end{bmatrix} \begin{array}{l} \\ \\ \text{\} } i\text{-th row} \\ \\ \\ \end{array}$$

$$\underbrace{}_{i\text{-th column}}$$

It has $n-1$ eigenvectors (in fact, canonical base vectors) with the eigenvalue 1, spanning the space of all vectors with zero in the i-th coordinate, and one eigenvector with the eigenvalue α, whose only nonzero coordinate is the i-th. The matrix $\mathbf{E}_2(\alpha)$ is therefore not defective.

Interchanging the i-th and the j-th rows corresponds to multiplying on the left by the matrix

$$\mathbf{E}_3 = \begin{bmatrix} 1 & & & \cdots & & & 0 \\ & \ddots & & & & \cdots & \\ & & 0 & & 1 & & \\ \vdots & & & \ddots & & & \vdots \\ & & 1 & & 0 & & \\ & \cdots & & & & \ddots & \\ 0 & & & \cdots & & & 1 \end{bmatrix} \begin{array}{l} \\ \\ \text{\} } j\text{-th row} \\ \\ \text{\} } i\text{-th row} \\ \\ \\ \end{array}$$

$$\underbrace{}_{j\text{-th}} \quad \underbrace{}_{i\text{-th}}$$

Matrix $\mathbf{E}_3$ has two eigenvectors with $+1$ and -1 in the i-th and j-th coordinates respectively, and 0 in all the other coordinates. The corresponding eigenvalue is -1. Furthermore, all the vectors with equal i-th and j-th coordinates are eigenvectors with eigenvalue 1. To summarize, $\mathbf{E}_3$ is not defective.

5. The characteristic polynomial of the given matrix is $p(r) = (-3 - r)$ $(-6 - r) + 2 = r^2 + 9r + 20$. Its roots are $r_1 = -5$, $r_2 = -4$. The eigenvectors corresponding to the first and the second eigenvalue are nontrivial solutions of the following homogeneous systems
$$\begin{bmatrix} 2 & 2 & \vdots & 0 \end{bmatrix} \text{ and } \begin{bmatrix} 1 & 2 & \vdots & 0 \end{bmatrix}$$
respectively. Hence $\mathbf{v}_1 = [1 \ -1]^T$ and $\mathbf{v}_2 = [2 \ -1]^T$.

7. The given matrix has characteristic polynomial $r(r-2)^2$. Its eigenvectors are $[3 \ 1 \ -1]^T$ with the eigenvalue 0 and $[-1 \ 1 \ 1]^T$ with the eigenvalue 2. The matrix is therefore defective.

9. Characteristic polynomial of a 2-by-2 matrix $\mathbf{A}$ is

$$p(r) = \begin{vmatrix} A_{11} - r & A_{12} \\ A_{21} & A_{22} - r \end{vmatrix}$$
$$= r^2 - (A_{11} + A_{22})r + (A_{11}A_{22} - A_{12}A_{21}).$$

The coefficient at r is the trace of $\mathbf{A}$ with a minus sign, and the constant term is the determinant of $\mathbf{A}$. Eigenvalues can therefore be found as the roots of $r^2 - 4r + 6 = 0$, that is $2 \pm i\sqrt{2}$.

11. The characteristic polynomial of this matrix is $p(r) = r^4 - 4r^2$. Eigenvectors for the eigenvalues -2 and 2 are $[1 \ -1 \ 1 \ -1]^T$ and $[1 \ 1 \ 1 \ 1]^T$ respectively. By Theorem 3 from Section 5.2, eigenvectors for different eigenvalues are orthogonal. It is not hard to find the two mutually orthogonal eigenvectors $[1 \ 0 \ -1 \ 0]^T$ and $[0 \ 1 \ 0 \ -1]^T$ for the eigenvalue 0 (alternatively, find any two eigenvectors for 0 and apply Gram-Schmidt algorithm).

In the problems 13–19 the matrix of the diagonalizing similarity transformation will contain eigenvectors of the original matrix as columns, see Theorem 1 in Section 6.1.

13. From Exercise 5, the matrix of eigenvectors in this case is $\mathbf{P} = \begin{bmatrix} 1 & 2 \\ -1 & -1 \end{bmatrix}$.

15. The given matrix is defective, so not similar to any diagonal matrix. Indeed, its characteristic polynomial is $r(r-2)^2$. Furthermore, for the

double root $r_{2,3} = 2$ the corresponding homogeneous system has a one-dimensional null space:

$$\begin{bmatrix} -1 & -2 & 1 & \vdots & 0 \\ -1 & 0 & -1 & \vdots & 0 \end{bmatrix}.$$

Thus there is only one eigenvector for this eigenvalue of multiplicity two.

17. This matrix has three distinct eigenvalues, namely, $\{1, 2, 3\}$. Applying the usual techniques, find the following matrix containing eigenvectors as columns:

$$\mathbf{P} = \begin{bmatrix} -1 & -2 & -1 \\ 1 & 1 & 1 \\ 2 & 4 & 4 \end{bmatrix}.$$

19. Observe that the given matrix is Hermitian, so it has 3 linearly independent eigenvectors. Its characteristic polynomial is $p(r) = r^3 - 5r^2 + 7r - 3$. The roots of $p(r)$ are: a double root $r_{1,2} = 1$, and $r_3 = 3$. Obtaining the matrix of eigenvectors as above, we have

$$\mathbf{P} = \begin{bmatrix} 1 & 1 & 0 \\ -i & i & 0 \\ 0 & 0 & 1 \end{bmatrix}.$$

21. According to Theorem 2 of Section 6.2 (principal axes theorem), eigenvectors of a real symmetric matrix generate an orthogonal change of coordinates that eliminates the cross terms in the corresponding quadratic form. The given form has the following matrix:

$$\mathbf{A} = \begin{bmatrix} 4 & 1 \\ 1 & 4 \end{bmatrix}.$$

It has two distinct eigenvalues, so the corresponding eigenvectors $[-1 \ 1]^T$ and $[1 \ 1]^T$ are orthogonal. Multiplying these vectors by an inverse to their norms, we obtain an orthogonal matrix representing the change of coordinates:

$$\begin{bmatrix} x_1 \\ x_2 \end{bmatrix} = \frac{1}{\sqrt{2}} \begin{bmatrix} -1 & 1 \\ 1 & 1 \end{bmatrix} \begin{bmatrix} y_1 \\ y_2 \end{bmatrix}.$$

The quadratic form becomes $3y_1^2 + 5y_2^2$ in the variables y_1, y_2, as can be seen from applying the similarity transform with the above matrix to $\mathbf{A}$.

23. The given quadratic form corresponds to the matrix

$$\mathbf{A} = \begin{bmatrix} 2 & 1 & 1 \\ 1 & 2 & -1 \\ 1 & -1 & 2 \end{bmatrix}.$$

Its eigenvalues are $r_1 = 0$ and $r_{2,3} = 3$. The eigenvector for r_1 is $\mathbf{v}_1 = [1 \ -1 \ -1]^T$. The other eigenvectors are found as solutions of the homogeneous system

$$\begin{bmatrix} -1 & 1 & 1 & \vdots & 0 \end{bmatrix}.$$

Hence, for example, $\mathbf{v}_2 = [1 \ 1 \ 0]^T$ and $\mathbf{v}_3 = [1 \ 0 \ 1]^T$. Applying the Gram-Schmidt algorithm to $\mathbf{v}_2$, $\mathbf{v}_3$ gives the matrix of orthogonal eigenvectors

$$[\mathbf{v}_1 \ \mathbf{v}_2 \ \mathbf{v}_3] = \begin{bmatrix} 1 & 1 & 1/2 \\ -1 & 1 & -1/2 \\ -1 & 0 & 1 \end{bmatrix}.$$

Recall that we seek a change of variables with an orthogonal matrix. It therefore remains to divide the columns (or rows) of this matrix by their respective norms to obtain the required change of coordinates (given below after rounding):

$$\begin{bmatrix} x_1 \\ x_2 \\ x_3 \end{bmatrix} = \begin{bmatrix} 0.5774 & 0.7071 & 0.4082 \\ -0.5774 & 0.7071 & -0.4082 \\ -0.5774 & 0 & 0.8165 \end{bmatrix} \begin{bmatrix} y_1 \\ y_2 \\ y_3 \end{bmatrix}.$$

25. The given quadratic form has the matrix

$$\mathbf{A} = \begin{bmatrix} 0 & 1/2 & 1/2 \\ 1/2 & 0 & 1/2 \\ 1/2 & 1/2 & 0 \end{bmatrix}.$$

Its eigenvalues are $r_{1,2} = -1/2$ and $r_3 = 1$. The corresponding eigenvectors are columns of the following matrix:

$$[\mathbf{v}_1 \ \mathbf{v}_2 \ \mathbf{v}_3] = \begin{bmatrix} 1 & 0 & 1 \\ 0 & 1 & 1 \\ -1 & -1 & 1 \end{bmatrix}.$$

Applying the Gram-Schmidt algorithm to the first two columns, we get three orthogonal eigenvectors:

$$\begin{bmatrix} 1 & -1/2 & 1 \\ 0 & 1 & 1 \\ -1 & -1/2 & 1 \end{bmatrix}.$$

Normalizing the column vectors of the above matrix gives the required change of variables:

$$\begin{bmatrix} x_1 \\ x_2 \\ x_3 \end{bmatrix} = \begin{bmatrix} 0.7071 & -0.4082 & 0.5774 \\ 0 & 0.8165 & 0.5774 \\ -0.7071 & -0.4082 & 0.5774 \end{bmatrix} \begin{bmatrix} y_1 \\ y_2 \\ y_3 \end{bmatrix}.$$

27. Let us compare the expressions for $\mathbf{AA}^H$ and $\mathbf{A}^H\mathbf{A}$:

$$\mathbf{AA}^H = \begin{bmatrix} 5 & 4 & 2 \\ 4 & a\bar{a}+4 & -2a \\ 2 & -2\bar{a} & 8 \end{bmatrix}$$

and

$$\mathbf{A}^H\mathbf{A} = \begin{bmatrix} a\bar{a}+4 & 4 & 2\bar{a} \\ 4 & 5 & -2 \\ 2a & -2 & 8 \end{bmatrix}.$$

Hence $\mathbf{AA}^H = \mathbf{A}^H\mathbf{A}$ is equivalent to $a = 1$. After substitution in $\mathbf{A}$, we have

$$\mathbf{A} = \begin{bmatrix} 0 & -1 & 2 \\ 1 & 0 & 2 \\ -2 & -2 & 0 \end{bmatrix}.$$

Its characteristic polynomial is $r^3 + 9r$ and eigenvectors are columns in the following display

$$\mathbf{P} = \begin{bmatrix} 1 & 1 & 1 \\ -1 & 0.8 + 0.6i & 0.8 - 0.6i \\ -0.5 & 0.4 - 1.2i & 0.4 + 1.2i \end{bmatrix}$$

with eigenvalues 0, $-3i$, $3i$ respectively. All the three eigenvalues are distinct, so the corresponding eigenvectors are orthogonal.

Printed in the USA
K019561SCI030716   01S29053000000001033